BIOMAT 2009

International Symposium on
Mathematical and Computational Biology

BIOMAT 2009

International Symposium on
Mathematical and Computational Biology

Brasilia, Brazil
1 – 6 August 2009

edited by

Rubem P Mondaini
Federal University of Rio de Janeiro, Brazil

World Scientific

NEW JERSEY • LONDON • SINGAPORE • BEIJING • SHANGHAI • HONG KONG • TAIPEI • CHENNAI

Published by

World Scientific Publishing Co. Pte. Ltd.
5 Toh Tuck Link, Singapore 596224
USA office: 27 Warren Street, Suite 401-402, Hackensack, NJ 07601
UK office: 57 Shelton Street, Covent Garden, London WC2H 9HE

British Library Cataloguing-in-Publication Data
A catalogue record for this book is available from the British Library.

BIOMAT 2009
International Symposium on Mathematical and Computational Biology

ISBN-13 978-981-4304-89-4
ISBN-10 981-4304-89-1

Printed in Singapore by World Scientific Printers

Preface

This is the volume of selected works of the 9th International Symposium on Mathematical and Computational Biology - BIOMAT 2009. The conference was held at the campus of University of Brasilia, in Brasilia, DF, Brazil, from 01st August to 06th August 2009. Ten invited scientists have been talking in the Plenary Sessions as Keynote Speakers. Some of them have also been lecturing in tutorials and mini-courses which is a tradition of the BIOMAT Symposia series. There were about sixty contributed papers in technical sessions - oral and poster. As a rule, we have invited successful authors to resubmit their accepted papers for presentation during the conference, to members of the BIOMAT Consortium Editorial Board and referees ad Hoc nominated by this Board. We received about 130 works and a third reviewing of the paper is now seen as strictly necessary in order to restrict the acceptance level of the BIOMAT published works around 22% as is usually made since the foundation of the BIOMAT Consortium. The organizers have done their best to provide the excellent level of communication among the participants in an informal scientific atmosphere. We also stress the opportunity of exchanging professional feedback with experts and research students of different scientific education but the same interest on a specific area of scientific research. This is the spirit of the BIOMAT Consortium and the BIOMAT Symposia series. We continue the offer of participation in the tutorials and mini-courses to young research students. The motivation of interdisciplinary work in the future career of prospective scientists being a special mission of the BIOMAT Consortium. In the present publication we can find contributions in the fields of Biomathematics, Biological Physics and Computational Biology as well as the generic Mathematical Modelling of Biosystems. In order to enhance the feedback mentioned above, we are always trying to invite professionals with a very different formation for lecturing in tutorials and mini-courses.

We are indebted to the Administrative Staff of two Brazilian Sponsoring agencies CAPES - Coordination for the Improvement of Higher Education Personnel and CNPq - National Research Council. Their Directors and Authorized Representatives for their continuous belief in the BIOMAT Symposia series. We thank the Foundation for Research Support of S. Paulo State and the University of S. Paulo, School of Medicine, Brazil for some financial help during the organization of the past symposium(BIOMAT 2008). Special thanks are due to Felipe Mondaini and Leonardo Mondaini for their collaboration with the Organizing Committee of the BIOMAT

2009 in Brasilia as well as Luiz A. S. de Oliveira and his LaTeX expertise for his help to the editorial work.
The organization of a scientific conference on an academic venue like a university or a research institute, has the favourable consequence of the possibility of using the facilities of the host institution. We thank the University of Brasilia and the Dean for Graduate Courses and Research for her collaboration with the offer of the auditorium of the Faculty of Technology and its electronic facilities. We thank also the members of the Administrative Staff of this institute for their invaluable help. On behalf of the BIOMAT Consortium, a non-profit association of scientific researchers from Universities and Research Institutes on many countries, we congratulate all the authors and participants for another opportunity of joint scientific work during the BIOMAT 2009 Symposium.

Rubem P. Mondaini
President of the BIOMAT Consortium

Rio de Janeiro, December 2009

Editorial Board of the BIOMAT Consortium

Contents

Population Dynamics

Computational Biology

Modelling Infectious Diseases

MACROPHAGES AND TUMOURS: FRIENDS OR FOE?

H. M. BYRNE AND M. R. OWEN
Centre for Mathematical Medicine and Biology,
School of Mathematical Sciences,
University of Nottingham, Nottingham NG7 2RD, UK
E-mail: helen.byrne@nottingham.ac.uk

Hypoxia or low oxygen is a characteristic feature of many solid tumours, associated with poor drug delivery and low rate of cell proliferation, factors that limit the efficacy of therapies designed to target proliferating cells. Since macrophages localise within hypoxic tumour regions, a promising way to target them involves engineering macrophages to express therapeutic genes under hypoxia. In this paper we present three mathematical models of increasing complexity that have been used to investigate the feasibility of using genetically-engineered macrophages to target hypoxic tumour regions. A robust prediction of the models is that the macrophage therapy is unable to eliminate the tumour. However it can reduce significantly the proportion of hypoxic tumour cells and, thereby, increase the sensitivity to standard chemotherapy. Thus we conclude that maximum therapeutic benefit will be achieved by using the macrophage-based therapy in combination with other drugs.

1. Introduction

Macrophages are white blood cells which perform diverse functions. At sites of infection they engulf foreign bodies (e.g. bacteria) and within wounds they secrete angiogenic factors that stimulate the ingrowth of new blood vessels [1,2]. The appearance of macrophages within solid tumours was originally viewed positively, the macrophages fighting to eradicate the foreign body or tumour. While in certain tumours macrophages are a good prognostic indicator there are cases for which the converse is true [3,2], the balance between pro- and anti-tumour functions being tumour-specific. Even so, it is clear that macrophages localise within low-oxygen (hypoxic) tumour regions[4], being recruited by chemoattractants which are produced by tumour cells (and, to a lesser degree, macrophages) under hypoxia[5].

Being characterised by reduced rates of cell proliferation and acting as primary sources of angiogenic factors, hypoxic regions are notoriously resistant to standard cancer treatments. Consequently, experimentalists are trying to exploit the tendency for macrophages to localise in hypoxic tumour regions by developing new anti-cancer therapies in which a patient's macrophages are genetically engineered to release, or indirectly to activate prodrug forms of, cytotoxic chemicals and/or antiangiogenic factors under hypoxia[4]. Laboratory results are promising, with reductions of up to 30% in the size of avascular tumour spheroids infiltrated with engineered macrophages. However many questions remain unanswered. For example, should the macrophages be engineered to kill tumour cells or targetted against angiogenesis? In which tumour regions are such therapies in fact effective? In this paper we review a series of increasingly detailed mathematical models developed to investigate the feasibility of using genetically-engineered macrophages to target hypoxic tumour regions.

While there is now a large literature devoted to modelling solid tumour growth [6,7,8,9,10,11,12] and the response of tumours to chemotherapy [13,14,15], with few exceptions, the role of macrophages has not been widely studied. Owen and Sherratt [16] studied the anti-tumour effect of normal, non-engineered macrophages caused by their cytolytic activity. The presence of macrophages was found to markedly alter the composition of the tissue but was unable to eliminate the tumour. In addition, only steady state solutions were observed. By extending their model to include spatial effects, such as chemotactic migration of the macrophages, Owen and Sherratt went on to show that macrophages may play a key role in generating spatial heterogeneity within solid tumours [17]. In [17] details of the tumour's spatial structure prior to macrophage infiltration were neglected, the tumour being assumed small enough that it was devoid of hypoxic and necrotic regions. These assumptions were relaxed in [18] where attention focused on the impact that macrophage chemotaxis and random motion and also tumour size had on the distribution of macrophages within well-developed avascular tumours that comprise necrotic and hypoxic regions.

The remainder of the chapter is structured as follows. In section 2 we study a time-dependent ODE model of tumour growth *in vivo* [19]. Attention focuses on interactions between macrophages, tumour cells and normal cells, with spatial effects neglected and nutrient assumed to be freely available. In section 3 we relax these assumptions and focus on a PDE model which describes the response to the therapy of a small, avascular tumour spheroid grown *in vitro*[20,21]. In section 4 we adapt an existing multiscale model of vascular tumour growth[22,23] to investigate the response of a developed tumour growing *in vivo* to the macrophage-based therapy. The chapter concludes in section 5 with a brief discussion of our results and suggestions for future work.

2. ODE models of macrophage-tumour dynamics *in vivo*

In this section we consider a model for the temporal evolution of a nutrient-rich, spatially-uniform tissue containing normal cells ($n_1(t)$), tumour cells ($n_2(t)$), genetically modified macrophages ($m(t)$)and a tumour cell-derived chemical ($\omega(t)$). We assume that all cells are undergoing mitosis and focus on tumour–macrophage interactions. The (dimensionless) equations that govern the dependent variables are stated below (for details, see Ref. [19]).

$$\frac{\mathrm{d}n_1}{\mathrm{d}t} = n_1[1 - (n_1 + n_2 + m)], \tag{1}$$

$$\frac{\mathrm{d}n_2}{\mathrm{d}t} = \xi n_2[1 - \phi(n_1 + n_2 + m)] - k_1 m n_2, \tag{2}$$

$$\frac{\mathrm{d}m}{\mathrm{d}t} = (m^* - m)\frac{\Delta\omega}{1 + \alpha\omega} - \sigma m(n_1 + n_2 + m), \tag{3}$$

$$\frac{\mathrm{d}\omega}{\mathrm{d}t} = n_2 - \lambda_0\omega, \tag{4}$$

with

$$n_1(0) = n_1^e, \quad n_2(0) = n_2^e, \quad m(0) = m^e, \quad \omega(0) = 0. \tag{5}$$

In (1) the normal cells undergo logistic growth (carrying capacity and growth rate normalised to unity by suitable scaling of the dimensional equations). We account for competition for space with tumour cells and macrophages by modifying the logistic growth term to include an additional death rate but assume that the rate at which they are destroyed by macrophages is negligible. Equation (2) is similar to (1), except that the tumour cells are assumed to proliferate at an elevated rate $\xi > 1$ and have a larger carrying capacity ($\phi^{-1} > 1$) than the normal cells. Additionally, the tumour cells are destroyed by macrophages at rate $k_1 n_2 m$. In (3) we neglect macrophage proliferation[24] and assume that macrophage levels are dominated by their rate of influx from the vasculature and death due to crowding effects. The rate of macrophage influx is an increasing and saturating function of the chemoattractant[25,26] that is also proportional to $m^* - m$ where m^* denotes the macrophage concentration in the blood vessels that perfuse the tissue. In (4) we assume that the chemoattractant is regulated by its rate of production by the tumour cells, with rate constant scaled to unity, and natural decay at rate λ_0. We close equations (1)–(4)

by assuming in (5) that initially the tissue comprises a mixture of normal cells, tumour cells and macrophages and is devoid of chemoattractant.

Numerical simulations were obtained by integrating (1)–(5) with a Runge–Kutta method. In the absence of detailed experimental data, we used the following parameter values (see Ref. [19] for details):

$$\lambda_0 = 1, \quad \xi = 100, \quad \phi = 0.7, \quad \Delta = 5, \quad \alpha = 30, \quad \sigma = 0.5, \qquad (6)$$

and focussed on the effect of varying the rate of tumour cell lysis (k_1) and the density of engineered macrophages in the vasculature (m^*).

Extensive numerical simulations indicate that the system typically evolves to a stable state involving several cell types. Depending on the values of k_1 and m^*, the stable solutions may be time-independent or oscillatory. Figure 1 illustrates the type of oscillatory solutions that can arise. If tumour cell lysis by the engineered macrophages is weak then the time-periodic solutions do not support normal cells (see figure 1(a)) whereas if k_1 is larger then normal cells may be present (see figure 1(b)). In both cases the peak macrophage density alternates with the peak tumour cell density and there are prolonged periods during which the tumour burden is minimal. The key differences between figures 1(a) and (b) are the appear-

Figure 1. Stable oscillatory solutions of the model equations (1)–(4), for the parameter values given in (6). (a) $m^* = 7$, $k_1 = 35$: coexistence of tumour cells and macrophages only. (b) $m^* = 5$, $k_1 = 50$: coexistence of normal cells, tumour cells and macrophages. Reproduced from Ref. [19], with permission.

ance of normal cells in figure 1(b), and the reduction in the peak tumour cell density. Figure 2 illustrates how indirect macrophage-tumour coupling, mediated by the chemoattractant, generates oscillatory solutions. In more detail, macrophages entering the tissue lyse the tumour cells, causing n_2 to decline. This causes the level of chemoattractant and, hence, the rate at which macrophages enter the tissue to decline. The number of tumour cells and the amount of chemoattractant then start to rise, leading to an increase in macrophage infiltration, and so the cycle repeats.

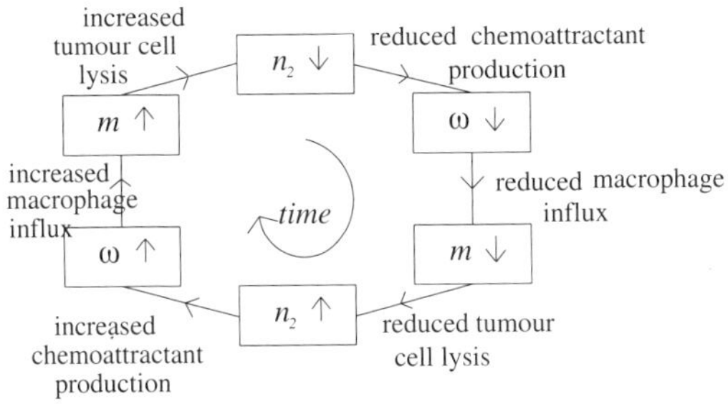

Figure 2. A schematic diagram explaining how oscillatory solutions are sustained. Reproduced from Ref. [19], with permission.

While the simulations presented in figure 1 depict stable solutions that are time-periodic, for other parameter values time-independent solutions may be stable. To illustrate this, figure 3 shows how the structure of the stable solution changes as k_1 and m^* vary. As k_1 increases, the average tumour cell density and the average macrophage cell density decline and the system passes through two bifurcation points at which the qualitative nature of the observed solutions changes. For small values of k_1 the stable solutions involve tumour cells and macrophages and are time-independent. As k_1 passes through the first bifurcation point ($k_1 \approx 20$ in figure 3(a)) this steady solution becomes oscillatory. At the second bifurcation point ($k_1 \approx 30$) this oscillatory solution loses stability to one for which all three cell types are present. Figure 3(b) shows how the structure of the stable solution changes as m^* varies. The mean tumour cell density falls as m^* increases whereas mean macrophage levels rise and eventually saturate. Two bifurcations occur: as m^* increases through $m^* \approx 2.5$ the tumour-macrophage steady state is replaced by one involving all three cell types; as m^* increases through $m^* \approx 4.2$ this stable steady state is superceded by an oscillatory solution of the type presented in figure 1(b). We remark that the details of the bifurcation structure depend on the path in parameter space that is selected. For example, in figure 3(a) none of the steady solutions involve all three cell types whereas such solutions arise in figure 3(b).

The above results may be explained by performing linear and weakly nonlinear analyses of the model equations (see Ref. [19] for details). In par-

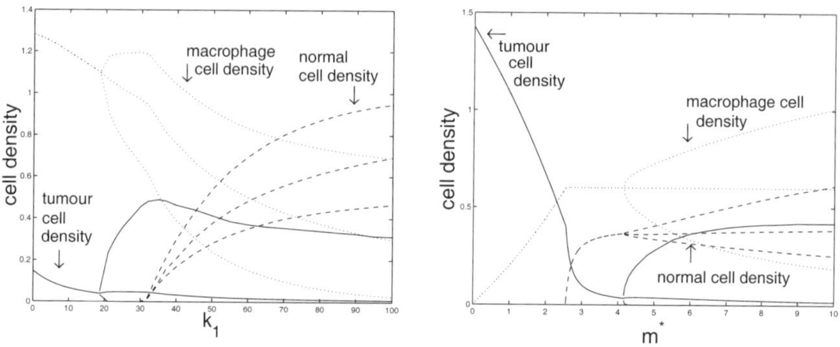

Figure 3. Changing the rate of tumour cell lysis (k_1) and the macrophage density in the vasculature (m^*) affects the nature of the stable solutions. (a) For k_1 small the stable solution is steady and involves tumour cells and macrophages. For k_1 large the stable solution is oscillatory and involves all three cell types. For intermediate values of k_1 time-periodic solutions involving tumour cells and macrophages are stable. (b) For m^* small the stable solution is steady and involves tumour cells and macrophages. For m^* large, time-periodic solutions involving all three cell types are stable. For intermediate values of m^* there is a stable steady solution involving all three cell types. Where oscillatory solutions arise the maximum, minimum and mean cell densities are plotted. Parameter values: as per equation (6) with $m^* = 8.0$ in (a) and $k_1 = 50$ in (b). Reproduced from Ref. [19], with permission.

ticular, it is possible to show that the system dynamics are organised about a well-defined point in (m^*, k_1) parameter space and to classify regions of parameter space according to whether the stable solutions that exist there are time-dependent or steady and whether they involve all three cell types or only macrophages and tumour cells.

In this section we have presented an ODE model that describes interactions between genetically-engineered macrophages, tumour cells and normal cells. For all choices of the model parameters, the macrophages were unable to eliminate the tumour, with time-independent solutions arising in some cases and time-periodic solutions in others. The mean tumour burden reduced as k_1 and/or m^* increased. However, large values of k_1 and m^* generally yielded oscillatory solutions (see figure 3). Such solutions may be undesirable for several reasons. Firstly, while the mean tumour cell density decreases as k_1 and/or m^* increased, the maximum tumour cell density does not. Further, since the macrophages do not eliminate the tumour, a second therapy must be used and its efficacy will be sensitive to the point in the cycle at which it is administered (the adjuvant therapy will be most effective when the tumour is smallest).

3. PDE models of therapeutic response

While the ODE model from section 2 generates useful insight, the tumour is treated as spatially-uniform and oxygen assumed to be plentiful. We now study a spatially-structured model, formulated as a system of PDEs, which includes oxygen as a dependent variable, enabling us to account for macrophage localisation in hypoxic regions and to describe more accurately their mode of action. In one mechanism that has been tested *in vitro*, the enzyme cytochrome P450 is transfected into the macrophages and expressed under hypoxia. The prodrug cyclophosphamide is injected separately and converted to active form on contact with the enzyme. The active drug is absorbed by the tumour cells, causing their death under mitosis[27]. As stated in section 1, preliminary results show reductions of up to 30% in the size of multicell spheroids treated with this macrophage-enzyme-prodrug approach.

Our PDE model describes the response of multicell spheroids to treatment with such genetically-engineered macrophages. We assume that the tumour undergoes spherically-symmetric growth and comprises tumour cells $m(r,t)$, extracellular material $n(r,t)$, and two types of genetically-engineered macrophages, those that express prodrug-activating enzyme, $l_e(r,t)$ and those that do not, $l_i(r,t)$. Transitions between $l_e(r,t)$ and $l_i(r,t)$ are regulated by the oxygen concentration $c(r,t)$, with activation occurring at a rate $f_u(c)$ which peaks under hypoxia, and inactivation occurring at a rate $f_d(c)$ that increases with c. Other model variables include a generic macrophage chemoattractant $a(r,t)$ which is produced by hypoxic tumour cells, the prodrug $\phi(r,t)$ and the active drug $\theta(r,t)$. If we assume that there are no voids within the tumour then, since the chemicals (c, a, θ, ϕ) make a negligible contribution to the spheroid volume, we have that

$$m + l_i + l_e + n = 1. \tag{7}$$

Each phase moves with a common advection velocity $v(r,t)$, modulated by diffusive and chemotactic terms. Combining the above concepts we arrive at the following system of PDEs (for details, see Ref. [21]):

$$\begin{aligned}\frac{\partial m}{\partial t} &= \frac{1}{r^2}\frac{\partial}{\partial r}\left[r^2\left(D_m\frac{\partial m}{\partial r} - mv\right)\right] + p_m(n,c)(1-2K(\theta))m \\ &\quad - d_m(c)m,\end{aligned} \tag{8}$$

$$\frac{\partial l_i}{\partial t} = \frac{1}{r^2}\frac{\partial}{\partial r}\left[r^2\left(D_l\frac{\partial l_i}{\partial r} - \chi_l l_i\frac{\partial a}{\partial r} - l_i v\right)\right] - d_l(c)l_i - f_u(c)l_i + f_d(c)l_e \tag{9}$$

$$\frac{\partial l_e}{\partial t} = \frac{1}{r^2}\frac{\partial}{\partial r}\left[r^2\left(D_l\frac{\partial l_e}{\partial r} - \chi_l l_e\frac{\partial a}{\partial r} - l_e v\right)\right] - d_l(c)l_e + f_u(c)l_i - f_d(c)l_e, \tag{10}$$

$$\frac{\partial n}{\partial t} = \frac{1}{r^2}\frac{\partial 1}{\partial r}\left[r^2\left(D_n\frac{\partial n}{\partial r} - vn\right)\right] - p_m(c,n)(1-2K(\theta))m + d_m(c)m + d_l(c)(l_i + l_e), \tag{11}$$

$$0 = \frac{D_c}{r^2}\frac{\partial}{\partial r}\left(r^2\frac{\partial c}{\partial r}\right) - d_c(m, l_i, l_e, c, n), \tag{12}$$

$$0 = \frac{D_a}{r^2}\frac{\partial}{\partial r}\left(r^2\frac{\partial a}{\partial r}\right) + p_a(m,c) - d_a a, \tag{13}$$

$$0 = \frac{D_\phi}{r^2}\frac{\partial}{\partial r}\left(r^2\frac{\partial \phi}{\partial r}\right) - d_\phi\phi - k_e l_e\phi, \tag{14}$$

$$0 = \frac{D_\theta}{r^2}\frac{\partial}{\partial r}\left(r^2\frac{\partial \theta}{\partial r}\right) + k_e l_e\phi - d_\theta\theta - k_i m p_m(n,c)K(\theta), \tag{15}$$

where $D_i\,(i = l, m, n)$ denote assumed constant random motility coefficients of the relevant phases, and χ_l represents the macrophage chemotaxis coefficient. In equations (8)–(11), $p_m(n, c)$ denotes the tumour cell proliferation rate and $d_m(c)$ denotes their death rate due to apoptosis and/or necrosis. $K(\theta)$ denotes the fraction of mitotic cells killed by the active drug. As in section 2, we neglect macrophage proliferation and assume that they die at a rate $d_l(c)$ which is a decreasing function of c.

In (12)–(15), $D_i(i = c, a, \theta, \phi)$ represent the diffusion coefficients of the chemicals, which we assume diffuse sufficiently fast, relative to the timescale of tumour growth, that their concentrations are quasi-steady [11,9]. Oxygen is consumed by tumour cells and both types of macrophages at rate $d_c(m, l_i, l_e, c, n)$. Macrophage chemoattractant is produced by tumour cells at a rate $p_a(m, c)$, which increases under hypoxia, and decays at a constant rate d_a. Active drug is produced at rate $k_e l_e \phi$. Pro- and active drug are assumed to decay linearly at rates d_ϕ and d_θ, and k_i represents the rate at which active-drug is depleted during tumour-cell lysis. Details of the kinetic terms and parameter values are given in Ref. [21].

The velocity v is specified by summing (8)–(11) and using (7):

$$v = D_l\left(\frac{\partial l_i}{\partial r} + \frac{\partial l_e}{\partial r}\right) - \chi_l(l_i + l_e)\frac{\partial a}{\partial r} + D_m\frac{\partial m}{\partial r} - D_n\left(\frac{\partial m}{\partial r} + \frac{\partial l_i}{\partial r} + \frac{\partial l_e}{\partial r}\right). \tag{16}$$

Eliminating $n = 1 - l_i - l_e - m$ supplies PDEs (8)–(10), (12)–(16) for $m, l_i, l_e, c, a, \theta, \phi, v$. We assume further that the spheroid occupies the region $0 < r < R(t)$ and that $R(t)$ moves with the tumour cell velocity at $r = R(t)$, so that, referring to (8), we have

$$\frac{dR}{dt} = \left(v - \frac{D_m}{m}\frac{\partial m}{\partial r}\right)\Bigg|_{r=R(t)}. \tag{17}$$

We close the model by imposing symmetry boundary conditions at $r = 0$ and mixed boundary conditions on $r = R(t)$ so that

$$0 = \frac{\partial l_i}{\partial r} = \frac{\partial l_e}{\partial r} = \frac{\partial m}{\partial r} = \frac{\partial c}{\partial r} = \frac{\partial a}{\partial r} = \frac{\partial \phi}{\partial r} = \frac{\partial \theta}{\partial r} \quad \text{at } r = 0, \tag{18}$$

and

$$\left.\begin{aligned} -D_l\frac{\partial l_i}{\partial r} + \chi_l l_i\frac{\partial a}{\partial r} + D_m\frac{l_i}{m}\frac{\partial m}{\partial r} &= h_l(l_i - l_\infty),\\ -D_l\frac{\partial l_e}{\partial r} + \chi_l l_e\frac{\partial a}{\partial r} + D_m\frac{l_e}{m}\frac{\partial m}{\partial r} &= h_l(l_e - l_{e\infty}),\\ -D_n\frac{\partial n}{\partial r} + D_m\frac{n}{m}\frac{\partial m}{\partial r} &= h_n(n - n_\infty),\\ c &= c_\infty,\\ D_a\frac{\partial a}{\partial r} &= h_a(a_\infty - a),\\ D_\phi\frac{\partial \phi}{\partial r} &= h_\phi(\phi_\infty - \phi),\\ D_\theta\frac{\partial \theta}{\partial r} &= h_\theta(\theta_\infty - \theta), \end{aligned}\right\} \quad \text{at } r = R(t), \tag{19}$$

with $n = 1 - l_i - l_e - m$. In (19) the macrophage and cellular material fluxes across $r = R(t)$ are proportional to $(l_\infty - l_i(R,t))$, $(l_{e\infty} - l_e(R,t))$ and $(n_\infty - n(R,t))$, respectively, l_∞, $l_{e\infty}$ and n_∞ denoting the levels of l_i, l_e and n exterior to the spheroid, while h_l, h_n are the constants of proportionality (permeabilities), which are assumed equal for inactive and active macrophages. Similarly, c_∞, a_∞, ϕ_∞ and θ_∞ denote the concentrations of c, a, ϕ and θ exterior to the spheroid and h_a, h_ϕ and h_θ represent the permeabilities of the tumour boundary to these variables. Henceforth we fix $l_{e\infty} = a_\infty = \theta_\infty = 0$ so that levels of enzyme-active macrophages, chemoattractant and active drug are negligible in the culture medium.

We assume that initially the tumour has radius R_0 and is devoid of macrophages, chemoattractant, prodrug and active drug so that

$$\begin{gathered} R(0) = R_0, \quad m(r,0) = m_0(r), \quad l_i(r,0) = l_e(r,0) = 0, \\ c(r,0) = c_0(r), \quad a(r,0) = \theta(r,0) = \phi(r,0) = 0, \end{gathered} \tag{20}$$

where $m_0(r)$ and $c_0(r)$ are prescribed functions of position r.

Numerical simulations were carried out using finite differences and the method of lines. Typical results are shown in figure 4. After an initial period of macrophage-free growth, the spheroid is continuously infused with prodrug and inactive macrophages from $t = 250$ onwards. The macrophages infiltrate by random motion and chemotaxis and accumulate in hypoxic regions where they express enzyme, giving an interior peak in active macrophages. The prodrug diffuses from the spheroid boundary and is activated by the active macrophages. Active drug that diffuses into the proliferating rim kills tumour cells when they attempt to divide, causing the spheroid to grow more slowly or shrink.

Figure 4 illustrates how increasing ϕ_∞, the prodrug concentration supplied to the tumour, can cause a transition from unbounded growth (a travelling wave) to a steady state. We characterise this transition in figure 5(a) which shows how the travelling wave velocity U and steady state spheroid radius R_∞ vary with ϕ_∞. Bifurcation curves were calculated by applying AUTO to steady state and travelling wave boundary value problems. As ϕ_∞ increases U decreases, until a bifurcation between travelling wave and steady state solutions occurs when $U = 0$ (at $\phi_\infty \approx 0.05$ in figure 5). As ϕ_∞ increases beyond this point, R_∞ decreases rapidly, before plateauing as ϕ_∞ (and hence the rate of tumour cell lysis) is increased further. The tumour is not eliminated as $\phi_\infty \to \infty$ because the spheroid becomes better oxygenated as R_∞ decreases and, in consequence, the rate of hypoxia-induced drug activation declines to zero. We note that for certain values of ϕ_∞ multiple steady solutions are predicted. Corresponding tumour cell profiles for selected values of ϕ_∞ are illustrated in figure 5(b). In particular, we show two profiles for coexisting steady states at $\phi_\infty = 0.2$. Direct simulations indicate that a middle branch of unstable steady states separates the pair of stable steady states.

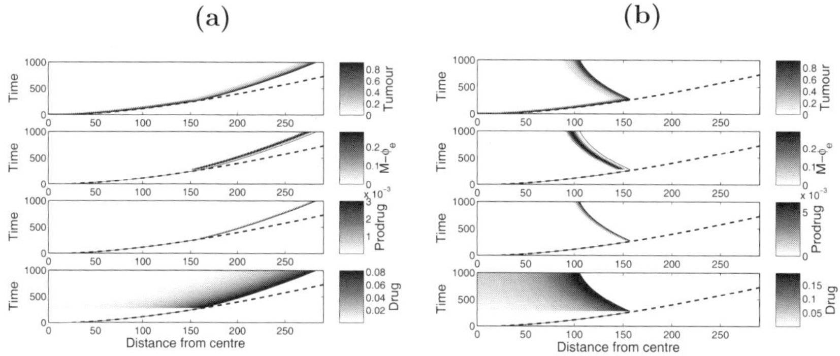

Figure 4. Typical simulations of (8)–(10),(12)–(19). We show how the spheroid composition changes over time, by plotting the tumour cell and enzyme-active macrophage volume fractions, and the concentrations of prodrug and active drug. The spheroid is grown in the absence of macrophages and prodrug (i.e. $l_\infty = 0$ and $\phi_\infty = 0$), until $t = 250$ when inactive macrophages and prodrug are continuously infused with $l_\infty = 0.2$ and (a) $\phi_\infty = 0.04$, and (b) $\phi_\infty = 0.1$. The dashed curves indicate the growth of the spheroid radius in the absence of treatment. (a) In this case, we observe a reduction in the tumour's growth rate but it still evolves to a travelling wave. (b) Higher levels of prodrug can cause tumour growth to be confined ($R(t)$ reaches a steady state $R_\infty \approx 95.9$). Reproduced from Ref. [21], with permission.

Figure 6(a) reveals how the solution structure changes when n_∞ and l_∞ vary: the steady state solutions may extend into the travelling wave region, so that steady state and travelling wave solutions coexist. This is borne out by the simulations presented in figure 6(b). When therapy is applied at $t = 6$ the spheroid shrinks and remains confined. Delaying therapy to $t = 7$ allows the tumour to increase in size and enter the basin of attraction of the travelling wave solution so the same level of treatment applied slightly later is unable to prevent unbounded growth.

Figure 7 shows where in (l_∞, ϕ_∞)-space travelling wave and steady state solutions exist for different values of n_∞. Travelling waves exist to the left of the bifurcation curves, these being paths in parameter space on which $U = 0$. To the left of each bifurcation curve there is inadequate prodrug to halt growth, whereas to the right there is sufficient drug activation to prevent wave-like growth. Regardless of whether macrophage chemotaxis is active, for n_∞ large the bifurcation curve is monotonic decreasing. This is because as the prodrug level (and hence the lysis rate) increases the number of infiltrating macrophages must decrease in order to maintain the balance necessary for zero speed waves. Thus for sufficiently large values of ϕ_∞ and l_∞ the system yields steady state solutions. Decreasing n_∞ moves the

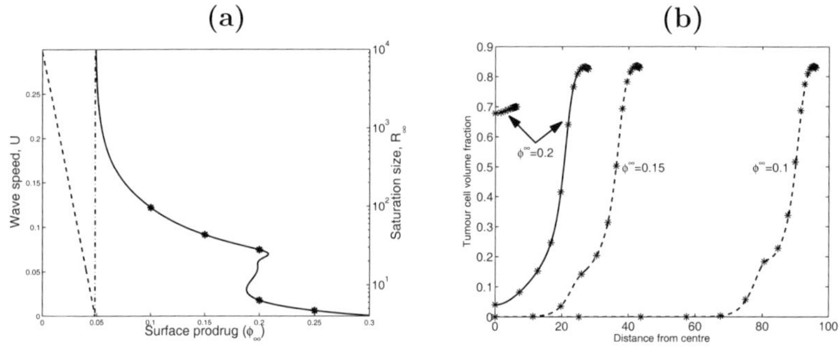

Figure 5. **(a)** Travelling wave velocity (U, dashed) and equilibrium tumour radii (R_∞, solid, log scale) against surface prodrug ϕ_∞. The dot-dashed vertical line indicates the steady-state/travelling wave bifurcation at $\phi_\infty \approx 0.05$. A comparison with solutions from the full PDE model is indicated by stars for $\phi_\infty = (0.1, 0.15, 0.2, 0.25)$. Corresponding tumour cell profiles are illustrated in part **(b)** — the curves are the solutions predicted by the large-time analysis and stars indicate those generated from numerical simulations of the full PDE. In each case the comparison is excellent. At $\phi_\infty = 0.2$ two stable steady state solutions arise (solid curves in part **(b)**). Reproduced from Ref. [21], with permission.

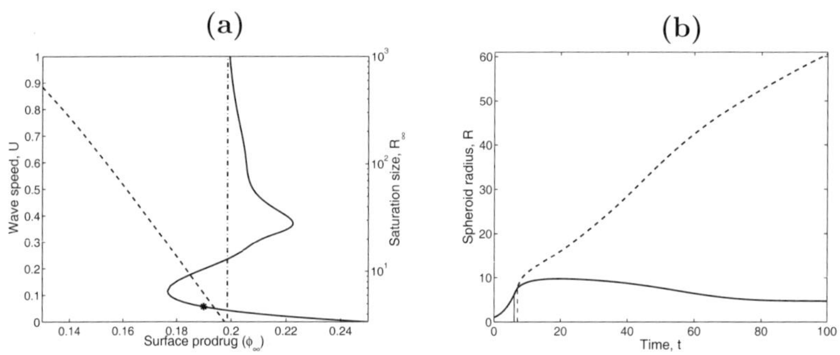

Figure 6. **(a)** Curves showing how U (dashed) and R_∞ (solid, log scale) vary with ϕ_∞ when $n_\infty = 0.2$ and $l_\infty = 0.4$. The dot-dashed vertical line indicates the steady-state/travelling wave bifurcation. For this choice of parameter values travelling wave and steady state solutions can coexist. **(b)** We fix $\phi_\infty = 0.19$ (i.e. the asterisk in part (a)) and show how the evolution of a tumour with $R(t = 0) = 1$ depends on the time at which therapy starts. If it starts at $t = 6$ (solid line) then the tumour regresses to a small equilibrium size ($R_\infty \approx 4.7$); if therapy starts at $t = 7$ (dashed line) then the tumour undergoes unbounded growth. Reproduced from Ref. [21], with permission.

bifurcation curve to the left so that steady state solutions occur for a wider range of values of ϕ_∞. This is because as n_∞ decreases the flux of cellular material across the tumour surface also falls.

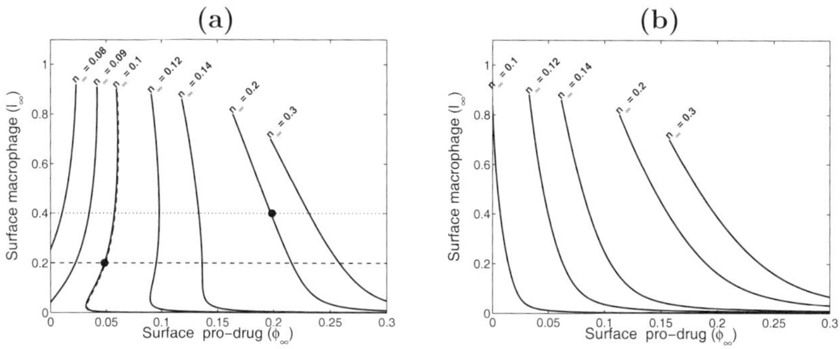

Figure 7. Bifurcation curves separating travelling wave and steady state solutions (solid curves) for different values of $0 < n_\infty < 1 - l_\infty$. In all cases, travelling wave solutions exist to the left of the relevant curve. In **(a)**, macrophages infiltrate by random motion and chemotaxis, in **(b)** they migrate by random motion alone ($\chi_l = 0$). Figure 5(a) corresponds to a slice through (l_∞, ϕ_∞)-space at $l_\infty = 0.2$, indicated by the dashed horizontal line in part (a) for $n_\infty = 0.1$. The marked point corresponds to the bifurcation at $\phi_\infty \approx 0.05$. Similarly the horizontal dotted line corresponds to the case where waves and steady states can coexist, shown in figure 6. Reproduced from Ref. [21], with permission.

Without macrophage chemotaxis ($\chi_l = 0$), the bifurcation curves are monotonic decreasing for all feasible values of n_∞ (see figure 7(b)). By contrast, with $\chi_l > 0$ and moderate values of n_∞ the bifurcation curve has a fold, and for smaller n_∞ it is monotonic increasing. If n_∞ is large (and extracellular material abundant) then travelling waves predominate. By contrast, when $\chi_l > 0$ and n_∞ is small the macrophages contribute significantly to the spheroid volume, and if ϕ_∞ is small then macrophage-derived tumour cell death can not counter volume increases due to macrophage infiltration so travelling wave solutions occur.

For moderate n_∞ there is a delicate balance between these effects: a fold in the bifurcation curve means that travelling waves exist if l_∞ is very small or large, with steady states for intermediate values of l_∞. Intuitively, we expect that increased macrophage infiltration should be of therapeutic benefit. However, our results suggest that increasing chemotactic infiltration into hypoxic regions may cause deleterious non-localised growth.

4. Multiscale models of macrophage delivery to vascularised tumours

By allowing spatial variation in oxygen levels, the PDE model developed in section 3 provided a realistic description of the response of avascular tumour spheroids cultured *in vitro* to macrophage-based gene therapy. Before this new approach can enter clinical practice, experimentalists need to establish whether the therapy is effective against vascularised tumours growing *in vivo*. As a first step towards this goal, in this section we modify an existing multiscale model in order to predict the efficacy of the prodrug-enzyme therapy for treating vascular tumours. We start by summarising the main features of the multiscale model (for more detailed information, including parameter values, see Ref. [22, 28]) and then explain how the macrophage-based therapy is included, before presenting some illustrative results.

Our model is formulated as a hybrid cellular automata [29,30] and integrates phenomena occurring on very different time and length scales. These inter-related features include blood flow and structural adaptation of the vascular network, transport into the tissue of blood-borne oxygen, competition between cancer and normal cells, cell division, apoptosis and VEGF (growth factor) release. We monitor diffusible species (here oxygen, VEGF and, for the macrophage-therapy, prodrug and active drug) and also account for intracellular and tissue-scale phenomena, and the coupling between them. To this end, our model is organised into three layers: vascular, cellular, and subcellular, which correspond, respectively, to the tissue, cellular and subcellular time and length scales (see figure 8).

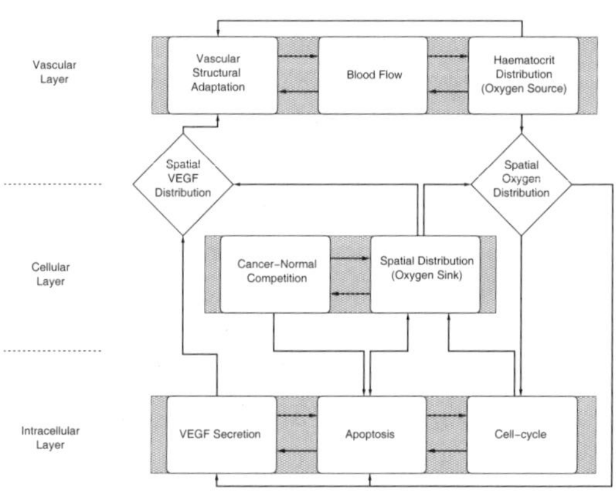

Figure 8. Basic structure of the model. Reproduced from Ref. [29], with permission.

In the vascular layer, we consider a hexagonal network in which each vessel undergoes structural adaptation (i.e. changes in radius) in response to a number of stimuli (see Ref. [29] for more details). We also compute the blood flow rate, the pressure drop and the haematocrit (i.e. relative volume of red blood cells) distribution in each vessel. Coupling between the vascular and cellular layers is mediated by the transport of blood-borne oxygen into the tissue. This process is modelled by a reaction-diffusion equation. The haematocrit acts as a distributed source of oxygen, whereas the cells act as spatially-distributed sinks of oxygen.

In the cellular layer, we focus on cell-cell interactions (competition) and the spatial distribution of the cells. We distinguish between normal and cancerous cells (and macrophages), and assume that there is a finite upper bound (typically, 1, 2 or 3) on the number of cells contained in each element. The different cell types compete for space and resources, cancerous cells usually out-performing the others. Competition between cell types is incorporated in simple rules, which connect the cellular and subcellular layers. Apoptosis (programmed cell death) is controlled by the expression of p53 (whose dynamics are dealt with in the subcellular layer): when the level of p53 in a cell exceeds a threshold value it undergoes apoptosis. However, this threshold depends on the distribution of cells in a given neighbourhood.

At the subcellular level, cell division, apoptosis, and VEGF production are modelled using ODEs. Since the distribution of oxygen depends on the spatial distribution of cells (cellular layer) and haematocrit (vascular layer), subcellular processes interact naturally with those at other layers: cell proliferation and apoptosis alter the spatial distribution of the cells (see Fig.8); the cellular and the intracellular layers modulate vascular adaptation through diffusion of VEGF in the tissue and its absorption by the endothelial cells lining the vessels.

We modify the multiscale model to investigate the impact of the macrophage-based therapy introduced in section 3 on vascularised tumours by introducing macrophages as a new cell type and monitoring levels of enzyme, prodrug and active drug. Since we monitor individual macrophages we need not distinguish inactive and active macrophages: their activity is controlled by the microenvironment (contrast this with the PDE model in section 3). Macrophages and prodrug injected into the vasculature are transported through the vessel network in the same way as oxygen (or haematocrit), extravasating at rates which are proportional to the differences between their vascular and tissue concentrations. The macrophages move by a combination of random motion and chemotaxis, the latter in response to spatial gradients in VEGF (since VEGF is expressed by hypoxic cells this leads to macrophage localisation in hypoxic regions). As in section 3, the macrophages release prodrug-activating enzyme under hypoxia.

Pro- and active drug levels are modelled using reaction-diffusion equations, similar to (14)-(15). Active drug absorbed by tumour cells causes cell death if present in sufficiently high levels when a cell attempts to divide.

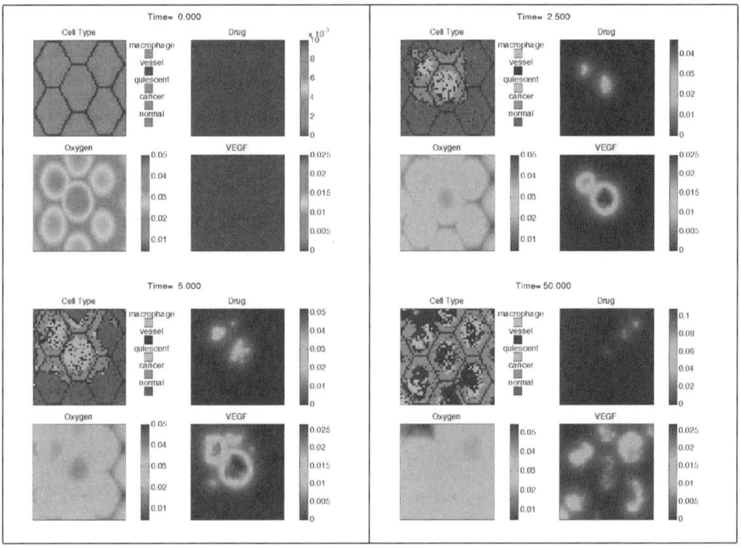

Figure 9. Results from a typical simulation, showing macrophage infiltration, prodrug activation and tumour cell kill when the active drug threshold for cell death is 0.01. The four frames show the state of the system at t=0.0, 2.5, 5.0 and 50 days.

Results from a typical simulation are presented in figure 9. Initially a small cluster of tumour cells is located towards the centre of a vascularised tissue. As the tumour grows, regions of hypoxia form. Macrophages injected into the blood supply extravasate and then localise in these hypoxic regions where they release prodrug-activating enzyme. Active drug diffuses throughout the tissue, being most effective when absorbed by tumour cells in oxygen-rich regions adjacent to blood vessels. The therapy reduces the number of quiescent (ie non-proliferating or hypoxic) tumour cells but is unable to eradicate the tumour.

The results from figure 9 are summarised in figure 10, along with results from two similar simulations. We plot the time-averaged macrophage and active drug concentrations, and the total number of tumour cells killed at each site within the tissue. In each case we note that while active drug levels peak under hypoxia, tumour cell kill is highest in the well-oxygenated regions that surround the blood vessels. The three simulations reveal how increasing the potency of the active drug (specifically lowering the threshold concentration required to stimulate tumour cell death on completion of the cell cycle) increases the effectiveness of the therapy.

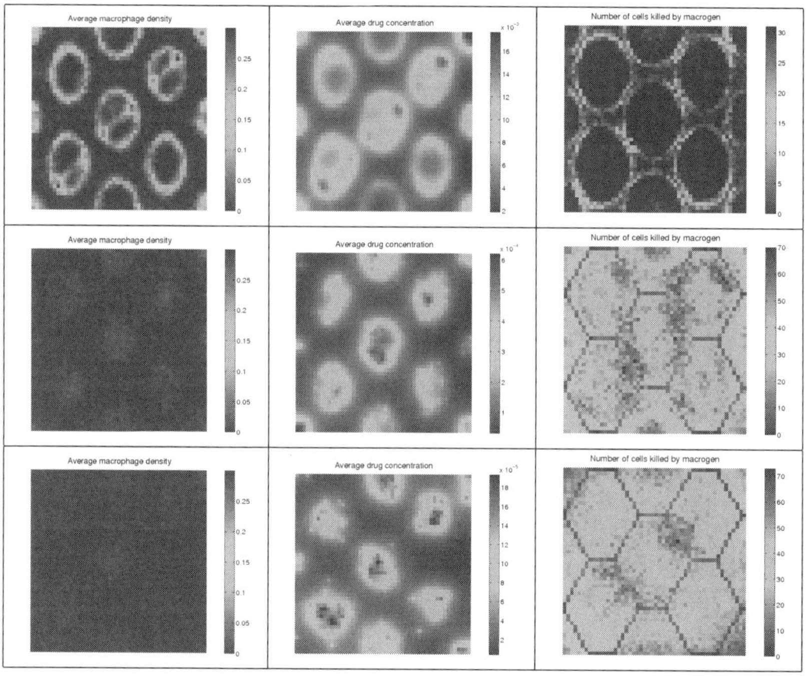

Figure 10. Series of plots showing how the time-averaged macrophage and active drug concentrations, and the total number of tumour cells killed vary across the tissue as the potency of the active drug increases. Key: drug threshold for tumour cell death on completion of cell cycle = 0.01 (top row), 0.001 (middle row), 0.0001 (bottom row).

In figure 11 we present summary time-course data for the simulations from figure 10. As the active drug becomes more effective (or, equivalently, the prodrug dose is increased), the tumour burden, the number of quiescent tumour cells and the number of macrophages fall, the latter considerably. However, the therapy fails to eliminate the tumour because it is self-limiting: hypoxia is necessary for the macrophages to produce prodrug-activating enzyme but when active drug kills proliferating tumour cells, levels of hypoxia within the tissue decrease, compromising the potency of the therapy. These results are consistent with those obtained from the PDE model of section 3 [21].

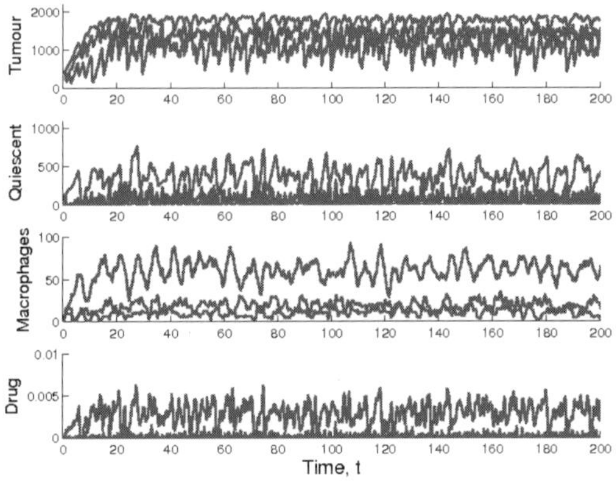

Figure 11. Summary plots showing how the tumour's composition changes over time in response to macrophage infiltration and prodrug activation. As the drug becomes more effective (i.e. the active drug threshold needed to stimulate tumour cell death on completion of the cell cycle, the tumour burden falls but it is not eliminated. Key: drug threshold for cell death = 0.01 (blue), = 0.001 (red), 0.0001 (green).

5. Conclusions

In this paper we have reviewed three alternative mathematical approaches that can be used to study the feasibility of using genetically-engineered macrophages to target hypoxic tumour regions. In spite of their differences, the models yield several common predictions. In particular, the macrophage-based therapy is unable to eliminate the tumour. However it is successful at driving down levels of hypoxic/quiescent tumour cells, rendering the tumour more vulnerable to standard chemotherapy that targets proliferating tumour cells.

When comparing the models, it is important to realise their relative strengths and weaknesses. For example, the simplicity of the ODE model of section 2 makes it amenable to analysis. Further the relatively small number of parameters involved should more easily be fit to experimental data than the larger number associated with the PDE and multiscale models. By contrast, the multiscale model of section 4 provides detailed spatial information about where the macrophages accumulate, where active drug is produced and the sites at which tumour cell kill occurs. While it is currently practically impossible to estimate reliably the parameters that appear in a detailed multiscale model, in the future, as imaging techniques improve, their estimation may become a reality, enabling the models to generate both qualitative and quantitative predictions.

Clearly much work remains to be done before this macrophage-based therapy realises its clinical potential. By extending the models presented in this chapter we believe that we can accelerate this process. Of particular interest will be investigating whether a better clinical outcome is achieved when the macrophages deliver cytotoxic drugs to the tumour (as considered here) or when they deliver anti-angiogenic drugs. Additionally, we should consider competition between the host's non-engineered macrophages and the modified ones, and investigate how the macrophage-based therapy should be coordinated with standard chemotherapy to maximise the therapeutic response. Work in progress involves introducing magnetic nanoparticles into the genetically-engineered macrophages and using magnetic fields to enhance their delivery to the tumour. Once the magnetically-loaded macrophages have localised in the tumour mass, this approach gives scope for applying an alternating magnetic field to heat the tissue and so using hyperthermia to kill the tumour cells[31].

In summary, we believe that mathematics has an increasingly important role to play in helping experimentalists to understand biological systems and to optimise the development of new therapies, such as the macrophage-based anti-cancer treatment that was the focus of this chapter. Additionally, biology represents a rich source of new mathematical challenges.

Acknowledgments

The authors gratefully acknowledge support from the BBSRC and EPSRC.

References

1. L.A. DiPietro, *Shock* **4**, 233 (1995).
2. C. O'Sullivan and C.E. Lewis, *J. Pathol.* **172**, 229 (1994).
3. R.D. Leek, C.E. Lewis, R. Whitehouse, M. Greenall, J. Clarke and A.L. Harris, *Cancer Res.* **56**, 4625 (1996).
4. J.S. Lewis, J.A. Lee, J.C.E. Underwood, A.L. Harris and C.E. Lewis, *J. Leukoc. Biol.* **66**, 889 (1999).
5. A. Mantovani, W.J. Ming, C. Balotta, B. Abdeljalil and B. Bottazzi, *Biochim Biophys Acta* **865**, 59 (1986).
6. C.J.W. Breward, H.M. Byrne and C.E. Lewis, *J. Math. Biol.* **45**, 125 (2002).
7. H.M. Byrne, T. Alarcon, M.R. Owen, S.D. Webb and P.K. Maini, *Phil. Trans. Roy. Soc. Ser. B.* **364**: 1563 (2006).
8. H.P. Greenspan, *Stud. Appl. Math.* **52**, 317 (1972).
9. J.P. Ward and J.R. King, *IMA J. Math. Appl. Med. Biol.* **14**, 39 (1997).
10. A.R.A. Anderson and M.A.J. Chaplain, *Bull. Math. Biol.* **60**, 857 (1998).
11. C.J.W. Breward, H.M. Byrne and C.E. Lewis, *Eur. J. Appl. Math.* **12**, 529 (2001).

12. C.J.W. Breward, H.M. Byrne and C.E. Lewis, *Bull. Math. Biol.* **65**, 609 (2003).
13. T.L. Jackson and H.M. Byrne, *Math. Biosci.* **164** 17 (2000).
14. C. Panetta and J.A. Adam, *Math. Comput. Modell.* **22**, 67 (1995).
15. J.C. Panetta, *Math. Biosci.* **146**, 89 (1997).
16. M.R. Owen and J.A. Sherratt, *IMA J. Math. Appl. Med. Biol.* **15**, 165 (1998).
17. M.R. Owen and J.A. Sherratt, *J. theor. Biol.* **189**, 63 (1997).
18. C.E. Kelly, R.D. Leek, H.M. Byrne, S.M. Cox, A.L. Harris and C.E. Lewis, *J. Theor. Med.* **4**, 21 (2002).
19. H.M. Byrne, S.M. Cox and C.E. Kelly, *Discrete Cont. Dyn.-B* **4**: 81 (2004).
20. M.R. Owen, H.M. Byrne and C.E. Lewis, *J. theor. Biol.* **226**, 377 (2004).
21. S.D. Webb, M.R. Owen, H.M. Byrne, C. Murdoch and C.E. Lewis, *Bull Math Biol.* 69, 1747 (2007).
22. T. Alarcón, **H.M. Byrne** and P.K. Maini, *SIAM J. Multiscale Model. Simul.* **3**, 440 (2005).
23. MR Owen, T Alarcon, PK Maini and HM Byrne, *J Math Biol* **58**, 689 (2009).
24. C.E. Lewis and J.O'D. McGee, "The Macrophage," Oxford University Press, 1992.
25. R.D. Leek, R.J. Landers, A.L. Harris and C.E. Lewis, *J. Pathol.* **190**, 430 (2000).
26. D.R. Senger, C.A. Peruzzi, J. Feder and H.F. Dvorak, *Cancer Res.* **46**, 5629 (1986).
27. L. Griffith, K. Binley, S. Iqball, O. Kin, P. Maxwell, P. Ratcliffe, C. Lewis, A. Harris, S. Kingsman and S. Naylor, *Gene Ther.* **7**, 255 (2000).
28. H.M. Byrne, T. Alarcon, M.R. Owen, J. Murphy and P.K. Maini, *Math. Mod. Meth. Appl. Sci.* **16**: 1219 Suppl S (2006).
29. T. Alarcón, H.M. Byrne, P.K. Maini, *J. theor. Biol.* **225**, 257 (2003).
30. A.R.A. Anderson, *Math. Med. Biol.* **22**, 163 (2005).
31. M Muthana, SD Scott, N Farrow, F Morrow, C Murdoch, S Grubb, N Brown, J Dobson, CE Lewis. *Gene Ther* **15**, 902 (2008).

TUMOR CELLS PROLIFERATION AND MIGRATION UNDER THE INFLUENCE OF THEIR MICROENVIRONMENT *

A. FRIEDMAN AND Y. KIM

Mathematical Biosciences Institute,
Ohio State University,
Columbus, OH 43210, USA
E-mail: afriedman@math.ohio-state.edu

It is well known that tumor microenvironment affects tumor growth and metastasis: Tumor cells may proliferate at different rates and migrate at different patterns depending on the microenvironment in which they are embedded. There is a huge literature that deals with mathematical models of tumor growth and proliferation, in both the avascular and vascular phases. In particular, a review of the literature of avascular tumor growth (up to 2006) can be found in Lolas [1]. In this article we report on some of our recent work. We consider two aspects, of proliferation and of migration, and describe mathematical models based on *in vitro* experiments. Simulations of the models show tight fit with experimental results. The models can be used to generate hypotheses regarding the development of drugs which will confine tumor growth.

1. Tumor cells and fibroblasts in a transwell

Significant evidence exists that fibroblasts and myofibroblasts residing in the tumor microenvironment affect tumor cell proliferation. Recently, Samoszuk *et al.* [2] demonstrated the ability of fibroblasts to enhance the growth of a relatively small number of breast cancer cells *in vitro*. Yashiro *et al.* [3] also demonstrated that tumor size is significantly increased in mice when breast cancer cells are co-inoculated with breast fibroblasts. In other experiments it was shown that fibroblasts cultured from normal tissue tend to have inhibitory effects on cell growth, whereas fibroblasts cultured from tumors stimulate the growth of several cell types, including muscle cells, mammary carcinoma cells and myofibroblasts [4,5]. A transwell kit that is

*A. Friedman and Y. Kim are supported by the national science foundation upon agreement 112050.

used to explore tumor cells proliferation under the influence of fibroblasts is shown in Figure 1. The interaction between the TECs and the fibroblasts

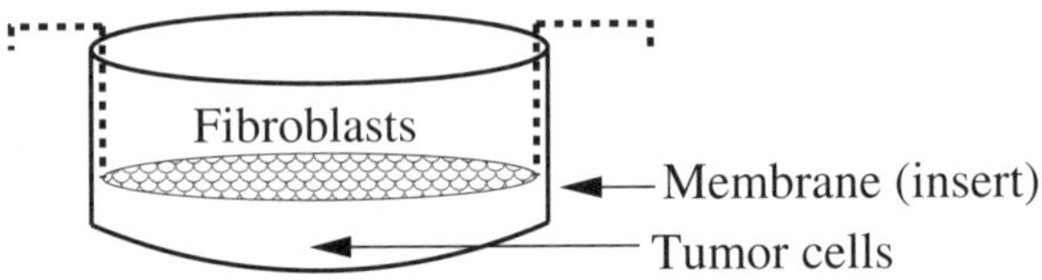

Figure 1. Structure of a transwell kit. Tumor epithelial cells (TECs) are deposited in the lower chamber and fibroblasts are deposited in the upper chamber. The two chambers are separated by a semi-permeable membrane.

is mediated by cytokines, namely, by epidermal growth factors (EGFs) produced by the TECs and transformed growth factor-β (TGF-β) produced by the fibroblasts. During a period of several days a large number of fibroblasts differentiate into myofibroblasts which secrete EGF at larger rates than fibroblasts. The cytokines can cross the membrane, but the cells cannot cross it. Figure 2 shows a schematic of the interaction between the cells and cytokines in a simple 2-d geometry. Experiments were conducted

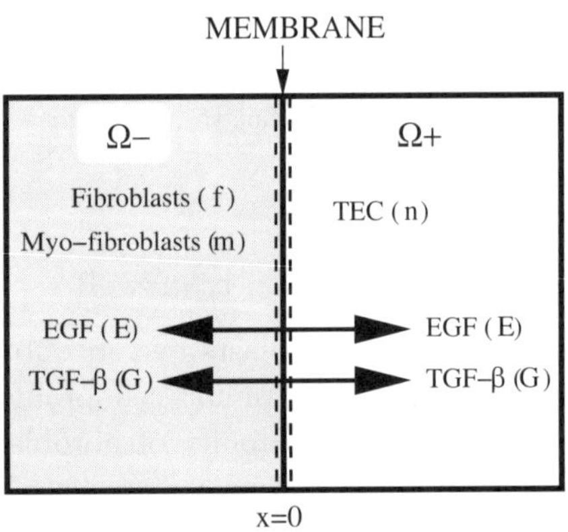

Figure 2. Schematics of the interactions across the membrane. While EGF and TGF-β can move across the semi-permeable membrane, tumor cells, fibroblasts, and myofibroblasts cannot cross the membrane.

by Kim *et al.* [6] using two kinds of fibroblasts: normal mammary fibroblasts and tumorigenic mammary fibroblasts. It was demonstrated that in the presence of tumorigenic fibroblasts, the TECs proliferated at a larger rate. A mathematical model, developed in Ref. [6], includes the following functions:

$$
\begin{aligned}
n &= \text{density of TEC,}\\
f &= \text{density of fibroblast,}\\
m &= \text{density of myofibroblast,}\\
E &= \text{concentration of EGF, and}\\
G &= \text{concentration of TGF-}\beta.
\end{aligned}
$$

Ignoring the vertical variable in Figure 2, these functions satisfy a system of partial differential equations in (x,t):

$$
\frac{\partial n}{\partial t} = \frac{\partial}{\partial x}\left(D_n \frac{\partial n}{\partial x}\right) - \underbrace{\frac{\partial}{\partial x}\left(\chi_n\, n\, \frac{\frac{\partial E}{\partial x}}{\sqrt{1+(\frac{\partial E}{\partial x}/\lambda_E)^2}}\right)}_{chemotaxis}
$$

$$
+ \underbrace{a_{11}\, \frac{E^4}{k_E^4 + E^4}\, n(1-n/\kappa)}_{proliferation}, \quad 0 < x < L/2, \tag{1}
$$

$$
\frac{\partial f}{\partial t} = \frac{\partial}{\partial x}\left(D_f \frac{\partial f}{\partial x}\right) - \underbrace{a_{21}\ G\ f}_{f\to m} + \underbrace{a_{22} f}_{proliferation}, \quad -L/2 < x < 0, \tag{2}
$$

$$
\frac{\partial m}{\partial t} = \frac{\partial}{\partial x}\left(D_m \frac{\partial m}{\partial x}\right) - \underbrace{\frac{\partial}{\partial x}\left(\chi_m\, m\, \frac{\frac{\partial G}{\partial x}}{\sqrt{1+(\frac{\partial G}{\partial x}/\lambda_G)^2}}\right)}_{chemotaxis} + \underbrace{a_{21}\ G\ f}_{f\to m}
$$

$$
+ \underbrace{a_{31}\ m}_{proliferation}, \quad -L/2 < x < 0, \tag{3}
$$

$$
\frac{\partial E}{\partial t} = \frac{\partial}{\partial x}\left(D_E \frac{\partial E}{\partial x}\right) + \underbrace{(a_{41}\ f + B a_{41}\ m)}_{production} - \underbrace{a_{43}\ E}_{decay},
$$

$$
-L/2 < x < L/2, \tag{4}
$$

$$
\frac{\partial G}{\partial t} = \frac{\partial}{\partial x}\left(D_G \frac{\partial G}{\partial x}\right) + \underbrace{a_{51}\ n}_{production} - \underbrace{a_{52}\ G}_{decay}, \quad -L/2 < x < L/2. \tag{5}
$$

The fact that the semi-permeable membrane allows EGF and TGF-β to cross over, but not cells, is represented mathematically by the following boundary conditions at the membrane x=0:

$$\frac{\partial n}{\partial x} = 0 \quad \text{at } x = 0+, \tag{6}$$
$$\frac{\partial f}{\partial x} = \frac{\partial f}{\partial x} = 0 \quad \text{at } x = 0-,$$

and

$$\frac{\partial E^+}{\partial x} = \frac{\partial E^-}{\partial x}, \quad -\frac{\partial E^+}{\partial x} + \gamma(E^+ - E^-) = 0, \tag{7}$$
$$\frac{\partial G^+}{\partial x} = \frac{\partial G^-}{\partial x}, \quad -\frac{\partial G^+}{\partial x} + \gamma(G^+ - G^-) = 0,$$

where

$$E(x) = \begin{cases} E^+(x) & \text{if } x > 0 \\ E^-(x) & \text{if } x < 0 \end{cases}, \qquad G(x) = \begin{cases} G^+(x) & \text{if } x > 0 \\ G^-(x) & \text{if } x < 0 \end{cases},$$

and γ is a positive parameter which is determined by the size and density of the holes in the membrane.

Simulations of the model show tight fit with experimental results, as seen, for example, in Figure 3.

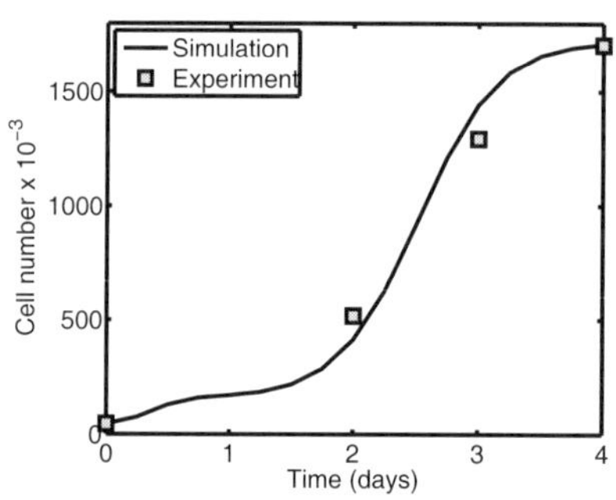

Figure 3. (a) Simulation results at day 4. Comparison of simulation results (solid curve) to experimental results (marked by squares) for F305 cell line.

2. Tumor cells and fibroblasts in invasion assay system

Invasion assay systems are used to study the influence of chemotactic and haptotactic forces on tumor cells. Figure 4 illustrates a Boyden chamber invasion assay that mimics tumor invasion *in vivo* [7]. A semi-permeable membrane separating the two chambers is coated with gel, or extracellular matrix (ECM), in order to represent *in vivo* situation of the basal membrane in mammary gland, for instance.

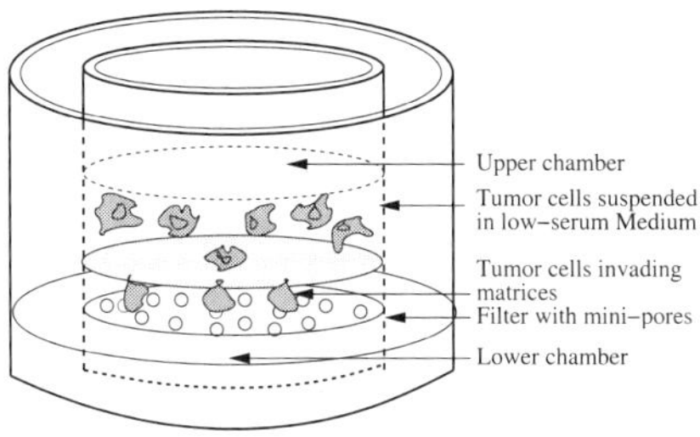

Figure 4. Illustration of a Boyden Chamber Invasion Assay that mimics tumor invasion *in vivo*

The mathematical model which describes tumor cells proliferation and migration now includes, in addition to the variables introduced in Section 1, also Matrix Metalloproteinase (MMP) secreted by the fibroblasts and myofibroblasts, and the ECM density. The model developed by Kim and Friedman [8] is based on the geometry shown in Figure 5. The ECM, which

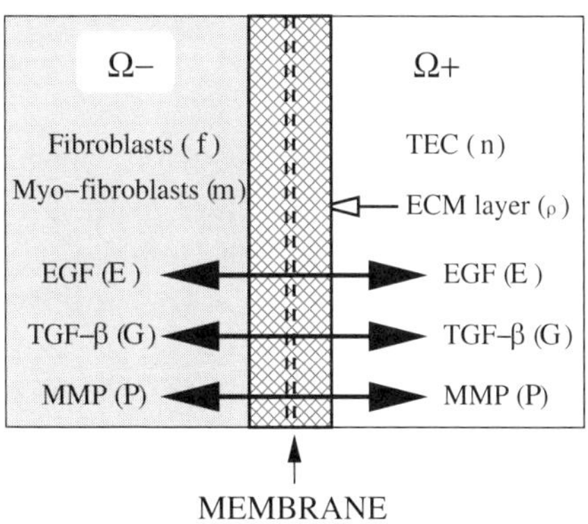

Figure 5. Schematics of an Invasion Assay System : EGF (E), TGF-β (G) and MMP (P) can cross the semi-permeable membrane, but the cells (TECs (n), fibroblasts (f), myofibroblasts (m)) may not cross it. Initially the TECs reside in the domain Ω_+ while fibroblasts and myofibroblasts are placed in the domain Ω_-. An ECM layer surrounds the semi-permeable membrane (filter).

is degraded by the TECs, gives rise to haptotaxis; chemotaxis arises from the gradient of the EGF and TGF-β. The mathematical model has also been extended, in Ref. [8], to the case where the membrane is permeable to cells.

Simulations of the model yield some interesting results:
(i) Figure 6(a) shows that when cells cannot cross the membrane, the TEC population (in the right chamber) has a biphasic dependence on the concentration of the gel. Figure 6(b) shows that when cells can cross the membrane and thus invade the left chamber, if the ECM concentration is increased in the left chamber Ω_-, then the population of TECs in the left chamber will increase. These results have been validated by experiments [9,10,11].

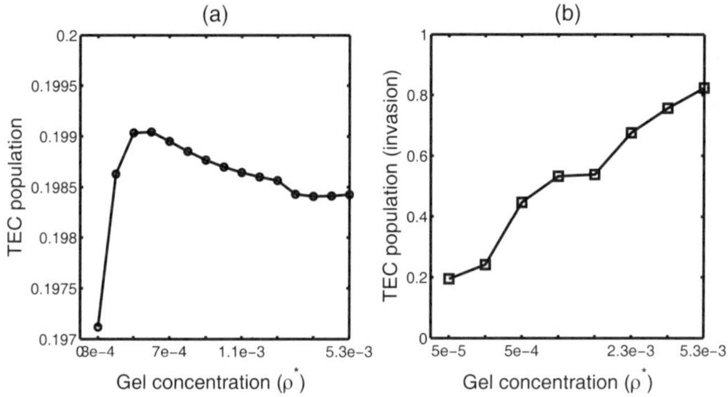

Figure 6. (a) Bifurcation diagram showing how total TEC population at $t = 13h$ depends on the gel concentration ρ^*. Biphasic dependence of the TEC population is seen. (b) Effect of gel concentration on TEC invasion: Total TEC population in the left chamber for different gel concentration in the left chamber and fixed gel concentration in the right chamber. As the gel concentration increases, the TEC population in the left chamber increases.

(ii) If we denote the width of the ECM layer by μ, then the total population of TECs is a decreasing function of μ for $\mu < \mu_0$ and an increasing function of μ for $\mu > \mu_0$, as shown in Figure 7 (Here cells can cross the membrane). This result, can be explained by the interaction between the haptotactic and chemotactic forces and the competition for space for TECs that are proliferating within the ECM. However, this result still needs to be validated by experiments.

The mathematical model can be used to make hypotheses regarding drugs that will block tumor growth. The model predicts that a drug which blocks the production of MMP by fibroblasts/myofibroblasts or the MMP

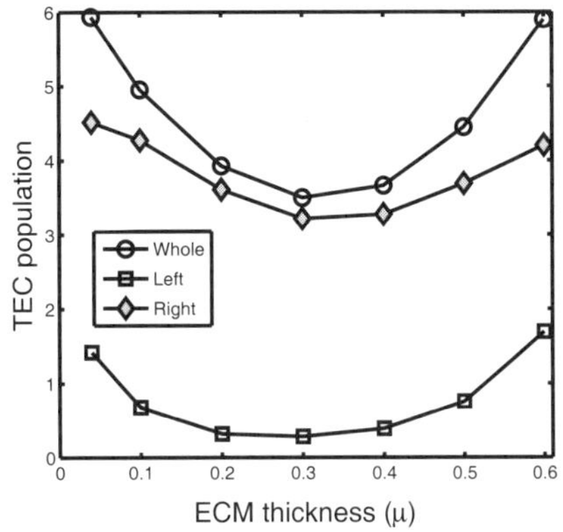

Figure 7. Effect of ECM coating (μ) on growth of TECs: Total population of TECs at day 4 in whole (circle), left (square), and right (filled diamond) chamber. The population of TECs has minimum values for the intermediate values of ECM thickness (μ).

activity will slow down tumor growth, as seen in Figure 8.

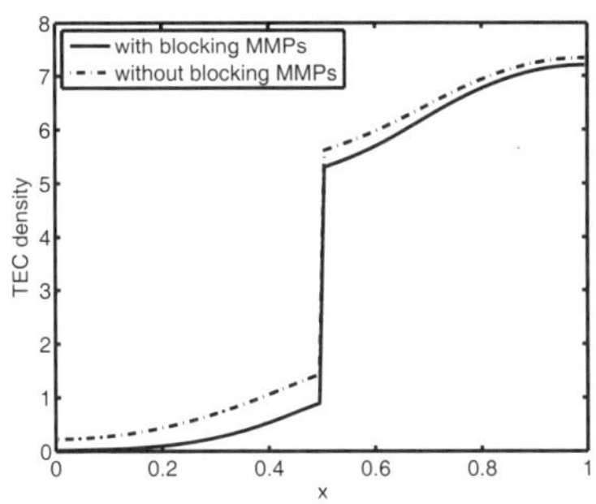

Figure 8. The growth of the TEC population where ECM degradation term is zero in comparison with its growth when MMP is not blocked. When proteolytic activity of TECs near membrane via MMPs is blocked, less cells are invading the left chamber. The filter is placed at x = 0.5.

Another strategy to stop or slow down tumor growth is to block the production of TGF-β by TECs. Figure 9 shows that proliferation and invasion are reduced by this procedure. The predictions described by Figures 8 and 9 need to be validated experimentally.

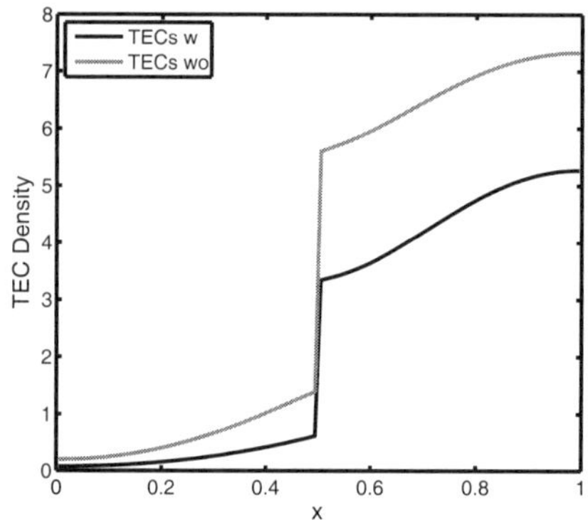

Figure 9. The growth of the TEC population at day 4 when TGF-β production is blocked or not blocked. Profile of TECs on domain x with (w) and without (wo) blocking TGF-β pathway.

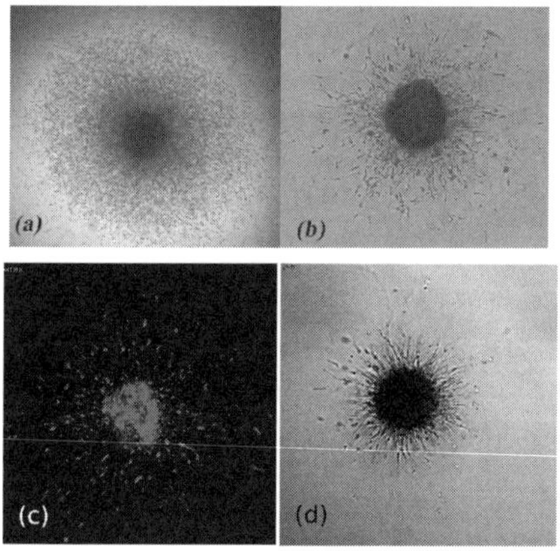

Figure 10. The four figures are taken from *in vitro* experiments with four different glioma lines: (a) U87, (b) U87ΔEGFR (a mutant of U87); (c) U87STM (d) X12RFP All cells shown were implanted into a type I collagen matrix and grown in DMEM containing 10% fetal calf serum and 4.5 g/liter glucose. Figures 10(a) and 10(b) were reprinted with permission from E. Khain and L.M. Sander, Dynamics and pattern formation in invasive tumor growth, Phys Rev Lett, 96, 188103 (2006). Copyright(2006) by the American Physical Society Figure 10(c)-(d) are replicated from experiments conducted by the co-authors Sean Lawler, Michal O. Nowicki of [17] in E. Antonio Chiocca's lab.

3. Patterns of migration

The pattern of cell migration is important for predicting metastasis. This is particularly important in the case of an aggressive cancer like glioblas-

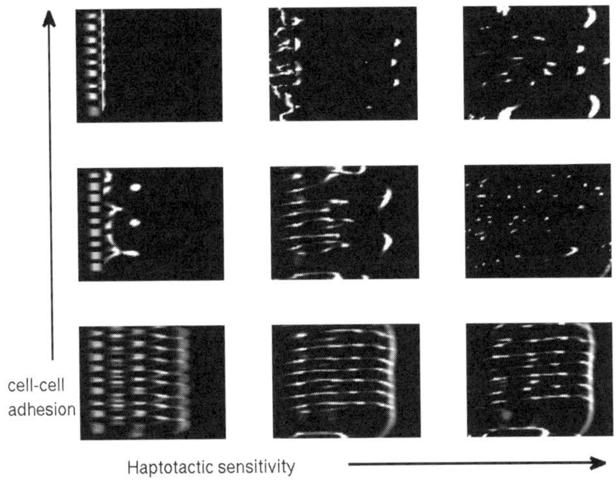

Figure 11. Effect of cell-cell adhesion (λ_a) and haptotactic sensitivity (χ_n^1) at a terminal time T. Cells are shedded from the left side of the frames in the first column. As cell-cell adhesion increases, migration is slowed down and, as it decreases, branching patterns appear.

toma. A major reason for treatment failure of glioblastoma is that by the time the disease is diagnosed tumor cells have already migrated from the primary tumor into other parts of the brain. It is therefore important to predict the migration pattern of the invasive cells; such detailed predictions are not known at this time. The migration of tumor cells from the primary tumor depends both on the tumor cell line and on the tumor microenvironment. Experimental results show various migration patterns of glioma, including isolated islands, branching, and dispersion near the tumor boundary as well as at some distance from the primary tumor. Some patterns of glioma cell migration seen in experiments are shown in Figure 10. Figures 10(b) and 10(d) show branching pattern whereas frame (a) shows a pattern of dispersion; (c) is a somewhat intermediate case between branching and dispersion. Sander and Deisboeck [12], Khain and Sander [13] and Stein *et al.* [14] developed mathematical models, based on PDEs, in attempts to capture such migration patterns. There is also some work based on a single diffusion equation with coefficients that depend on geometric and physical features of the brain[15,16].

Here we report on recent work by Kim *et al.* [17] which includes the chemotactic and haptotactic forces as well as cell-to-cell adhesion.

The model consists of a system of PDEs for the following variables: density of glioma cells and concentrations of ECM, MMP and nutrients (glucose). The model assumes that glioma cells shed off from the surface of the tumor, and tracks down the density of the cells for a period of time, for any particular choice of the following three parameters: chemotactic sensitivity (χ_n) of cells moving in the direction of the gradient of glucose, haptotactic sensitivity (χ_n^1) of cells moving in the direction of gradient of ECM concentration, and cell-cell adhesion force λ_a. The simulation of the model for a simple rectangular geometry for different choices of χ_n^1 and λ_a are shown in Figure 11.

We can use the model to develop testable hypotheses for slowing down glioma cells migration. One hypothesis is that increase in cell-cell adhesion will slow down tumor migration. This, in fact, is also suggested by experiments conducted *in vitro* and *in vivo* by Asano *et al.* [18]. Another way to slow migration is by blocking the effect of MMP, which can be achieved, for example, by viral transduction of siRNA, or some chemical inhibitors. We note that blocking the activity of MMP in order to slow tumor proliferation and migration was also suggested by the simulations in Section 2.

The model in [17] assumes that the tumor is spherical and that its microenvironment is initially homogeneous. However any inhomogeneities, which occur naturally in the brain, may result in significant changes in the pattern of cell migration. Modeling of migration patterns in the presence of such inhomogeneities remains an interesting and important problem.

Acknowledgments

This work was supported by NSF/DMS upon agreement 112050.

References

1. G. Lolas, *Lecture Notes in Mathematics, Springer Berlin / Heidelberg*, **1872**, 77 (2006).
2. M. Samoszuk, J. Tan and G. Chorn, *Breast Cancer Res*, **7**, R274 (2005).
3. M. Yashiro, K. Ikeda, M. Tendo, T. Ishikawa and K. Hirakawa, *Breast Cancer Res Treat*, **90**, 307 (2005).
4. N.A. Bhowmick, E.G. Neilson and H.L. Moses, *Nature*, **432**, 332 (2004).
5. M.M. Mueller and N.E. Fusenig, *Nat Rev Cancer*, **4**, 839 (2004).
6. Y. Kim, A. Friedman, J. Wallace, F. Li and M. Ostrowski, *J Math Biol*, submitted (2009).
7. K. Yuan, R.K. Singh, G. Rezonzew and G.P. Siegal, *Cell Motility in Cancer Invasion and Metastasis, Springer, Netherlands*, **25**, 25 (2006).
8. Y. Kim and A. Friedman, *Bull Math Biol*, submitted, (2009).

9. S. Aznavoorian, M.L. Stracke, H. Krutzsch, E. Schiffmann and L.A. Liotta, *J Cell Biol*, **110**, 1427 (1990).
10. A.J. Perumpanani and H.M. Byrne, *Eur J Cancer*, **35**, 1274 (1999).
11. J. A. Sherratt and J. D. Murray, *Proc.R. Soc.Lond.*, **B241**, 29 (1990).
12. L.M. Sander and T.S. Deisboeck, *Phys. Rev. E*, **66**, 051901 (2002).
13. E. Khain and L.M. Sander, *Phys Rev Lett*, **96**, 188103 (2006).
14. A.M. Stein, T. Demuth, D. Mobley, M. Berens and L.M. Sander, *Biophys J*, **92**, 356 (2007).
15. K.R. Swanson, E.C. Alvord and J.D. Murray, *Cell Prolif*, **33**, 317 (2000).
16. K.R. Swanson, E.C. Alvord and J.D. Murray, *Math Comp Modelling*, **37**, 1177 (2003).
17. Y. Kim, S. Lawler, M.O. Nowicki, E.A Chiocca and A. Friedman, *J Theo Biol*, submitted (2009).
18. K. Asano, C.D. Duntsch, Q. Zhou, J.D. Weimar, D. Bordelon, J.H. Robertson and T. Pourmotabbed, *J Neurooncol*, **70**, 3 (2004).

PHENOMENOLOGICAL STUDY OF THE GROWTH RATE OF TRANSFORMED VERO CELLS, CHANGES IN THE GROWTH MODE AND FRACTAL BEHAVIOUR OF COLONY CONTOURS

M. A. C. HUERGO, M. A. PASQUALE, A. E. BOLZÁN, A. J. ARVIA

Instituto de Investigaciones Fisicoquímicas Teóricas y Aplicadas (INIFTA), (UNLP, CONICET), Sucursal 4, Casilla de Correo 16, (1900) La Plata, Argentina

E-mail: `mac.huergo@gmail.com`

P. H. GONZÁLEZ

Cátedra de Patología, Facultad de Ciencias Médicas, Universidad Nacional de La Plata, La Plata, Argentina

The 2D growth of Vero cell colonies grown from both quasi-straight-edges and radial expanding fronts is investigated. The latter are followed starting from either a 3D cluster of cells or from a few basal cells. The quasi-straight-edge strategy is first selected for the dynamic scaling of rough growth fronts to derive the critical exponents related to roughness characteristics and dynamics of the biological processes at the interface. In the range $0 \leq t \leq 12000$ min, the average unidirectional colony front propagation velocity is $v_h = 0.24 \pm 0.04$ $\mu\mathrm{m\,s}^{-1}$. The fractal dimension of propagation front, determined by the box-counting method, is $D_F = 1.3 \pm 0.1$, and the growth exponent from the dynamic scaling analysis is $\beta = 0.50 \pm 0.05$ without roughness saturation. From the dynamic behaviour of growth patterns, changes in the order of cell domains related to cell size and shape modifications are observed. At constant growth front velocity, comparable results are obtained from radially expanding growth fronts of confluent cell colonies. The Vero cell colony growth dynamics is likely determined by the random birth of cells at the interface, a process that is compatible with a random distribution of cells at different cell-cycle stage, yielding non-equilibrium colony contours.

1. Introduction

In recent years a number of papers dealing with neoplasic tumour growth and their therapies have been published[1,2]. Some of them have been devoted to attempting to classify tumour growth patterns in terms of generic mechanisms of single cell dynamics at the growing interface. This type of work has been approached utilising experimental procedures and model simulations of typical phenomena in which the moving interface can be characterised by the interface fluctuations. This approach involves obtaining data about the interface front features, and to process them applying dynamic scaling to obtain the critical exponents involved in the equation[3,4,5],

$$w(L,t) = t^{\alpha/z} f(L/t^{1/z}) \tag{1}$$

$f(y)$ being the scaling function and $y = L/t^{1/z}$. For $y \ll 1$ it results in $y = y^{\alpha}$, and for $y \gg 1$, $f(y)$ is a constant. The global roughness exponent α and the dynamic exponent $z = \alpha/\beta$, β being the growth exponent, determine the universality class of the growing interface dynamics. Equation (1) can also be utilised to describe the local scaling behaviour of the interface width,

$$w(\Delta,t) = \left[\langle [h(x_i) - \langle h \rangle_\Delta]^2 \rangle_\Delta\right]^{1/2} = t^{\zeta/z} f(\Delta, t^{1/2}) \tag{2}$$

where $\langle \cdots \rangle_\Delta$ corresponds to an average value over x in a size window Δ. When the system exhibits no characteristic length, the local and global behaviours of the interface are characterised by the same scaling exponents. However, the local roughness exponent ζ is either less or equal to the global one, i.e. $\zeta \leq \alpha$[3,4,5,6].

On the basis of the critical exponents, a mechanism of cell colony growth, both in vitro and in vivo, has been proposed and extensively discussed [7,8]. This situation encouraged us to develop a new method to follow cell colony growth from a quasi-straight-line front in order to facilitate the interpretation of quantitative data derived from the dynamic scaling analysis utilising standard growth models. In doing so, we also found that the growth process involves changes in the shape and size of cells in the growing colony, particularly at the interfacial region. Our study is made on quasi-2D Vero cell patterns from quasi-straight-line and radial fronts to demonstrate the evolution of both the cell colony growth front and the cell colony structure. These cells were chosen because, unlike primary cells, they continue growing and dividing indefinitely in vitro as long as the adequate culture conditions are maintained. Vero cells, although not being tumourigenic, exhibit either a null or almost null contact inhibition property, i.e. they behave like cells of typical neoplasic tumoural tissues.

The Vero cell colony quasi-2D growth in a constant composition medium and under constant linear and radial growth front velocity is characterised by critical exponents derived from the dynamic scaling analysis indicating that the cell colony growth at the moving interface would be related to the universality class of stochastic processes[3], i.e., it would involve the stochastic duplication of cells at the interface that would result from the random distribution of cells at different cell-cycle stage.

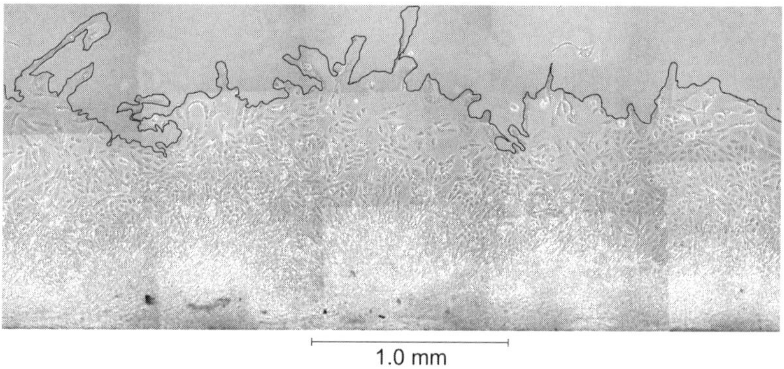

Figure 1. Pattern of a cell colony growth that started from a quasi-straight-line edge. The colony contour is shown by the continuous black trace. The total image of the colony is made by stitching partial images.

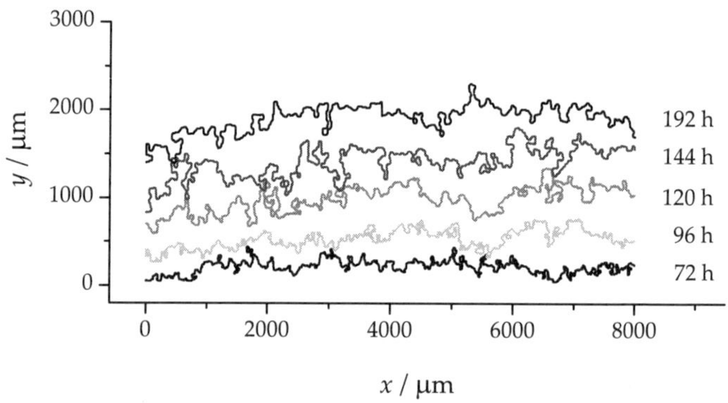

Figure 2. Evolution of a cell colony contour starting from a quasi-straight-line edge. Culture times are indicated.

2. Materials and Methods

2.1. *Cultures preparation*

Two types of quasi-2D colony cell growth experiments were carried out, utilising either initially quasi-straight-line or radial expanding fronts. Vero cell colonies are cultured in a RPMI 1640 medium containing 10 % fetal bovine serum (FBS). They are maintained in a 5 % carbon dioxide controlled atmosphere at 37 C changing one half of the medium every two days. An inverted phase-contrast Nikon TS 100 microscope coupled to a Canon Powershot G5 camera is utilised to follow the evolution of cell colonies. The distance covered by the growing front at each point of the contour, their average value, the roughness, and their fractal dimension (D_F) are obtained by computational analysis of each image employing conventional programs. For large cell colonies partial images of each growth pattern are stitched to obtain the entire propagation front.

2.2. *Growth from initially quasi-straight-line edges*

The growth of initially quasi-straight-line 2D cultures is followed by two experimental procedures. Procedure I consists in: i) seeding the cells on a 10×5 mm^2 rectangular glass plate 100 μm thick located on Petri dishes; ii) after reaching 100 percent cell confluence, the glass is transferred to a second Petri dish and 2 ml fresh medium is added; iii) the cell colony starts to grow from the thin glass plate edges and each colony contour is photographed daily. Procedure II involves a glass with 100 percent cell confluence as in Procedure I that is placed upside down in a Petri dish with fresh medium, so that the quasi-straight-line colony growth starts with cells at the same dish base level. The snapshots at the μm scale of colonies grown according to Procedures I and II at different times are considered to follow the growth front displacement. For colonies prepared according to Procedure I, lag times of several days are obtained, in contrast to lag time of a few hours of those from Procedure II. The large lag times in Procedure I are related to the time required for cells to slide down the glass plate thickness to reach its growth edge at the base of the Petri dish.

2.3. *Growth from radially expanding regimes*

Two different procedures are employed to obtain cell colonies expanding in a quasi-radial geometry. Procedure III: colonies are formed from 2 ml of medium containing 3000-5000 disaggregated cells/ml, poured into 3.6

cm-diameter Petri dishes waiting for a day to ensure cell sticking to the dish base. Two types of cell colonies are distinguished: those consisting of a small number of cells (less than 20) with a poor confluence, and those with a large number of cells in a confluent colony. Procedure IV: 3D-cell aggregates are grown on a culture initially made on a glass plate for two weeks. Then, aggregates are taken with a micro-pipette and placed into a Petri dish and covered with fresh medium. After 24 h, the radial 2D growth developed around the aggregate is followed as described above.

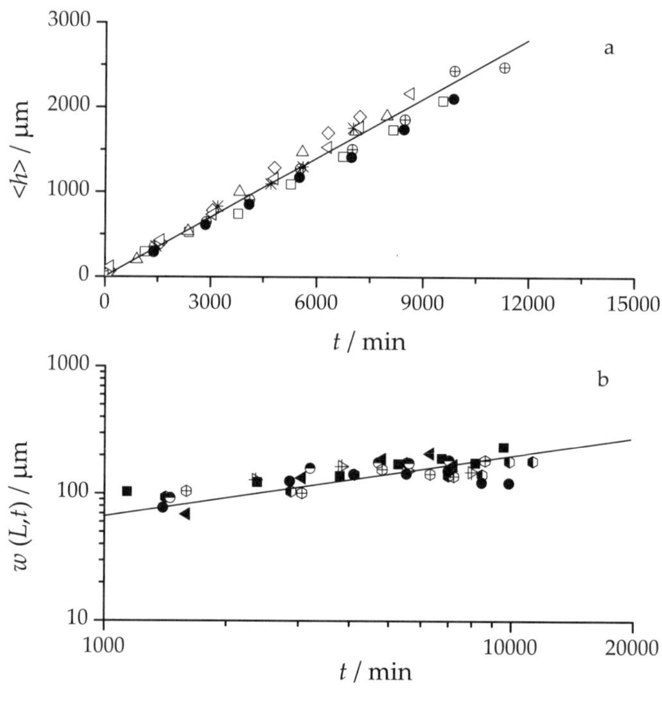

Figure 3. (a) Plot of $\langle h \rangle$ versus culture time (t) for different Vero cell colonies grown from a quasi-straight-line edge. (b) Log $w(L,t)$ versus log t plot for different Vero cell colonies grown from a quasi-straight-line edge.

3. Results and discussion

3.1. *Quasi-straight-line growth fronts*

The growth of Vero cell colonies from a quasi-straight-line edge (Figure 1) is followed in the range $0 \leq t \leq 12000$ min and the corresponding growth front is evaluated by processing the images data. The evolution of the rough growth front is depicted in Figure 2.

A linear dependence of the average distance covered by the growing front $\langle h \rangle$ versus t is obtained (Figure 3a). Data plotted in this Figure is corrected for the lag time. From the slope of this plot, the growth front velocity results in $v_h = 0.24 \pm 0.02$ μm min^{-1}, irrespective of the value of L.

Fractal analysis of the colony growth fronts utilising the box-counting method yields a fractal dimension $D_F = 1.3 \pm 0.1$. Dynamic scaling applied to the rough contours depicted in Figure 2 gives rise to a linear log $w(L,t)$ versus log t plot with a slope $\beta = 0.50 \pm 0.05$ (Figure 3b). Despite data scattering, no saturation roughness could be determined, at least under the conditions of our work, i.e., $\alpha \approx 0$.

3.2. *Radially expanding growth fronts*

Photographs of quasi-2D radial Vero cell colonies consisting of a small ($\leq$ 20) and a large (about 100) number of cells are shown in Figure 4. The evolution of rough quasi-radial 2D colony growth front (Figure 5) is followed by plotting the mean radius of the colony ($\langle r \rangle$) versus time (Figure 6a). These plots approach an exponential increase of $\langle r \rangle$ with time over the time range $0 \leq t \leq 12000$ min. The equation of these plots is of the form

$$r = Ae^{kt} \tag{3}$$

where $k = 1.2 \times 10^{-4}$ min^{-1}, irrespective of the colony size. This figure would express the rate constant of a relaxation process at the colony growth front, and the value of A would represent the radius of the colony above which the front can be reasonably assumed as a circle of mean radius A.

Fractal analysis of quasi-radial colony fronts utilising the box-counting method yields $D_F = 1.30 \pm 0.05$. Furthermore, dynamic scaling applied to rough contours gives rise to a linear log $w(L,t)$ versus log t plots. For aggregates initially made with a small number of cells, it results in $\beta = 0.70 \pm 0.05$ (Figure 6b), whereas for those large aggregates, $\beta = 0.50 \pm 0.05$ (Figure 6c). Concerning roughness saturation these plots are consistent with results

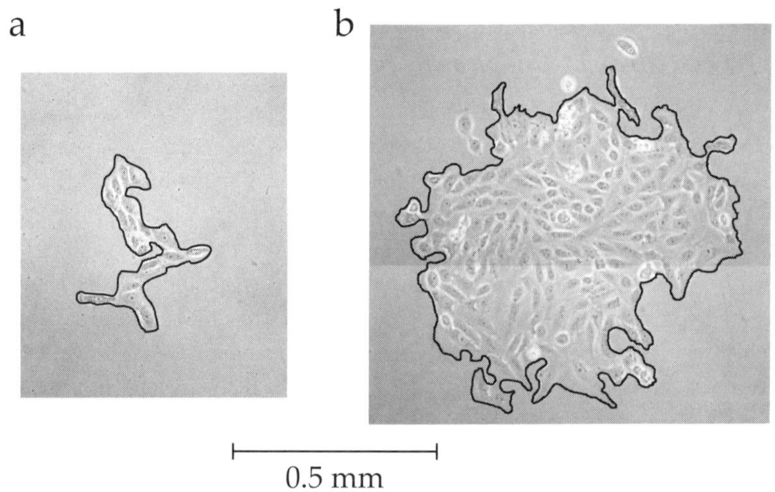

Figure 4. Pattern of cell colonies consisting of a few cells (a) and about 100 cells (b). The radially grown colony contour is shown by the continuous black trace. The total image of the colony is made by stitching partial images.

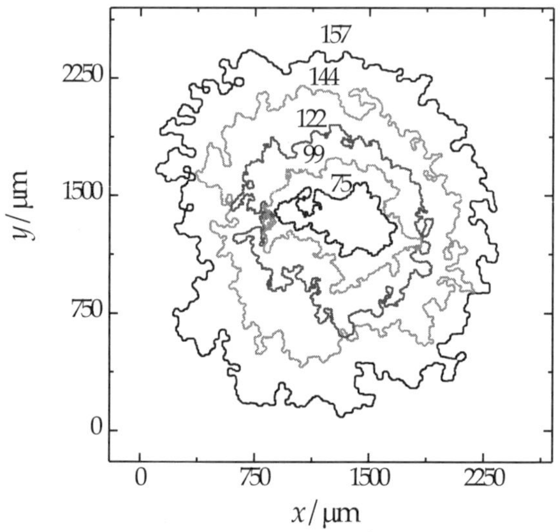

Figure 5. Evolution of a radial Vero cell colony contour. The numbers in the countour indicates the growth time in hours.

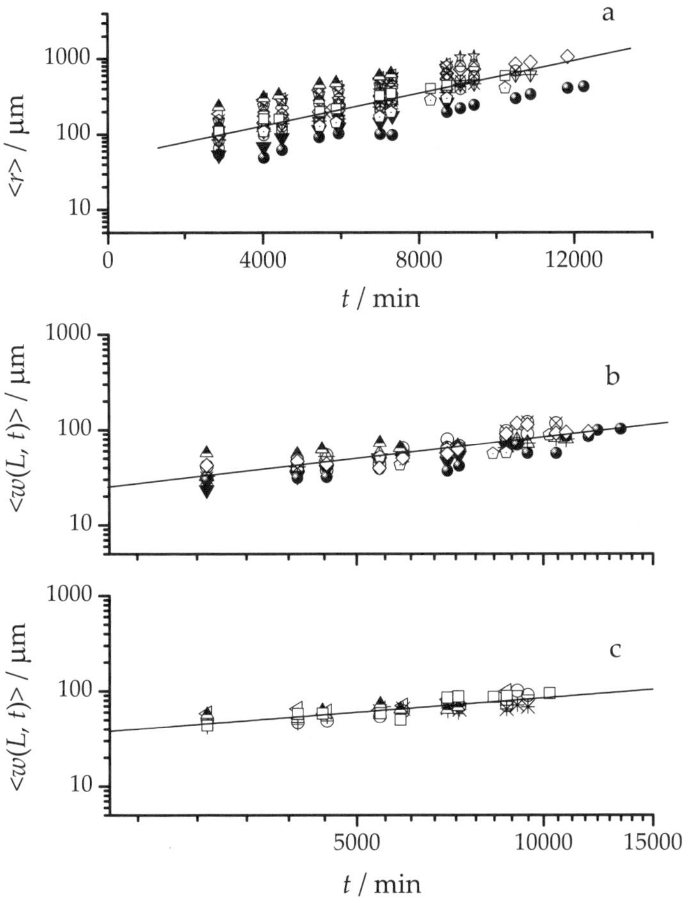

Figure 6. (a) Plot of $\langle r \rangle$ versus culture time (t) from radially expanding fronts of different Vero cell colonies. (b,c) Log $w(L, t)$ versus log t plots from radially expanding fronts for different Vero cell colonies. (b) Initially cell clusters containing less than 20 cells; (c) cell colonies consisting of more than 30 cells.

above described in 3.1. These results contrast with those reported in the literature from radially grown colonies[1,2], but agree with our above-described results from quasi-straight-line edge experiments.

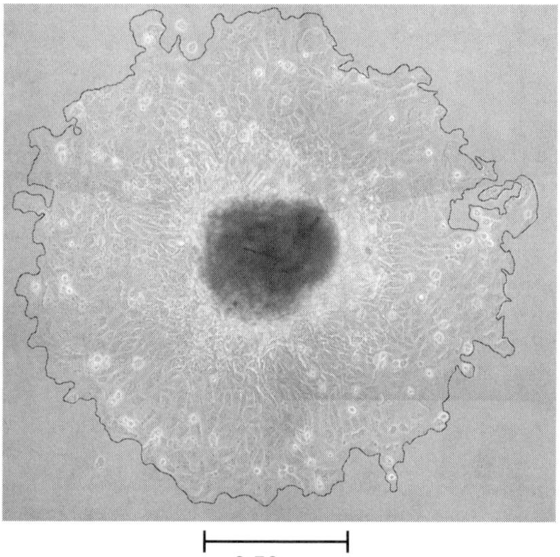

Figure 7. Pattern of a cell colony growth starting from a 3D cluster of Vero cells. The colony contour is shown by the continuous black trace. The total image of the colony is made by stitching partial images.

3.3. *Radially expanding quasi-2D growth fronts from 3D cell clusters*

The quasi-2D colony growth is also studied from a 3D cell cluster grown on glass plate for at least two weeks. This cluster is placed in a Petri dish to follow its 2D growth pattern (Figures 7 and 8). In this case the cell colony expansion produces a quasi-2D layer, and the evolution of the radial growth front yields a linear increase of the $\langle r \rangle$ with time in the range $1000 \leq t \leq 10000$ min (Figure 9a). The slope of this plot is the same as that observed in Figure 3, i.e., a constant $v_r = 0.24 \pm 0.04$ is established. The fractal dimension of the rough front is $D_F = 1.3 \pm 0.1$ and the dynamic scaling applied to the radial colony contours yields $\beta = 0.50 \pm 0.05$ (Figure 9b) and $\alpha = 0$, as for those runs started from a quasi-straight-line edge. The application of the dynamic scaling to the radial object has been justified in references[9,10,11,12].

Above results contrast with those reported in the literature for the same biological system in which a saturation roughness and a global value $\alpha = 1.50$ have been reported[2]. It should be noted that despite this important difference, the value of v_r, as indicated above, and the fractal dimension

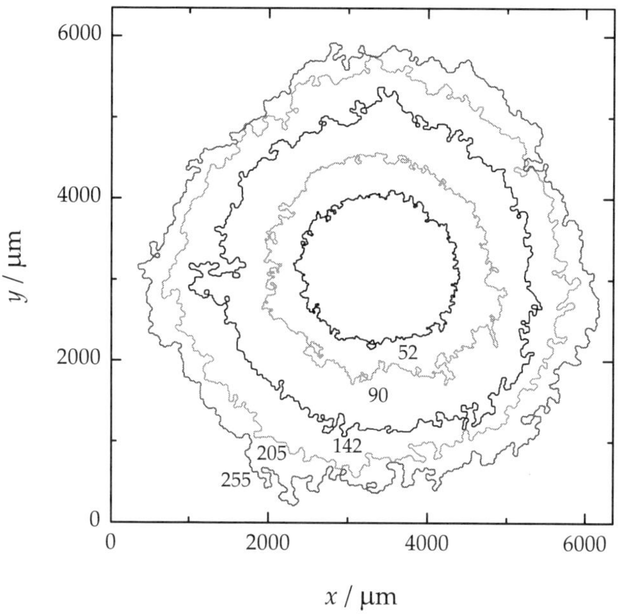

Figure 8. Evolution of a cell colony contour starting from a 3D cluster of Vero cells. The numbers in the countour indicates the growth time in hours.

are close to those given in previous work[2].

The fact that no influence of the geometry on the experiment data is observed suggests that the dynamics of the biological system is dominated by a random process that takes place at the level of individual cells at the interface, as has recently been proposed in a mesoscopic model for tumour growth[13].

3.4. *Changes in cell morphology*

For $t < 50$ h, the colony fronts consist of a quasi-2D layer of cells with almost the same average size and shape (Figure 10a). At longer time, i.e. $t > 96$ h, a relatively small amount of larger, multinuclear cells with philopodia located close to the colony border are formed (Figure 10b). At this stage, due to the change in the cell size and shape, the overall colony be-

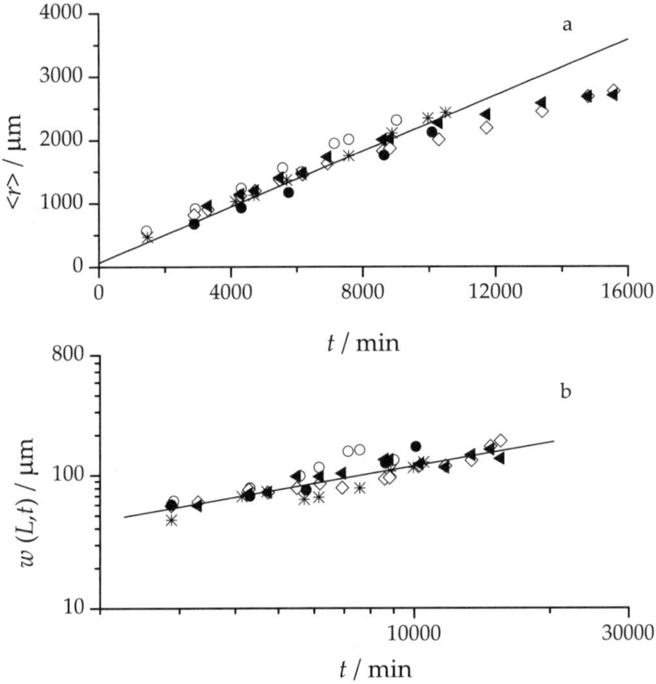

Figure 9. (a) Plot of $\langle r \rangle$ versus culture time (t) from radially expanding 2D fronts for different Vero cell colonies grown from 3D cluster. The departure of data from the linear relationship for $t >$ 10000 min can be related to the increasing density number of large polynuclear cells at the interface region. The appearance of these cells hinders the growth front velocity. (b) Log $w(r, t)$ versus log t plot from radially expanding 2D fronts for different Vero cell colonies grown from 3D clusters.

comes more heterogeneous. The cell size distribution of the colony changes as the colony grows, as concluded from the ordered-disordered cell domains ratio. This size and shape anisotropy of constituents becomes a drawback for making a reliable interpretation of experimental data considering standard growth model predictions in which a single constituent is involved. This makes that those results obtained from dynamic scaling of biological systems be handled with caution[8]. On the other hand, Equation (1) has been derived for a strictly self-affine geometry without overhangs, a situation that seldom occurs in cell cultures where μm size single cell overhangs

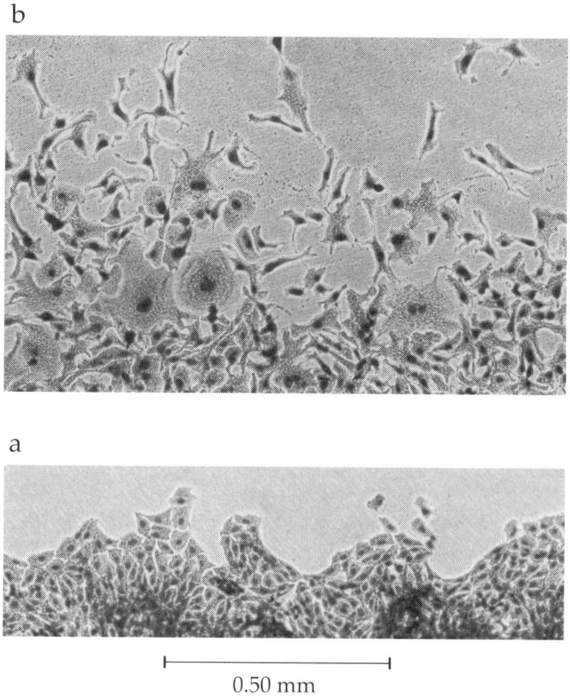

Figure 10. (a) Domain of a Vero cell growth pattern at the initial stage of the process constituted by cells with uniform size and shape. (b) Domain of a growth pattern in which the formation of polynuclear cells with some filopodia close to the interface are observed.

are produced. However, despite these observation, the dynamic scaling approach applied to Vero cell colony growth reveal that the evaluation of $w(L, t)$ is scarcely influenced by the presence of overhangs.

4. Conclusions

- Experimental data resulting from both the initially quasi-straight-line edge and the confluent radial system are comparable. For both systems $\alpha \approx 0$ and $\beta = 0.50 \pm 0.05$ are obtained. These figures are consistent with a stochastic growth process at the mobile interface under non-equilibrium roughness.
- The tentative explanation of the above data obtained by the dynamic scaling of interfacial contours is compatible with a random distribution of cellular division there. This interpretation could be

related to the random distribution of cells at different cell-cycle stage at the colony interface.

- The occurrence of changes in the cell morphology along the colony growth, particularly for domains close to the interface, constitutes a new variable to be considered in revising growth models applicable to biological systems.
- As a consequence of the present work, different transformed cell lines are currently being studied to evaluate the universality of the present approach.

Acknowledgments

This work was financially supported by PICT 34530 from Agencia de Promoción Científica y Técnica. Authors thank the Consejo Nacional de Investigaciones Científicas y Técnicas (CONICET, Argentina) and the Comisión de Investigaciones Científicas de la Provincia de Buenos Aires (CIC).

References

1. A. Brú, J. M. Pastor, I. Fernaud, I. Brú, S. Melle, C. Berenguer, *Phys. Rev. Lett.* **81**, 4008 (1998).
2. A. Brú, S. Albertos, J. L. Subiza, J. L. García-Asenjo, I. Brú, Biophys. J. **85**, 2948 (2003).
3. A. L. Barabasi, H. E. Stanley, *Fractal Concepts in Surface Growth*, Cambridge University Press, 1995.
4. P. Meakin, *Fractals Scaling and Growth Far From Equilibrium*, Cambridge University Press, London, 1998.
5. J. J. Ramasco, J. M. López, M. A. Rodríguez, *Phys. Rev. Lett.* **84**, 2199 (2000).
6. C. Guiot, P. P. Delsanto, A. Carpinteri, N. Pugno, Y. Mansury, T. S. Deisboeck, *J. Theor. Biol.* **240**, 459 (2006).
7. J. Buceta, J. Galeano, *Biophys. J.* **88**, 3734 (2005).
8. M. Block, E. Schöll, D. Drasdo, *Phys. Rev. Lett.* **99**, 248101 (2007) .
9. J. Krug, H. Spohen, in *Solids Far From Equilibrium*, C. Godrèche (editor), Cambridge University Press, Cambridge (England), 1991.
10. C. Escudero, *Phys. Rev. Lett.* **100**, 116101 (2008).
11. J. Krug, *Phys. Rev. Lett.* **102**, 139601 (2009).
12. C. Escudero, *Phys. Rev. Lett.* **102**, 139602 (2009).
13. E. Izquierdo-Kulich, J. M. Nieto-Villar, *Math. Biosci. Engin.* **4**, 687 (2007).

EVIDENCE OF DETERMINISTIC EVOLUTION IN THE IMMUNOLOGICAL MEMORY PROCESS *

A. DE CASTRO[†‡]

C. F. FRONZA[‡]

[†]*Embrapa Informática Agropecuária, Empresa Brasileira de Pesquisa Agropecuária - EMBRAPA, Campinas, 13083-886, Brazil*

[‡]*Departamento de Informática em Saúde, Universidade Federal de São Paulo - UNIFESP São Paulo, 04023-062, Brazil*
E-mail: acastro@cnptia.embrapa.br
E-mail: c.fronza@unifesp.br

R. H. HERAI

Instituto de Biologia, Universidade Estadual de Campinas - UNICAMP, Campinas, 13083-970, Brazil
E-mail: rherai@yahoo.com

D. ALVES

Departamento de Medicina Social, Faculdade de Medicina de Ribeirão Preto, Universidade de São Paulo - USP, Ribeirão Preto, 14049-900, Brazil
E-mail: quiron@fmrp.usp.br

In this paper, we study the behavior of immune memory against antigenic mutation. Using a dynamic model proposed by one of the authors in a previous study, we have performed simulations of several inoculations, where in each virtual sample the viral population undergo mutations. Our results suggest that the sustainability of the immunizations is dependent on viral variability and that the memory lifetimes are not random, what condradicts what was suggested by Tarlinton *et al.*. We show that what may cause an apparent random behavior of the immune memory is the antigenic variability.

*This work is supported by brazilian agricultural research corporation (embrapa).

1. Introduction

In recent decades, models have been widely used to describe biological systems, mainly to investigate global behaviors generated by cooperative and collective behavior of the components of these systems. More recently, several models were developed to study the dynamics of immune responses, with the purpose of comparing the results obtained by simulations, with the experimental results, so the connection between the existing theories on the functioning of the system and the available results can be adequately established[1,2,3,4,5].

From an immunological point of view, these models contribute to a better understanding of cooperative phenomena, as well as they lead to better understanding of the dynamics of systems out of equilibrium. Currently, the existing models to simulate the behavior of the immune system have been mainly based on differential equations, cellular automata and coupled maps. Processes and mechanisms of the natural immune system are being increasingly used for the development of new computational tools. However, the mathematical formalization of the functioning of the immune system is essential to reproduce, with computer systems, some of its key biological characteristics and skills, such as the ability of pattern recognition, information processing, adaptation, learning, memory, self-organization and cognition . Researchers, inspired by the intelligent techniques of recognition and elimination used by white blood cells, are planning a new generation of antivirus, with the purpose of searching in the biological systems the solutions to carry out strategic attacks, which may be the transforming elements of future technologies.

Within this scenario, in 2000, Lagreca *et al.*[6] proposed a model that uses techniques of multi-spin coding and iterative solution of equations of evolution (coupled maps), allowing the global processing of a system of higher dimensions. The authors showed that the model is capable of storing the information of foreign antigens to which the immune system has been previously exposed. However, the results obtained by these authors for the temporal evolution of clones, include only the B cells, not taking into account the antibodies population soluble in the blood. In 2006, one of the present authors has proposed[7,8,9] an extension of the Lagreca model, including the populations of antibodies. With this assumption, we considered not only the immunoglobulins attached to the surfaces of the B cells, but also the antibodies scattered in serum, that is, the temporal evolution of the populations of secreted antibodies is considered, to simulate the role in mediation of the global control of cell differentiation and of immunological memory[6,7,8,9]. In that work, our approach showed that the soluble antibodies alter the global properties of the network, diminishing the memory capacity of the system. According to our model, the absence or reduction of the production of antibodies favors the global maintenance of the immunizations. This result contrasts with the results obtained by Johansson and Lycke[10], who stated that the antibodies do not affect the maintenance of immunological memory.

Without considering terms of antigen mutation, this same extension[7,8,9] also led us to suggest a total randomicity for the memory lifetime, in order to maintain homeostasis of the system[11,12,13,14]. This random behavior was also recently proposed by Tarlinton *et al.*[14]. In our earlier work[7,8,9], the results indicate that, in order to keep the equilibrium of the immune system, some populations of memory B cells must vanish so that others are raised and this process seemed completely random. However, the results shown in this study suggest that the durability of immunological memory and the *raised-vanished* process is strongly dependent on the variability of viral populations. Thus, the lifetimes of immune memory populations are not random, only the antigenic variability from which they depend upon is a random feature, resulting in an apparent randomicity to the lifespan of B memory cells.

2. Extended model and methodology

To perform the simulations, we use the same mathematical model presented in BIOMAT 2008 – International Symposium on Mathematical and Computational Biology[15]. In that model, the molecular receptors of B cells are represented by bit-strings with diversity of 2^B, where B is the number of bits in the string[16,17]. The individual components of the immune system represented in the model are the B cells, the antibodies and the antigens. The B cells (clones) are characterized by its surface receptor and modeled by a binary string. The epitopes –portions of an antigen that can be connected by the B cell receptor (BCR) – are also represented by bit-strings. The antibodies have receptors (paratopes) that are represented by the same bit-string that models the BCR of the B cell that produced them[17,18,19,20,21,22,23,24,25,26].

Each string shape is associated with an integer σ ($0 \leq \sigma \leq M = 2^B - 1$) that represents each of the clones, antigens or antibodies. The neighbors for a given σ are expressed by the Boolean function $\sigma_i = (2^i xor \sigma)$. The complementary form of σ is obtained by $\overline{\sigma} = M - \sigma$, and the temporal evolution of the concentrations of different populations of cells, antigens and antibodies is obtained as a function of the integer variables σ and t, by direct iteration.

The equations that describe the behavior of clonal populations $y(\sigma, t)$ are calculated using an iterative process, for different initial parameters and conditions:

$$\begin{aligned} y(\sigma, t+1) &= (1 - y(\sigma, t)) \times \\ &\left\{ m + (1-d)y(\sigma, t) + b\frac{y(\sigma, t)}{y_{tot}(t)}\zeta_{a_h}(\overline{\sigma}, t) \right\} \end{aligned} \tag{1}$$

and all the complementary shapes included in the term $\zeta_{a_h}(\overline{\sigma}, t)$

$$\begin{aligned} \zeta_{a_h}(\overline{\sigma}, t) &= (1 - a_h)[y(\overline{\sigma}, t) + y_F(\overline{\sigma}, t) + y_A(\overline{\sigma}, t)] + \\ &a_h \sum_{i=1}^{B} [y(\overline{\sigma}_i, t) + y_F(\overline{\sigma}_i, t) + y_A(\overline{\sigma}_i, t)]. \end{aligned}$$

In these equations, $y_A(\sigma, t)$ and $y_F(\sigma, t)$ are, respectively, the populations of antibodies and antigens, b is the rate of proliferation of B cells; $\overline{\sigma}$ and $\overline{\sigma}_i$ are the complementary forms of σ, and of the B nearest neighbors in the hypercube (with the ith bit inverted). The first term (m), within the brackets in equation (1) represents the production of cells by the bone marrow and is a stochastic variable. This term is small, but non zero. The second term in the bracket describes the populations that have survived natural death (d), and the third term represents the clonal proliferation due to interaction with complementary forms (other clones, antigens or antibodies). The parameter a_h is the relative connectivity between a certain bit-string and the neighborhood of its image or a complementary form. When $a_h = 0.0$, only perfectly complementary forms are allowed. When $a_h = 0.5$, a string can equally recognize its image and its first neighbors. The factor $y_{tot}(t)$ is given by

$$y_{tot}(t) = \sum_{\sigma} [y(\sigma, t) + y_F(\sigma, t) + y_A(\sigma, t)]. \tag{2}$$

The temporal evolution of the antigens is determined by:

$$\begin{aligned} y_F(\sigma, t+1) &= y_F(\sigma, t) - k\frac{y_F(\sigma, t)}{y_{tot}(t)} \times \\ &\{(1 - a_h)[y(\overline{\sigma}, t) + y_A(\overline{\sigma}, t)] + \\ &a_h \sum_{i=1}^{B} [y(\overline{\sigma}_i, t) + y_A(\overline{\sigma}_i, t)]\}, \end{aligned} \tag{3}$$

where k is the rate with which populations of antigens or antibodies decay to zero. The population of antibodies is described by a group of 2^B variables, defined in a B-dimensional hypercube, interacting with the antigenic populations:

$$y_A(\sigma, t+1) = y_A(\sigma, t) + b_A \frac{y(\sigma, t)}{y_{tot}(t)} \times$$
$$\left[(1 - a_h) y_F(\overline{\sigma}, t) + a_h \sum_{i=1}^{B} y_F(\overline{\sigma}_i, t)\right] -$$
$$k \frac{y_A(\sigma, t)}{y_{tot}(t)} \zeta_{a_h}(\overline{\sigma}, t), \tag{4}$$

where the contribution of the complementary forms $\zeta_{a_h}(\overline{\sigma}, t)$ is again included in the final term, b_A is the rate of proliferation of antibodies, and k is the rate of removal of antibodies, which measures their interactions with other populations. The antibody populations $y_A(\sigma, t)$ (which represent the total number of antibodies) depend on the inoculated antigen dose. The factors $\frac{y_F(\sigma,t)}{y_{tot}(t)}$ and $\frac{y_A(\sigma,t)}{y_{tot}(t)}$ are responsible for the control and decay of the antigens and antibodies populations, while the factor $\frac{y(\sigma,t)}{y_{tot}(t)}$ is the corresponding accumulation factor for the populations of clones, in the formation of the immunological memory. The clonal population $y(\sigma, t)$ (normalized total number of clones) may vary from the value produced by bone marrow (m) to its maximum value (in our model, the unit), since the Verhulst factor limits its growth[27,28].

Eqs. (1)-(4) form a set of equations that describe the main interactions in the immune system between entities that interact through key-lock match, i.e., that specifically recognize each other. This set of equations was solved iteratively, considering the viral mutations as initial conditions.

Using virtual samples – representing hypothetically 10 individuals (mammals) with the same initial conditions – we inoculated *in silico* each one of the 10 samples with different viral populations (110, 250 and 350) with fixed concentration, where the virus strains occur at intervals of 1000 time steps, that is, at each 1000 time steps a new viral population, different from the preceding one, is injected in the sample. When a new antigen is introduced *in maquina*, its connections with all other entities of the system are obtained by a random number generator[7,8,9].

Changing the seed of the random number generator, the bits in the bit-strings are flipped and, taking into account that in our approach the bit-strings represent the antigenic variability, the bits changes represent, therefore, the corresponding viral mutations. Thus, in order to simulate the influence of viral mutation on the duration of immunological memory, the seed of the number generator is altered for each of the 10 samples. Figure 1 shows the design of the experiments.

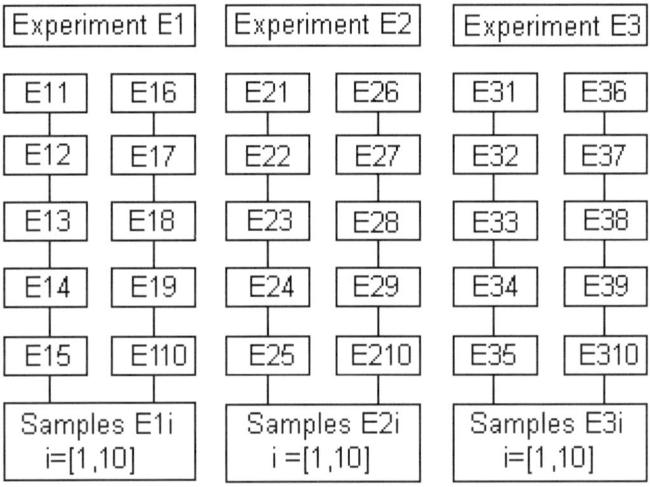

Figure 1. In each experiment were considered different lifetimes for the individuals. The lifetimes for E1, E2 and E3 are respectivelly 110000, 250000 and 350000.

In this model, we use for the rate of apoptosis (d), or natural cell death, the value of 0.99, the proliferation rate of the clones is equal to 2.0, and of the antibodies is 100. The connectivity parameter a_h was considered equal to 0.01 and the term of bone marrow m was set to 10^{-7}. The value 0.1 was set to represent each virus strains (the antigen dose) and the length of the bit-string B was set to 12, corresponding to a potential repertoire of 4096 distinct cells and receptors. Injections of different viral populations were administrated at time intervals corresponding to a period of the life of the individual.

3. Results

To investigate the relationship between the viral variability and the memory time of the population of lymphocytes that recognized a certain species of virus, three experiments *in silico* (E1, E2 and E3) were performed and organized as follows: in the first (E1) a lifetime equal to 110000 was chosen for the individuals, in the second (E2), lifetime of 250000 and in the third (E3), 350000 time steps. In each of the experiments were used sets of 10 virtual samples (E1i, E2i and E3i), representing 10 identical individuals what, in our approach, corresponds to keep the amount of interaction of the coupled maps with the same initial conditions in all samples –30 samples, taking into account the 3 experiments (Fig.1).

To simulate the viral strains, at each 1000 time steps, a new dose of virus was administered. Therefore, in the first experiment were injected *in maquina* in the samples (individuals) 110, in the second, 250 and in the third 350 *distinct* viral populations and to represent the inoculation of *mutated* viral populations in each individual, the seed for the random number generator was changed for each one of the 30 samples. It is important to clarify that in our approach, *distinct* viral populations are populations of different species and *mutated* viral populations are genetic variations of the same population.

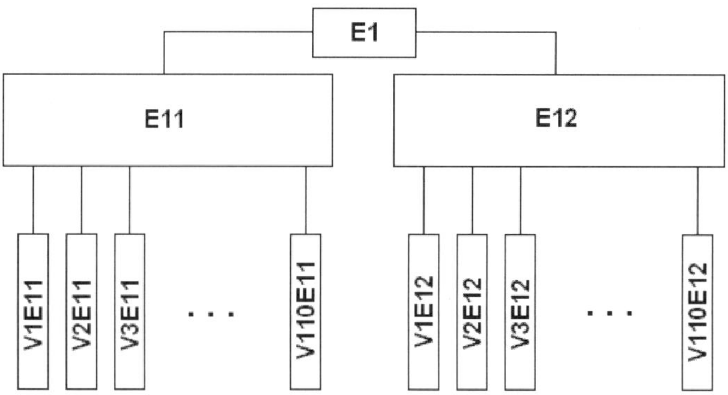

Figure 2. Design of inoculations of viral populations in the samples for the experiment E1.

Figure 2 shows more clearly the entities used to simulate the behavior of memory against the antigenic variability. In the scheme, for example, the virus identified by V1E12 is a mutation of the virus V1E11 (both belong to the same original population, who suffered mutation) and the virus V2E11 is distinct of virus V1E11 (belonging to viral populations of different species).

Figure 3 shows the average lifetime of the populations of lymphocytes that specifically recognized the antigens in each of 3 experiments (E1, E2 and E3). In the experiments, the average lifetime of each population, ex-

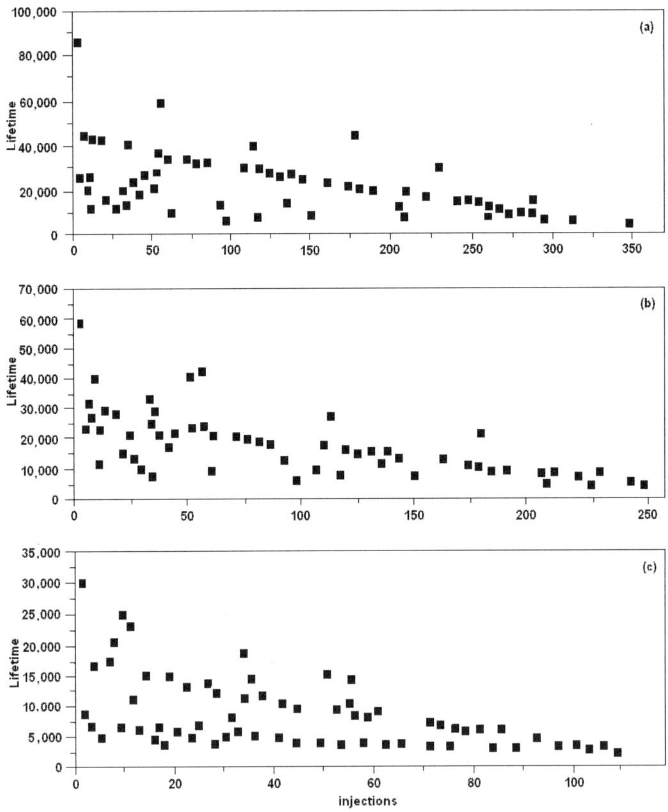

Figure 3. Lifetime of the populations that recognize the antigens, for (a) 110, (b) 250 and (c) 350 inoculations.

cited by a kind of virus, is calculated over 10 samples (E1i, E2i and E3i). The difference in the average behavior of the memory in the three experiments (Fig. 3 (a), (b) and (c)) was expected, and is due to the fact that the samples were inoculated with virus of high genetic variability (in our approach, the random number generator seed was changed for each new sample to represent the genetic variability).

Even considering an uneven evolution for the memory in the three experiments, it is possible to see that there is a tendency of the first populations, on average, to survive longer than the subsequent. This result is consistent with the current hypothesis that, on average, the first vaccines administered in naïve individuals tend to have longer useful life than the vaccines administered in adulthood[29,30,31].

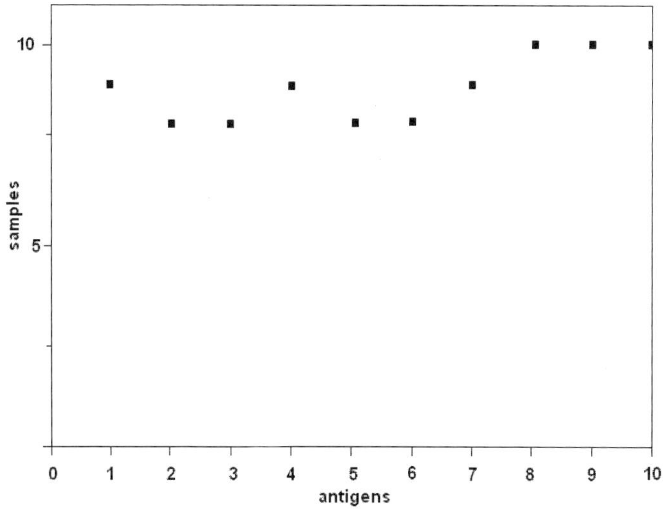

Figure 4. Number of samples (N_S) of the experiment E2 with up to the tenth excited (live) clonal population, after 10000 time steps have passed.

Figure 4 shows that for the experiment E2, the first clonal population (B cells population) remains alive in 9 samples, when the lifetime of individuals is 10000. Similar situation was observed in experiment E3. However, in Figure 5(a) - (j) it is possible to visualize separately the behavior of each one of the 10 samples, when we administer, *in maquina*, 110 injections (experiment E1). From Figure 5, it is clearly noticed that we can not say that the first clonal populations persist longer than the populations who later recognized other antigens, since the simulations indicate that only in two samples the first clonal population survived for a long period of time (Fig. 5 (b) and (h)).

It is important to point out that this apparent discrepancy between the results of Figures 3 and 5 can be explained by the fact that in two samples the lifetime of the first excited clonal population was long, what determined that the simple arithmetic average was high – even if in the other samples the first clonal population has not survived for a long period of time. The latter result suggests that it is not safe to predict that the first excited populations tend to last longer – only on average these first immunizations tend to last longer than the others.

In specific situations, depending on the viral variability, our model shows that you can not say that vaccines used in the first years of life of the in-

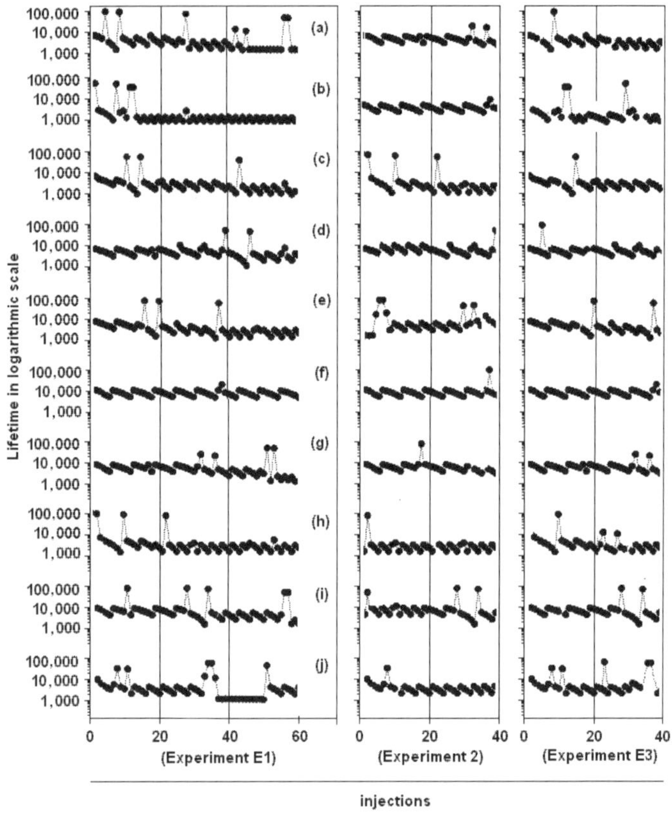

Figure 5. Lifetime of the clonal populations in each sample (experiment E1 until 60 time steps). Similar behaviors were obtained for the experiments E2 and E3, until 40 time steps.

dividual provide with certainty immunization for long periods. Our results also suggest that the sustainability of immunization is dependent of the on viral variability and that the lifetimes of the memory populations are not completely random, but that the antigenic variability of which they depend on causes an apparent randomicity to the lifespan of memory lymphocytes

4. Conclusion

We have performed simulations using a model of bit-strings that considers the temporal evolution of B cells, antigens and antibodies, and have studied the behavior of immune memory and which main factors are influencing the durability of vaccines or immunizations[7,8,9]. In previous studies, we have suggested that it is not possible to accurately determine the duration of memory[10], however, our results suggested that decreasing the production of antibodies, we can achieve greater durability for the immunizations[9]. Those simulations indicated that it is not possible to accurately determine the clonal populations that will survive for long periods, but those that survive, will have higher durability if there is a reduction in the production of soluble antibodies. This result may be biologically important, as it suggests a strategy to give more durability to the vaccines by inhibiting the production of antibodies harmful to the memory of the system.

In this article we present results that indicate that, besides the influence of populations of soluble antibodies[9], another factor that may be decisive for the durability of immunological memory is the antigenic mutation of the viral population, which brings on a reaction of the system. In this work we show that the lifetimes of the memory clones are not random, but the antigenic variability from which they depend originates an apparent randomicity to the lifespan of the B memory cells. As a consequence, our results indicate that the mainteinance of the immune memory and its relation with the mutating antigens can mistakely induce the *wrong deduction* of an stochasticity hypothesis for the sustainability of the immunizations, as proposed by Tarlinton *et al.*[14]. Our results show that what presents a random aspect is the mutation of the viral species, resulting in an *apparent* unpredictable duration for the lifetime of the memory clones.

Acknowledgements

The authors wish to thank Dr. V.B. Campos for valuable discussions.

References

1. L.C. Ribeiro, R. Dickman, A.T. Bernardes, 387 (2008) 6137.
2. F. Castiglione, B. Piccoli, J. Theor. Biol., 247 (2007) 723.
3. M.N. Davies, D.R. Flower, Drug Discov. Today, 12 (2007) 389.
4. J. Sollner, J. Mol. Recognit., 19 (2006) 209.
5. N. Murugan, Y. Dai, Immunome Res., 1 (2005) 6.
6. M.C. Lagreca, R.M..C. de Almeida, R.M.Z. dos Santos, Physica A 289 (2000) 42.
7. A. de Castro, Simul. Mod. Pract. Theory. 15 (2007) 831.
8. A. de Castro, Eur. Phys. J. Appl. Phys. 33 (2006) 147.
9. A. de Castro, Physica A 355 (2005) 408.
10. M. Johansson, N. Lycke, Immunology 102 (2001) 199.
11. I. Roitt, J. Brostoff, D. Male, Immunology, fifth ed., Mosby, New York, 1998.

12. A.K. Abbas, A.H. Lichtman, J.S. Pober, Cellular and Molecular Immunology, second ed., W.B. Saunders Co., 2000.
13. B. Alberts, D. Bray, J. Lewis, M. Raff, K. Roberts, J.D. Watson, Molecular Biology of the Cell, fourth ed., Garland Publishing, 1997.
14. D. Tarlinton, A. Radbruch, F. Hiepe, T. Dorner, Curr. Opin. Immunol. 20 (2008) 162.
15. A. de Castro, C.F. Fronza, R.H. Herai, D. Alves, In: BIOMAT 2008 International Symposium on Mathematical and Computational Biology. World Scientific Publishing Co. (2008) 351.
16. A.S. Perelson, G. Weisbush, Rev. Mod. Phys. 69 (1997) 1219.
17. A.S. Perelson, R. Hightower, S. Forrest, Res. Immunol. 147 (1996) 202.
18. J.H.L. Playfair, Immunology at a Glance, sixth ed., Blackwell Scientific Publications, Oxford, 1996.
19. O. Levy, Eur. J. Haematol. 56 (1996) 263.
20. M. Reth, Immunol. Today 16 (1995) 310.
21. G. Moller, Immunol. Rev. 153 (1996).
22. N.K. Jerne, Clonal Selection in a Lymphocyte Network, in: G.M. Edelman (Ed.), Cellular Selection and Regulation in the Immune Response, Raven Press, 1974, 39-48.
23. F.M. Burnet, The Clonal Selection Theory of Acquired Immunity .Vanderbuilt University, Nashville, TN, 1959.
24. N.K. Jerne, Ann. Immunol. 125 (1974) 373.
25. F. Celada, P.E. Seiden, Immunol. Today 13 (1992) 53.
26. F.Celada, P.E. Seiden, J. Theoret. Biol. 158 (1992) 329.
27. P.F. Verhulst, Correspondance mathématique et physique 10 (1838) 113.
28. P.F. Verhulst, Nouveaux Memoires de l'Academie Royale des Sciences et Belles-Lettres de Bruxelles, 18, (1845) 1.
29. D. Gray, M. Kosko, B. Stockinger, Int. Immunol. 3 (1991) 141.
30. D. Gray, Ann. Rev. Immunol. 11 (1993) 49.
31. C.R. Mackay, Adv. Immunol. 53 (1993) 217.

MATHEMATICAL AND COMPUTER MODELLING CONTROL MECHANISMS OF HIERARCHICAL MOLECULAR-GENETIC SYSTEMS *

H. B. NABIEVICH, S. MAHRUY, H. M. BAHROMOVNA

Institute of Mathematics and Information Technologies,
Durmon yuli, 29
100125, Tashkent, Uzbekistan
E-mail: regulatorika@yahoo.com

A. B. RAHIMBERDIEVICH

Uzbek Institute of Virology,
Head of Department Uzbek Institute of Virology
Murodov, 29
100194, Tashkent, Uzbekistan

The progress achieved in biology during last century has led to development of mathematical and computer modelling control mechanisms in molecular-genetic processes in living systems. In this work we consider one of the possible methods for mathematical and computer modelling control mechanisms of hierarchical molecular-genetic systems. The corresponding equations (in the class of delay functional-differential equations) are developed, using the approaches by B.Goodwin, Bl. Sendov, M. Eigen, B.A. Ratner, and taking into account temporary relations, presence of multifunctional feedback and cooperative nature of processes in cell's regulatory loops. Results of using the considered method for analyzing cell's molecular-genetic systems in the presence of the alien genes (on the example of hepatic cell infections by hepatitis B virus) are given.

*This work is partially supported by RUz grants (FA-F1-F011, FA-A17-F009)

1. Introduction

The analysis of main protein-ferments, participating in the specific functioning of cells in a multicellular organism shows, that each cell contains different complex of protein molecules, though all cells have the same set of chromosome and, consequently, identical genetic information[1,2,3]. It happens because in eukaryotic cells there is a mechanism of hierarchical organizations of molecular-genetic system, regulating genes activity: which groups of genes at the moment must be active and which genes should be in the inactive condition. Existence of biochemical properties and processes, inherent to all cells of the organism (syntheses of energy proteins, amino acid, to maintain and build membrane, etc) shows that there are universal groups of genes functioning in all cells regardless of specialization. The remaining part of genes consists of groups of general genes (its operation depends on activity conditions in universal groups of genes), containing information on general functions characteristic for many, but not all cells and groups of specific genes (its functioning depends on activity conditions of universal and general genes groups), keeping information on strictly specific functions of concrete cells[4]. Under such approach, the whole cell genome of multicellular organism works as evolutionary established hierarchical system of universal, general and specific genes groups (Figure 1).

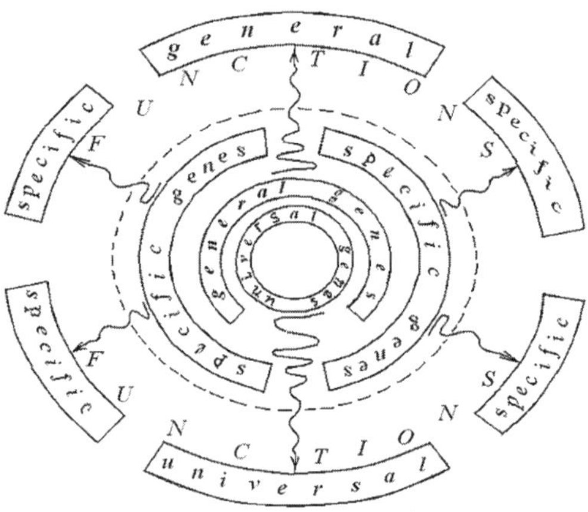

Figure 1. Structural-functional organization of cell genome.

Hierarchical organization of molecular-genetic systems is of the utmost importance in functioning control mechanisms of intracellular processes and cellular functions. Their account makes it possible to carry out studying mechanisms of origin, existence and development of living systems when quantitatively analyzing activity patterns of molecular-genetic systems. In this work we develop the mathematical modeling of genes activity (section 2), possible equations control mechanisms of molecular-genetic systems (sections 3, 4) and certain problems concerning the mathematical and computer modeling of hierarchical molecular-genetic systems (sections 5, 6).

2. Development of modeling control mechanisms molecular-genetic systems by ordinary differential equations (ODE).

The intensive development of quantitative methods for the analysis of regulation mechanisms of genetic systems[5,6,7,8], observed in the last decade, is basically conditioned by outstanding achievements in studying structural-functional organization of processes which occur in a cell's molecular-genetic system[9,10,11], as well as by a wide intrusion of cybernetic concepts, mathematical ideas and methods into molecular biology. Objects formalization in mathematical and computer modeling control mechanisms of molecular-genetic system is realized within the bounds of mathematical biology, biocybernetics and bioinformatics. There are many different approaches for quantitatively studying the functioning regularities of molecular-genetic systems[12,13,14]. The mathematical and computer modeling control mechanisms of molecular-genetic systems using ODE is probably by now a routine method. B. Goodwin's work[15] is one of the first works, based on the operon idea by F.Jakob and J.Monod[9] and contains elementary differential equations of cellular functions regulatory system (Figure 2, see B. Goodwin Ref. [15]). In Figure 2 L_i denotes genetic locus, where there occurs the synthesis of RNA molecules (X_i), which are in the cellular organoid R - ribosome, where there occurs the specific protein (Y_i) synthesis. On a certain section C in the cell this protein regulates metabolism degree, acting as ferment. Under ferment activity the metabolite M_i, which closes the feedback loop, is produced, as its parts return to the genetic locus L_i and operate there as a repressor[15].

Based on ODE, M.Eigen, B.A.Ratner and their followers[16,17] describe the dynamics of informational macromolecules community in the course of evolution - hypercycles and sysers. These models propose that the macromolecules synthesis is far from steady state and occurs only in complex

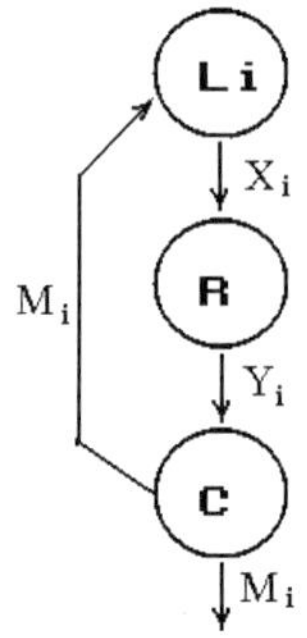

$$\frac{dX_i}{dt} = \frac{a_i}{A_i + k_i M_i} - b_i X_i ;$$

$$\frac{dY_i}{dt} = \alpha_i X_i - \beta_i Y_i ; \qquad (1)$$

$$\frac{dM_i}{dt} = r_i Y_i - s_i M_i .$$

Figure 2. Diagram of elementary regulatory system of cellular functions - systems of protein synthesis - and its simple equations

of matrix and ferments carrying out this synthesis. In general view the dynamics of informational macromolecules community can be described by the following system [16]:

$$\frac{dX_i}{dt} = (A_i Q_i - D_i) X_i + \sum_{k \neq i} \omega_{ik} X_k + F_i, \tag{2}$$

$$\sum_k X_k = const \equiv C.$$

In Eq.(2) X_i is a concentration of i-th informational macromolecules community; $A_i Q_i X_i$ and $D_i X_i$ express formation and decay of i-th informational macromolecules community, accordingly; ω_{ik} denotes individual rate of mutations; F_i determines outflow velocity of reactions products $(i, k = 1, 2, .., N)$.

The mathematical models of molecular-genetic systems based on ODE were repeatedly improved[18,19,20,21,22,23] in the sequel. Bl. Sendov and R. Tsanev[18] have shown a possibility for making a model, which imitate acting regulatory system of cellular functions, tissue's cells groups using systems of nonlinear ODE like Eq.(1). J. Smith has modified the Eq.(1) by taking into account the delay in regulation loop of cell biosynthetic processes[21]. Using the considered models the mechanisms of malignant growth in liver [19], regulation of biosynthetic functions[20], metabolic systems[24], rhythmical processes[25] have been quantitatively investigated, mechanisms of probionts, phages's molecular-genetic systems have been studied and similar evolutionary problems have been considered[16,17]. Improved versions of Eq.(1)

taking into account the cooperativity, end product inhibition and temporal mutual relations in cell control mechanisms system were applied for quantitative analyzing gene control mechanisms[21], cell's malignant growth[26], cell function control mechanisms and cellular communities[26,27,28,29]. In the following sections we consider a possible method for modelling control mechanisms molecular-genetic system in the class of functional-differential equations and its application for quantitative studying control mechanisms of hierarchical molecular-genetic systems.

3. Functional-differential equations for control mechanisms of molecular-genetic systems.

The main equations of molecular-genetic systems, in the most general aspects, has the following form

$$\frac{dx_i(t)}{dt} = a_i F_i(x_1, x_2, ..., x_n) - b_i x_i, \tag{3}$$

$$i = 1, 2, ..., n$$

where $x = (x_1, x_2, ..., x_n)$ is a vector of molecular-genetic system products concentrations; $F_i = (x_1, x_2, ..., x_n) : R^n \to R$ are functions which express activity degree by i-th molecular-genetic elements; a_i, b_i are non-negative constants of "synthesis" and "decay" for i-th molecular-genetic element products, $i = 1, 2, ..., n$.

Since the equations of type Eq.(3) were appeared in B.Goodwin's work[15], the functions F_i $(i = 1, 2, ..., n)$ are the most subject to improvement. In choosing the functions F_i, starting from functioning regularities for the considered class of molecular-genetic systems, the influence from stimulating and inhibitory factors on molecular-genetic systems functioning in different cases are taken into account[18,19,22,25,26,28,30]. One of the possible variants Eq.(3)-like-equations, applying for mathematical and computer modelling control mechanisms of molecular-genetic systems, can be general regulatorika equations[22,26,28,30]. In this case, with taking into account the following characteristics for molecular-genetic systems:

- the biologically reasonable genetic text's appearance in the cells, without the participating living system's genetic apparatus, is absent;
- the genetic processes in the cells are subordinated to a united regulation system: initiations, transcriptions and translations of genes;

forming the active protein -ferments and their operation; forming effector complexes and repressors;
- genetic processes regulation, occurring in cells, can be realized autonomously

the mathematical and computer models for control mechanisms of molecular-genetic systems can be based (in accordance with Eq.(3)) on the following equations:

$$\frac{\theta_i}{h}\frac{dx_i(t)}{dt} = A_i^N(X(t-1))exp\left(-\sum_{k=1}^{N}\delta_{ik}x_k(t-1)\right) - x_i(t) \qquad (4)$$

with

$$A_i^N(X(t-1)) = \sum_{j=1}^{N}\left(\sum_{k_1,\ldots,k_j=1}^{N}\gamma_{i_{k_1,\ldots,k_j}}\prod_{m=1}^{j}x_{k_m}(t-1)\right),$$

where $x_i(t)$ is the value which characterizes m-RNA count, transcribed from i-th gene at time t; θ_i express average time of genes products vital activity; h is the feedback time in considered molecular-genetic systems; $\gamma_{i_{k_1,\ldots,k_j}}$ are the induction matrix elements and δ_{ik} are repression matrix elements of intergenic mutual relations; $i_{k_1,\ldots,k_j}, i, j, k_j = 1, 2, ..., N$.

Elements of the vector $M_c(C_1, C_2, ..., C_N)$ of the molecular-genetic systems relationships with external medium are calculated by the formula

$$C_i = \int_0^\infty \ldots \int_0^\infty A_i^N(S)exp\left(-\sum_{j=1}^{N}\delta_{ik}S_j\right)dS_1 \ldots dS_N - 1. \qquad (5)$$

This vector outlines the boundares of admissible values for equation coefficients of control mechanisms of genetic processes. System Eq.(4) belongs to the class of delayed functional-differential equations and if we have continuous functions on initial time segment, then its continuous solution can be obtained by consequent integration method[31,32].

4. Control mechanisms equations of homogeneous molecular-genetic systems.

As expected, the evolutionary development control mechanisms of molecular-genetic system began from homogeneous molecular-genetic systems. Let us consider the equations of a possible homogeneous (associative, inter-conjugate and self-conjugate) molecular-genetic system based on

Eq.(4) and formula (5). For qualitative, studying the functioning of the associative molecular-genetic system we offer the following equations on the basis of Eq.(4)

$$\frac{\theta_i}{h}\frac{dX_i(t)}{dt} = \sum_{j=1}^{n} a_{ij}X_j(t-1)exp\left(-\sum_{k=1}^{n} X_k(t-1)\right) - X_i(t) \tag{6}$$

with vector elements M_c (see (5))

$$C_i = \sum_{j=1}^{n} a_{ij} - 1.$$

$$i = 1, 2, ..., n$$

In Eq.(6), θ_i expresses the average vital activity time of gene products ; h is the feedback time control mechanisms of molecular-genetic systems; a_{ij} are parameters describing the i-th product formation by j-th gene.

For functioning inter-conjugated molecular-genetic systems it is necessary to have all the genes products. We have

$$\Lambda_i^n(X(t-h)) = a_i \prod_{j=1}^{n} X_i(t-h);$$

$$\frac{\theta_i}{h}\frac{dX_i(t)}{dt} = a_i\left(\prod_{j=1}^{n} X_i(t-h)\right) exp\left(-\sum_{k=1}^{n} X_k(t-h)\right) - X_i(t); \tag{7}$$

$$C_i = a_i - 1.$$

$$i = 1, 2, ..., n$$

If there is balance with the external medium ($C_i = 0; i = 1, 2, ..., n$) then the right part of Eq.(7) does not have arbitrary constants. In some cases for functioning the molecular-genetic systems it is necessary to have n products of the same gene (here n can be associated with Hill coefficient). Then all genes have identical activity and

$$\frac{\theta_i}{h}\frac{dX_i(t)}{dt} = a_i X_j^n(t-1)exp\left(-\sum_{k=1}^{n} X_k(t-1)\right) - X_i(t);$$

$$C_i = a_i \int_0^{\infty} S_j^n exp\left(-\sum_{k=1}^{n} S_k\right) dS_1...dS_n - 1,$$

$$i, j = 1, 2, ..., n$$

This leads to the following equations for control mechanisms of self-conjugated molecular-genetic system (n is self-conjugate degree)

$$\frac{\theta}{h}\frac{dX(t)}{dt} = aX^n(t-1)e^{-nX(t-1)} - X(t); \tag{8}$$

$$C = a\int_0^\infty S^n e^{-nS} dS - 1.$$

Here θ, h are the parameters of average vital time for molecular-genetic system products and feedback system; a, n are the parameters of resource-provision and self-conjugate of molecular-genetic system. The Eq.(6), Eq.(7) and Eq.(8) were used for modelling the concrete molecular-genetic systems which accordingly have the following properties: they are additive, inter-conjugate and self-conjugate[21,26,27,28,29,30,33]. In the following sections possible applicability of the Eq.(7) and Eq.(8) for analyzing main regularities in control mechanisms of uniform and hierarchical molecular-genetic systems will be considered.

5. Qualitative analyzing uniform self-conjugated molecular-genetic systems.

Based on the results of the analysis of control mechanisms in biological systems[28,33] we suppose that initial development of molecular-genetic systems is realized from associative to inter-conjugate and hereinafter to self-conjugate systems. In this event the study of general regularities of further evolutionary developments of molecular-genetic systems can be realized using Eq.(8).

Within the framework of the given problem the value h is defined by means of feedback common time in the considered biosystem population with given type of molecular-genetic system and consequently $\theta \ll h$. Then, qualitative analysis of characteristic solutions Eq.(8) can be realized by the following functional equations[28,29,33]

$$X(t) = aX^n(t-1)e^{n(1-X(t-1))} \tag{9}$$

and its discrete analogue

$$X_{k+1} = aX_k^n e^{n(1-X_k)}. \tag{10}$$

If the analyzed molecular-genetic systems are in equilibrium with the external medium, we have

$$X_{k+1} = (n^n/(n-1)!)X_k^n e^{n(1-X_k)}.$$

When $n \gg 1$ using Stirling's approximation[33] we have

$$X_{k+1} = \sqrt{n/2\pi} X_k^n e^{n(1-X_k)}, \tag{11}$$

where X_k expresses product count, outputting by molecular-genetic system on k-th step. Analyzing Eq.(11) solutions behavior we show[28,33] that the solutions are in the first quadrant of phase space when parameters values and initial conditions are non-negative; infinitely distant points are unstable; it is possible that there exist stable trivial steady state - trivial attractor (for all $n \geq 1$), unstable positive steady state (α) and positive attractors (β) if $n \geq 6$. The basin β-attractor is area of possible control mechanisms for hierarchical molecular-genetic systems. Results of the qualitative studying solutions Eq.(9)-Eq.(11) in the β-attractor basin have shown that there is sufficiently complex behavior of the considered models for control mechanisms of self-conjugated molecular-genetic systems (Figure 3). Besides periodic functioning regimes there can exist irregular oscillations (where Lyapunov exponent $L(\beta) > 0$; see Table 1 and "black hole" effect[28,29,33]. In this case, the last effect consists of appearance of the destructive changes in the system and is expressed by oscillations failure[34] - solutions tend into the trivial attractor.

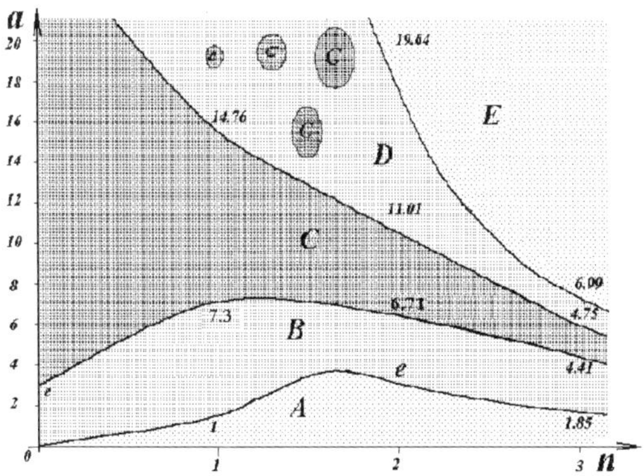

Figure 3. Parametric portrait Eq.(10). A is the trivial attractor area, B is the area of stationary modes, C is the auto-oscillations area, D is the area of irregular oscillations, E is the "black hole" area.

Table 1. Positive steady states and Lyapunov exponent self-conjugated molecular-genetic systems calculated based on Eq.(11).

n	6	7	8	9	10	11	12
α	0.719	0.699	0.694	0.694	0.695	0.699	0.702
β	0.964	1.038	1.086	1.118	1.142	1.158	1.171
$L(\beta)$	-1.395	-1.326	-0.375	-0.161	-0.111	+0.379	+0.681

The solutions given by Eq.(11) and conducted on a PC using Lamerey diagrams construction, calculations of Lyapunov exponent, Hausdorff and high dimensions (based on Ref. [34]) show the existence of irregular oscillations when $n \geq 11$ and the appearance of a "black hole" effect when $n > 12$. Consequently, stable self-conjugated molecular-genetic systems, which are in the equilibrium with the external medium, can exist under $6 \leq n \leq 12$ only, moreover under $6 \leq n < 9$ their control mechanisms are characterized by stable stationary behavior and under $9 \leq n < 11$ control mechanisms are characterized by stable regular oscillations, if $11 \leq n \leq 12$ there are irregular oscillations. In the last case so-called "orasta power" is formed [21,26,28,33] leading to failure of normal functioning in cell's molecular-genetic system (activation of silent genes, intensification of genes mutations and failure of cellular regulatorika).

Thereby, within the framework of the accepted suggestions, based on the study results we can get the following effects of comparatively evolutionary stable molecular-genetic systems: they have genes from six to twelve in number, if genes in number are greater than twelve then the systems are organized by the hierarchical type, if genes in number are less than eight then the systems basically have stable stationary activity, if genes in number are greater than eight then we have behavior in the manner of irregular oscillations.

6. Modelling control mechanisms of hierarchical molecular-genetic systems.

In general cases the control mechanisms of hierarchical molecular-genetic systems can be investigated using Eq.(4). The equations of inter-conjugate molecular-genetic systems type Eq.(7) are elementary equations and available for modelling hierarchical molecular-genetic systems. When constructing the control mechanisms equations of hierarchical molecular-genetic systems it is necessary to take into account the consequent dependence of molecular-genetic systems activity of primitive hierarchical level on molecular-genetic systems activity of higher hierarchical level. Let us consider this on the example of constructing elementary equations control mechanisms of eukaryotic cells genome by taking into account common hierarchical levels (universal, general and specific genetic systems). For the simplicity we present each level as one genetic system. Then as the simplest equations control mechanisms for considered molecular-genetic system we have

$$\begin{aligned}
\frac{\theta_1}{h}\frac{dX(t)}{dt} &= aX(t-1)e^{-\omega(t)} - X(t);\\
\frac{\theta_2}{h}\frac{dY(t)}{dt} &= bX(t-1)Y(t-1)e^{-\omega(t)} - Y(t);\\
\frac{\theta_3}{h}\frac{dZ(t)}{dt} &= cX(t-1)Y(t-1)Z(t-1)e^{-\omega(t)} - Z(t);\\
\omega(t) &= X(t-1) + Y(t-1) + Z(t-1),
\end{aligned} \tag{12}$$

where $X(t), Y(t), Z(t)$ are values, characterizing common m-RNA count for universal, general and specific genetic systems products at time t; θ_1, θ_2, θ_3 are parameters, expressing "lifetime" for molecular-genetic systems products; h is the time, necessary for fulfilling feedback in cell; a, b, c are rate constants of product formation for considered genetic systems.

Let us consider functional-differential equations for control mechanisms of inter-conjugate molecular-genetic systems Eq.(7) and Eq.(12) for investigating the hepatitis B development mechanism, which is one of the actual medical problems in viral hepatology[35]. The analysis of possible equations variants and preliminary studying its behavior using mathematical and virology viewpoints shows[36] that it is necessary to take into account specific aspects of interconnected activity between hepatocyte and hepatitis virus molecular-genetic systems, i.e., we must take into consideration independent functioning hepatocyte, hepatitis B virus genome adoption ability into hepatocyte genome and dependence of hepatitis B virus activity on conditions of hepatocyte intracellular medium.

Subject to aforesaid, the elementary system of control mechanisms equations for interconnected activity between hepatocyte and hepatitis B viruses molecular-genetic systems on the basis of Eq.(7) has the form

$$\frac{\theta_1}{h}\frac{dX(t)}{dt} = aX^2(t-1)e^{-X(t-1)-cY(t-1)} - X(t); \tag{13}$$

$$\frac{\theta_2}{h}\frac{dY(t)}{dt} = bX(t-1)Y(t-1)e^{-dX(t-1)-Y(t-1)} - Y(t),$$

where $X(t)$, $Y(t)$ are values, characterizing hepatocyte and hepatitis B virus molecular-genetic systems activity, accordingly; θ_1, θ_2 are parameters, expressing products "lifetime" of hepatocyte and hepatitis B virus molecular-genetic systems; h is the time necessary for feedback fulfilment in the considered systems; a, b are velocity constants of product formation in the considered genetic systems; c, d are parameters, expressing repression degree in hepatocyte and hepatitis B virus molecular-genetic systems; all parameters are positive. Qualitative study of Eq.(13) shows that there are stable trivial and complex positive attractors. Analysis of character solutions Eq.(13), using the methods of qualitative analysis of functional-differential equations and results of computer studies, shows that there is an ensemble of parameters values under which there exists hepatocyte genome domination. The most interesting case correspond to joint activity regime between hepatocyte and hepatitis B virus genomes (chronic hepatitis B) (Figure 4). Under certain values of the functional-differential equations system parameters Eq.(13) the positive attractor is stable (Figure 4) and its stability loss is realized by Hopf bifurcations by beginnings of stable, regular oscillations. Results of computer studies have shown that in the interconnected symbiotic activity between hepatocyte and hepatitis B virus there are irregular oscillations regimes (dynamic chaos) and "black hole" effect (Figure 5). In the last case, there exist oscillations failure and solutions Eq.(13) tend to a trivial attractor (Figure 5 c, d).

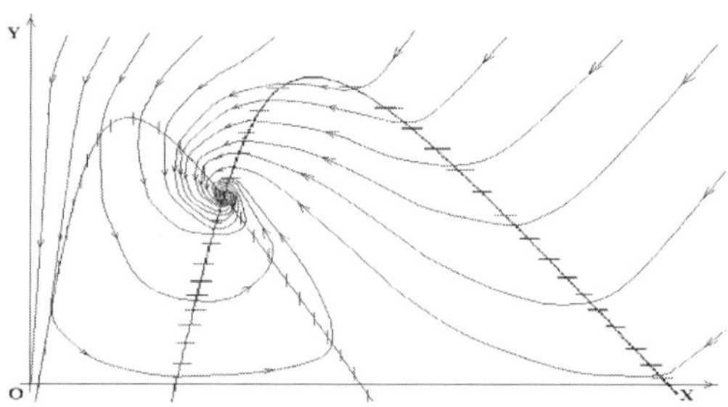

Figure 4. Symbiotic regime in "hepatocyte-hepatitis B virus" system Eq.(13).

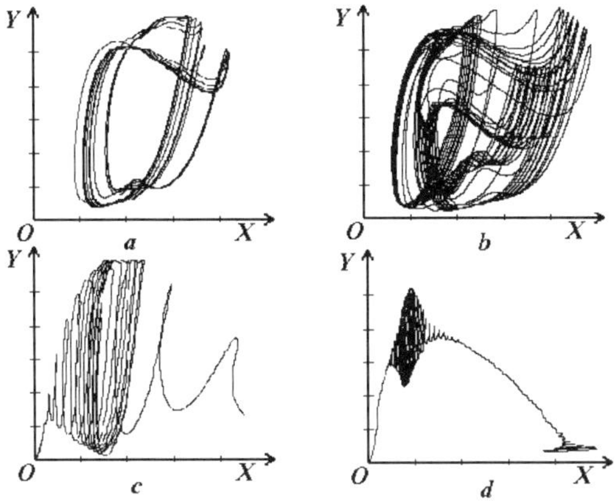

Figure 5. Characteristic phase trajectories Eq.(13) are obtained using PC (A is the auto-oscillation, b is the irregular oscillation (chaos), c, d are the "black hole" variants (trajectories go right to left)).

Our research using a PC showed that in dynamic chaos area there are small regions (r-windows), in which the behavior of the solutions of Eq.(13) have regular character. This indicates that there exists possibility for effective controlling hepatocyte molecular-genetic systems for the purpose of getting its normal functioning. The entry in the area of irregular oscillations can be forecasted: splash series of Lyapunov exponents (Hopf bifurcations by Feigenbaum scenario) precede this occurrence[34]. Splash can be fixed by analyzing solutions using PC. This allows to forecast the coming of the destructive changes in hepatocyte under the hepatitis B virus influence and to realize a diagnostics and forecasting specific stages in disease current under infection by hepatitis B virus.

7. Conclusion

Mathematical and computer modeling control mechanisms of cell's molecular-genetic systems of multicellular organisms supposes accounting for their hierarchical organizations. This can be reached by constructing the equations of control mechanisms of molecular-genetic systems based on the general functional-differential equations of regulatorika[22,26,28,30].

The equations obtained allow for conducting the quantitative study of the origin and development of hierarchical molecular-genetic systems and the mathematical and computer modeling control mechanisms of concrete molecular-genetic systems at the norm and at the interaction with alien genes.

Mathematical and computer modeling control mechanisms of interconnected activity between hepatocytes and hepatitis B virus molecular-genetic systems shows the possibility of quantitatively diagnosing the consequent regimes of the development of this hierarchical process.

References

1. E.B. Popov, *Agropromizdat*, (1991)(in Russian).
2. M. Ptashne, *Novosibirsk, Mir*, (1966)(in Russian).
3. E.H. Davidson, *3rd ed. New York*, (1986).
4. B.N. Hidirov, *"Mathematical theory of biological processes", Kaliningrad*, 166 (1976)(in Russian).
5. H. Salgado, A. Santos and et all. *Nucl.Acids Res*, **27(1)**,59 (2000).
6. F.A Kolpakov, E.A. Ananko, G.B. Kolesov, and N.A. Kolchanov. *Bioinformatics*, **14(6)**,529 (1999).
7. M.G. Samsonova, E.G. Savostyanova, V.N. Serov, A.V. Spirov, and J.Reinitz, *First Int. Conf. Bioinformatics Genome Regul. Struct., BGRS'98. Novosibirsk*, (1998).
8. N. Friedman, M. Linial, I. Nachman, and D. Pe'er, *J. Comput. Biol.*, **7**, 601 (2000).
9. J.D. Watson and F.H.C. Crick, *Cold.Spring Harb Symp*, (1953).
10. F. Jacob and J. Monod, *J. Molec. Biol.*, **3**, 318 (1961).
11. A. A. Neyfax, *Regeneratsiya i kletochnoe delenie*, (1968) (in Russian).
12. L. Glass, *Plenum Press, New York*, (1977).
13. H.M. McAdams and A. Arkin, *Proc. Natl. Acad. Sci. USA*, **94**,814 (1997).
14. Endy D. and R. Brent, *Nature*, **409**,391 (2001).
15. B.C. Goodwin, *AcademicPress, London and New York*, (1963).
16. M. Eigen, P. Schuster, *J. Mol. Evol.*, **19**, 34 (1982).
17. V.A. Ratner, V.V. Shamin, *Novosibirsk, ITSiGSO AN SSSR*, 60 (1980)(in Russian).
18. R. Tzanev, Bl. Sendov, *J. Theoret. Biol.*, **12**,327 (1966).
19. Bl. Sendov, R. Tzanev, *J. Theoret. Biol.*, **18**,90 (1968).
20. J. Smith, *Cambridge*, Cambridge Univ. Press, (1968).
21. B.N. Hidirov, *Problems of informatics and energetics*, **2**, 39 (2003) (in Russian).
22. M. Saidalieva, *Scientiae Mathematicae Japonicae*, **58(2)**,415 (2003).
23. J.D. Murray, *Clarendon Press, Oxford* (1977).
24. J. J. Tyson, H. G. Othmer, *Prog. Theor. Biol.*, **5**, 1 (1978).
25. P. Ruoff, M. Vinsjevik, C. Monnerjahn, L. Rensing, *J. Theoret. Biol.*,**209**,29 (2001).
26. B.N. Hidirov, *International Scientific and Practical Conference "Innovation-2001"*,156 (2001).
27. M. Saidalieva, *Scientiae Mathematicae Japonicae*, **64(2)**, 469 (2006).
28. B.N. Hidirov, *Doclady ANRUz*, **3**, 26 (1998) (in Russian).

29. B.N. Hidirov, *Scientiae Mathematicae Japonicae*, **58(2)**, 419 (2003).
30. B.N. Hidirov, M. Saidalieva, *World Conference on Intelligent Systems "WCIS 2000"*,138 (2003).
31. R. Bellman, K.L. Cooke, *Academic Press, London* (1963).
32. J.K. Hale, *Introduction to functional differential equations, Springer-Verlag* (1993).
33. B.N. Hidirov, *Voprosy kibernetiki*, **128**, 41 (1984) (in Russian).
34. L. Glass, M. Mackey, *Princeton University Press* (1988).
35. B.R. Aliev, Z.A. Kushimov, *Pathology*, **58(2)**, 415 (2003) (in Russian).
36. Sh.Kh. Khodjaev, B.R. Aliev, B.N. Hidirov, M. Saidalieva, *Problems of informatics and energetics*, **1**, 7 (2005) (in Russian).

MONTE CARLO SIMULATIONS OF PROTEIN MODELS: AT THE INTERFACE BETWEEN STATISTICAL PHYSICS AND BIOLOGY *

T. WÜST AND D. P. LANDAU

Center for Simulational Physics,
The University of Georgia,
Athens, GA 30602 U.S.A.
E-mail: dlandau@hal.physast.uga.edu

C. GERVAIS AND YING XU

Institute of Bioinformatics,
The University of Georgia,
Athens, GA 30622 U.S.A.

Systems of atoms or molecules with complex free energy landscapes are common for quite diverse systems in nature ranging from magnetic "glasses" to proteins undergoing folding. Although Monte Carlo methods often represent the best approach to the study of suitable models for such systems, the complexity of the resultant rough energy landscape presents particular problems for "standard" Monte Carlo algorithms because of the long time scales that result at low temperatures where behavior is "interesting". We shall first review several inventive sampling algorithms that have proven to be useful for such systems and attempt to describe the advantages and disadvantages of each. Then we shall prevent results for wide ranges of temperature, obtained primarily using Wang-Landau sampling, for three models that are physically quite distinct. For pedagogical reasons we begin with spin glasses in condensed matter physics and then consider HP "lattice proteins" in which interest comes from disciplines as diverse as statistical mechanics, statistics, and biology. We shall then close with a "realistic" model for membrane protein dimerization in the continuum. All of these results will demonstrate advances in our understanding of the behavior of diverse systems that possess rough free energy landscapes.

*This work is partially supported by grants NIH-1R01GM075331 and NSF-DMR-0810223.

1. Introduction

Protein folding is one of the great frontiers of the early 21st century science and methods of statistical physics are finding their way into investigation of these systems. Proteins are long linear polymers with different amino acids along the backbone and rather complicated interactions that result in rough free energy landscapes which are in many ways similar to those found in spin glass models of statistical physics. At high temperatures protein molecules are distended, but below some characteristic temperature they fold into a "native state" of low free energy. Proteins are sufficiently complicated that attempts to study them numerically rely upon simplifying the problem to one of manageable proportions yet retaining the fundamental features of the protein. Even then, investigation is non-trivial.

In this presentation we shall describe the advances that have been made using a new form of Monte Carlo simulation for two different systems: the HP (hydrophobic-polar) model, a very simplified lattice protein model that was introduced by biochemists over 20 years ago, and a continuum model for glycophorin A dimerization. In spite of its simplicity, the HP model is quite challenging to study, particularly as it attempts to fold into the native state. Because of their small size, these lattice proteins are at the intersection of biology, nanoscience, and statistical physics. In this manuscript we shall review numerical studies of lattice proteins with different sequences and introduce a new Monte Carlo method that performs a random walk in energy space that is highly efficient for such problems. We shall then consider our large scale Monte Carlo studies of several different HP proteins. Finally, we shall describe the application of Wang-Landau sampling to the problem of dimerization of glycophorin A in a membrane which is even more challenging in some ways because of the continuous degrees of freedom.

1.1. *"Wang-Landau" sampling*

While the "classic" Metropolis Monte Carlo algorithm continues to be widely used, a quite different Monte Carlo algorithm offers substantial advantages [1] in simplicity, broad applicability, and performance, particularly for systems with rough energy landscapes. (Originally termed the "random walk in energy space with a flat histogram" method, the technique is now referred to in the simulational physics community simply as "Wang-Landau sampling".) Unlike "traditional" Monte Carlo methods that generate canonical distributions at a given temperature $P(E) \sim g(E)e^{-E/k_BT}$, where $g(E)$ is the density of states, this method estimates $g(E)$ directly (iteratively) and then uses it for the estimate of canonical averages for thermodynamic quantities at any temperature, e.g. the free energy

$$F(T) = -k_B T \ln(Z) = -k_B T \ln(\sum_E g(E) e^{-\beta E}). \quad (1)$$

and its derivatives.

If a random walk in energy space is performed with a probability proportional to $\frac{1}{g(E)}$, then a flat histogram will be generated for the energy distribution. This is done by modifying the estimated $g(E)$ systematically to produce a "flat" histogram over the allowed range of energy and simultaneously making it converge to the correct value from some initial estimate, e.g. $g(E) = 1$ for all E. The random walk proceeds by randomly applying a trial move (e.g. flipping spins in a spin glass model) with transition probability

$$p(E_1 \to E_2) = \min(\frac{g(E_1)}{g(E_2)}, 1). \quad (2)$$

where E_1 and E_2 are energies before and after having performed the move. Each time an energy level E is visited, $g(E)$ is updated by multiplying the existing value by a *modification factor* $f > 1$, i.e. $g(E) \to g(E) * f$. A histogram of the energies that are "visited" is also updated; and when $H(E)$ is approximately "flat", f is reduced, e.g. $f_1 = \sqrt{f_0}$, the histogram is reset to $H(E) = 0$ for all E, and a new random walk is begun. This process is iterated n times, until f_n is smaller than some predefined final value (e.g. $f_{\text{final}} = \exp(10^{-8}) \simeq 1.00000001$). The final results are normalized, and if multiple walks are performed within different energy ranges, they must be matched up at the boundaries in energy. During the early stages of iteration the algorithm does not satisfy detailed balance since $g(E)$ is modified continuously; however, detailed balance is recovered to high precision after many iterations.

The final accuracy of $g(E)$ is controlled by two parameters: the final modification factor f_{final} and the flatness criterion p. Whereas Wang-Landau studies of some polymeric systems reported that $\ln(f_{\text{final}}) \simeq 10^{-6}$ is sufficient, we found that for the HP model, reliable DOS estimates over the entire energy range (including **the** lowest energies) required $\ln(f_{\text{final}}) \leq 10^{-7}$. We, thus, used a very stringent parameter set for our simulations of the HP model: $\ln(f_{\text{final}}) = 10^{-8}$ and $p = 0.8$. Knowledge of the exact energy range is needed for the examination of the flatness of the histogram, but the energy boundaries for polymer/protein models are *a* *priori* unknown. (For this reason, there has been substantial use of ground state search algorithms, e.g. for the HP model). To overcome this difficulty, the following procedure was used: Each time a new energy level $E_{\rm new}$ was found, it was marked as "visited" and $g(E_{\rm new})$ was set to $g_{\rm min}$, i.e. the minimum of g among all previously visited energy levels. The flatness of the histogram is checked only for those energy levels which have been visited. With this self-adaptive procedure, new regions of conformational space can be explored simultaneously as the current estimate of the density of states is further refined.

Wang-Landau sampling is a very flexible, highly efficient **and robust** Monte Carlo algorithm for the determination of the density of states of quite diverse statistical physical systems [1,2,3,4]. To demonstrate the effectiveness of the algorithm, in Fig.1 we show the dramatic variations in the canonical probabilities of states in the Edwards-Anderson (EA) spin glass as determined from a **single simulation**. (This system has been a great challenge in statistical physics for decades.)

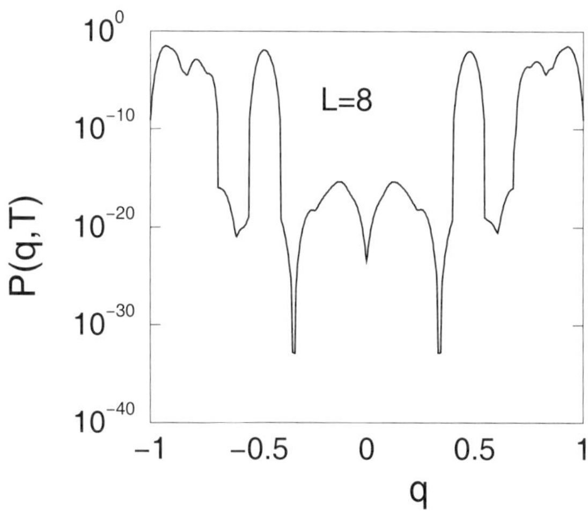

Figure 1. Canonical probability, extracted using the density of states, for the Edwards-Anderson spin glass in three dimensions. The result is for a temperature of $T/J = 0.1$ and the order parameter for this model is q.

Several different methods with a computer science flavor have been applied to the HP model with varying degrees of success. In addition, the configurations of an HP polymer may be studied using the traditional Metropolis method, but the complexity of the resultant free energy surface at low temperature renders the method extremely inefficient. Multicanonical Monte Carlo (i.e. sampling with a modified, fixed probability) can also be applied to the system and is more efficient than Metropolis. In our Wang-Landau sampling we have determined that traditional trial moves, e.g. end-flips, kink-flips, crankshaft, and pivots are inadequate in the collapsed state, so we have implemented a new set of trial moves that turns out to be very efficient, namely a combination between pull moves [5] and bond-rebridging moves [6]. These moves complement each other extremely well, as pull moves allow the polymer chain to naturally and fastly fold/unfold, while bond-rebridging moves change the polymer's configuration even at high densities. Our approach has been to implement Wang-Landau sampling together with these new types of trial moves to determine the density of states over the entire range of energies in a single run [3]. In addition to allowing us to find the ground state (or "native state" in biological language) we can determine thermodynamic properties as a function of temperature.

2. What have we already learned from simulations?

2.1. *The HP model*

In the hydrophobic-polar (HP) lattice model [7] the protein is represented as a self-avoiding chain of beads (i.e. coarse grained representations of the amino acid residues) on a rigid lattice. Only two types of amino acids - hydrophobic (H) and polar (P) - are included and an attractive interaction ϵ acts only between non-bonded neighboring H residues (i.e. $\epsilon_{HH} = -1, \epsilon_{HP} = \epsilon_{PP} = 0$). Different sequences of H- and P- monomers are used to "match" different proteins, and several sequences which have been deemed "benchmarks" have been studied extensively by a variety of methods. At low temperatures the HP chains fold into compact structures, but there are many different folds with different energies which are separated by free energy barriers. For this reason we have the same kinds of difficulties that one encounters with spin glass models of magnetism, and the location of the true ground state, or native state, becomes a difficult optimization problem.

We simulated various HP benchmark sequences found in the literature, emphasizing longer sequences like a 103mer in three dimensions (3D103) or a 100mer in two dimensions (2D100b). (Two dimensional models are of substantial significance within the context of statistical physics and could be physically realized if they were placed on a strongly attractive surface.) A number of different HP sequences are defined elsewhere[8]. The ground states of sequence 2D100b are believed to have an energy $E = -50$ [9], and various methods have confirmed this result [10,5,8]. However, previous attempts to obtain $g(E)$ over the entire energy range $[-50, 0]$ within a single simulation have failed [9,11]. In contrast, with our approach we were able to achieve this with high accuracy and Fig. 2 shows the resulting specific heat $C_V(T)/N$, depicting a peak at $T \approx 0.48$ (coil-globule transition) and a very weak shoulder at $T \approx 0.23$ (folding transition)[3].

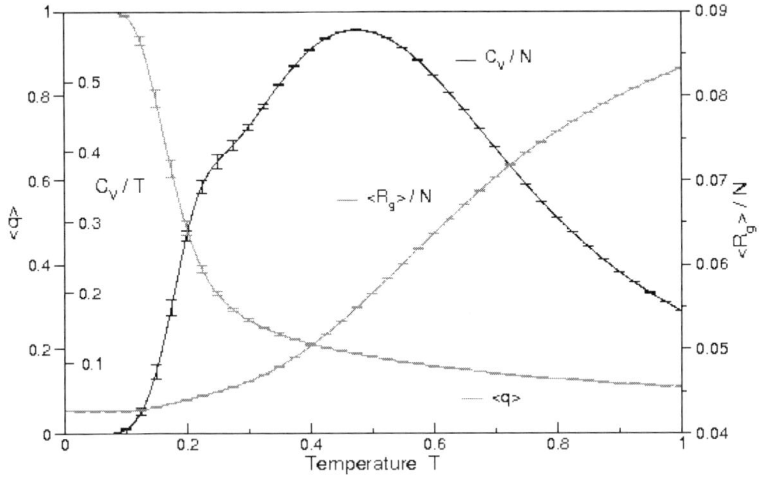

Figure 2. Specific heat C_V/N, mean radius of gyration $\langle R_g \rangle/N$ (N, chain length) and mean Jaccard index $\langle q \rangle$ as a function of temperature T for the two dimensional HP sequence 2D100b. Statistical errors were calculated by a Jackknife analysis from 15 independent Wang-Landau (C_V) and multicanonical production runs ($\langle R_g \rangle$ and $\langle q \rangle$) for each sequence.

As we shall see later, such two-step acquisition of the ground (native) state has been observed in studies of realistic protein models, e.g. glychophorin A, and is not restricted to lattice models. For different HP chains the relative locations of the peak and shoulder differ, but the general features do not.

For sequence 3D103, the lowest energy found by fragment regrowth Monte Carlo via energy-guided sequential sampling (FRESS) [8] was $E = -57$, but with our approach, we discovered an even lower state with energy -58 [3]. It was also possible to determine the density of states in the energy range $[-57, 0]$, from a single simulation, and with very high accuracy, although it was not possible to determine the relative magnitudes of the ground state and 1st excited state with high precision. Figure 3 displays the specific heat for 3D103, showing a peak at $T \approx 0.51$ and a shoulder at $T \approx 0.27$[3].

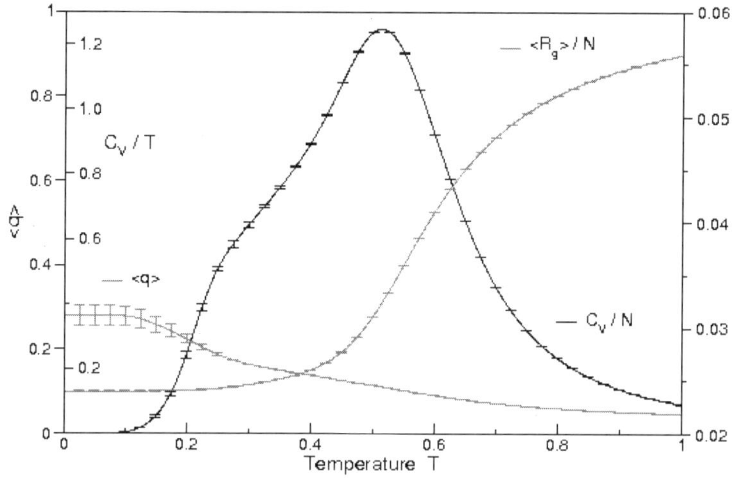

Figure 3. Specific heat C_V/N, mean radius of gyration $\langle R_g\rangle/N$ (N, chain length) and mean Jaccard index $\langle q\rangle$ as a function of temperature T for the three dimensional HP sequence 3D103. Statistical errors were calculated by a Jackknife analysis from 15 independent Wang-Landau (C_V) and multicanonical production runs ($\langle R_g\rangle$ and $\langle q\rangle$) for each sequence.

After determining $g(E)$ with high resolution we used it in a multicanonical run to measure the radius of gyration R_g [12] and the Jaccard index q [13], i.e. the similarity between a conformation and the ground state of a HP sequence:

$$q = \max\left\{ \frac{C_{s,g}}{C_{s,g} + C_s + C_g} \mid E_g = E_{\min} \right\}. \tag{3}$$

$C_{s,g}$ denotes the number of common (native) H-H contacts between a conformation s and the ground state g, and C_s, C_g are the numbers of H-H contacts found only in s and g, respectively, (the maximum stems from the degeneracy of ground states). Figures 2 and 3 also show the temperature averages $\langle R_g\rangle$ and $\langle q\rangle$ for sequences 2D100b and 3D103 and illustrates the complementary information provided by these two quantities. While $\langle R_g\rangle$ indicates the coil-to-globule collapse, $\langle q\rangle$ identifies the folding transition to the native state and may thus serve as a structural order parameter. In the case of the sequence 3D103, the ground state ($E = -58$) was excluded (due to the difficulty of finding the relative density for this state) which results in only a rather small Jaccard index for $T \to 0$. This shows that there are still large structural differences between conformations with $E = -57$ and the ground state with $E = -58$.

TABLE 1 compares results obtained using various methods, and, if available, the $g(E)$ for common benchmark HP sequences[3]. We also include results from methods which were focused on the low temperature range only, i.e. FRESS [8] and the variants of PERM (pruned-enriched Rosenbluth method) [10] and do not provide the entire density of states.

Table 1. Energy minima ($E_{\min}$) found by several sophisticated numerical methods for benchmark HP sequences in 2D and 3D. The first column names the sequence (dimension and length).

	WLS	EES^{11}	$MCCG^{14}$	$MSOE^{9}$	$FRESS^{8a}$	$PERM^{10a}$
$2D100a$	-48	-48	-	-	-48	-48
$2D100b$	-50	-49	-	-50[b]	-50	-50
$3D88$	-72	-	-	-	-72	-69
$3D103$	-58	-	-56	-	-57	-55

Note: [a] Ground state search only.
[b]g(E) not found.

Figure 4 shows the ground state structure for 3D103. The hydrophobic core is easy to see and the overall structure is quite symmetric. Observation shows that "antiphase" domains form near the shoulder so that portions of the native state are shifted with respect to other portions, and at the higher temperature peak the protein structure becomes distended.

2.2. *Glycophorin A*

We have used Wang-Landau sampling to investigate the behavior of glycophorin A [15]. The system contains two identical α-helices, A and B, of 22 residues each, running from $E72$ to $Y93$ (EITLIIFGVMAGVIGTILLISY). In order to decrease the large number of degrees of freedom inherent to this system, the backbone was kept fixed during the simulations and a unified atom representation was employed, in which only the heavy atoms and those polar hydrogen atoms susceptible to being involved in hydrogen bonding are explicitly modelled (a total of 378 atoms). Besides, the

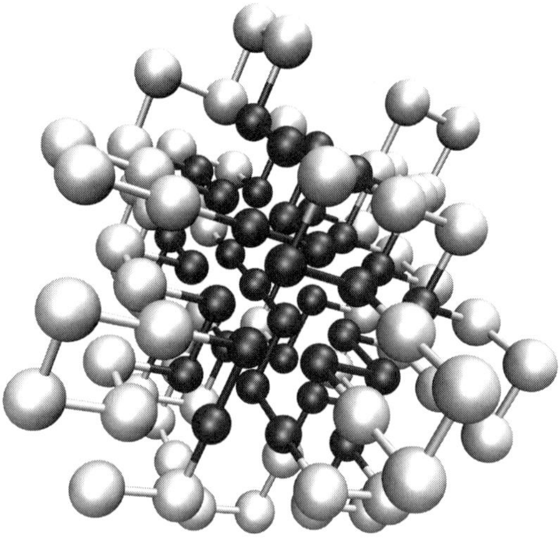

Figure 4. Native state structure for the 3D103 HP model. Hydrophobic residues are the small, dark balls and the large grey balls represent polar residues.

membrane was represented implicitly, i. e. the interaction between membrane and protein was treated by a mean-field (see E_{lipid} below). Like for the HP model, designing appropriate moves that allow the algorithm to search the entire energy landscape efficiently is crucial for the success of the simulation. Here, a total of seven trial moves were employed, designed to allow either global modifications of the protein or local changes in the conformation[16].

The energy is based on the CHARMM19 force-field [17] and a knowledge-based potential designed to take into account the membrane environment implicitly [18,19]. The energy is then given by

$$E = E^{A,B}_{inter} + E^{A}_{intra} + E^{B}_{intra} + E^{A}_{lipid} + E^{B}_{lipid}, \tag{4}$$

where $E^{A,B}_{inter}$ is the sum of the van der Waals and electrostatic energies between atoms of the helix A and those in helix B; E^{A}_{intra} and E^{B}_{intra} define the sum of the van der Waals, electrostatic and dihedral energies within the helix A and B, respectively, see [17]. Finally, E^{A}_{lipid} and E^{B}_{lipid} are the sums of the lipid-residue interactions of helix A and B, respectively. The lipid energy is a function of the z-coordinates of the C_α atoms of the residues,

$$E_{lipid} = \sum_{residues} \mu_{type}(|z|). \tag{5}$$

It reflects the propensity of an amino-acid to be located in different regions of the membrane. Because of this simplified representation of the membrane-protein interaction the values for transition temperatures obtained below should not be taken too seriously as the flexibility and deformation of the membrane at high temperature are not included in our model. We can, however, still draw conclusions about the characteristic behavior of the dimer, especially at low and medium temperatures where membrane-protein interactions are likely to play a limited role on the thermodynamics of the system [20].

The computed density of states illustrates the complexity of the system as a large range of $g(E)$, spanning nearly 110 orders of magnitude, was obtained. Such complexity requires making some compromises to reduce the computation time while keeping good thermodynamic resolution. In our case, one way to achieve this is to neglect the few bins with a very low energy. Indeed, the density of states for these bins is likely to have only a negligible influence on the thermodynamic quantities at temperatures of interest (i.e. $T > 200$K). For the glycophorin A system, the lowest energy found during a simulation was -666.7 kcal/mol. Restricting the energy range from -665 to -300 kcal/mol with a bin width $\Delta E = 1$kcal/mol, allowed all runs to converge within about 100 CPU hours per run (AMD Opteron processors) yet still provided good thermodynamic resolution. Once the density of states is obtained, one can also gather thermodynamic information for all types of observables, both energetic and structural. For that purpose, it suffices to run a second simulation with Wang-Landau sampling, however without updating $g(E)$ anymore. This so-called production run enables efficient sampling of the observables, including in conformational regions with low energies[16]. In our case, we sampled the inter-helix interaction ($E_{inter}^{A,B}$) and studied global structural changes by looking at the distance between the center of mass of the two helices ($d^{A,B}$) and the root-mean square deviation ($RMSD$) of the C_{alpha} atoms with respect to the experimental reference structure (model 1 of the NMR structure with PDB code 1AFO [15]). All results presented below were obtained by running and averaging five production runs of 8×10^8 MC moves each (about 100 CPU hours per run), a simulation length found sufficient to ensure the reliability of our results.

Figure 5 shows the specific heat of the glycophorin A (C_V), the thermodynamic average of $\langle E_{inter}^{A,B} \rangle$ and its derivative as a function of temperature. The specific heat shows a clear, rounded peak at $\sim$ 800K followed by a shoulder around $\sim$ 300K. While there is little effective interaction between the helices at 1100K, the system undergoes a dimerization transition at $\sim$ 800K characterized by a significant peak in the derivative. However, $d\langle E_{inter}^{A,B}\rangle/dT$ does not show another peak at lower temperatures, indicating that the inter-helix interaction is not responsible for the shoulder observed in the specific heat at 300K. The two structural observables

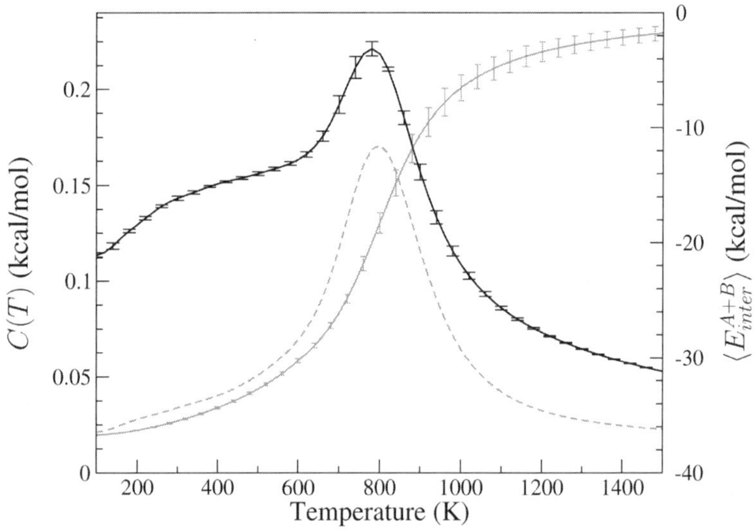

Figure 5. Black: Specific heat $C(T)$ of glycophorin A. Statistical errors were estimated using a Jackknife analysis from ten runs. Note that only the temperature range for which results are reliable (i.e. T > 100K) is shown. Gray: Variation of the inter-helix energy $\langle E_{inter}^{A,B}\rangle$ (solid line) and its derivative (dashed line) as a function of temperature.

$\langle d^{A,B}\rangle$ and $\langle RMSD\rangle$ show a similar behavior at high temperatures (results not shown). A large structural transition slightly above 800K is observed, indicating that close contact between the helices has established. However, contrary to $\langle E_{inter}^{A,B}\rangle$, an additional peak is observed around 300K. At this temperature, the RMSD has fallen below 2Å and the distribution of $d^{A,B}$ shows a preference for $d^{A,B} \approx 6.7$Å, a value close to the native distance of 6.64Å[16]. These observations clearly suggest that this second peak corresponds to the association of the helices into a native-like conformation. Indeed, representative structures at this transition temperature (300K) were found to be similar to the experimental structure (see Fig. 6).

The motif assumed to be responsible for the dimerization of the homodimer glycophorin A is composed of 7 residues `LIxxGVxxGVxxT` [21]. The

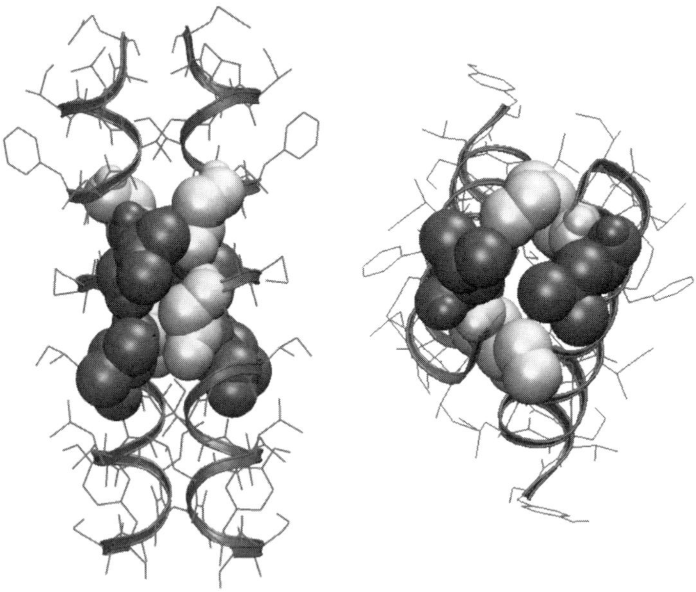

Figure 6. Typical structure for glycophorin A at 300K: *(Left)* Side view. Glycine residues G79 and G83 (white) pack densely with valine residues V80 and V84 (gray); *(Right)* Top view. Isoleucine I76 (white) and leucine L75 (gray) form a hook which stabilizes the dimer interface via hydrophobic interactions and close contact packing between branched residues.

motif `GxxxG`, well known to promote dimerization in membrane proteins and to favor helix-helix interactions in soluble proteins [22], acts as an anchor point (Fig. 6, left). The glycine residues G79 and G83 facilitate the approach of the two helices because of their small size and their minimal entropic contribution (absence of side chain). Hence, they allow a dense packing in a "groove and ridge" fashion with the two neighboring valines V80 and V84 [15]. Above this anchor point, leucine L75 and isoleucine I76 are necessary to stabilize the dimer. At the other end of the α-helix, threonine T87 stabilizes the dimer by forming an inter-helical hydrogen bonding [23,24].

To investigate the relative importance of these three strategic points at the dimer interface, during a production run we sampled the temperature dependence of the average energies $\langle E_{\rm motif} \rangle$ of the three following *motifs*: the motif *leucine*, composed of the interactions of residues L75 and I76; the motif *glycine*, composed of the interactions of residues G79, V80 and

G83; and the motif *threonine*, composed of the interactions of the residue T87. All three motifs show qualitative thermodynamic behavior similar to that observed above, i.e. their derivatives (dE/dT) exhibit two peaks near $\sim$ 800K and $\sim$ 300K, respectively. However, the significantly different magnitudes of the peaks at the two transitions clearly indicate the difference in thermal stability of the three motifs. Whereas $E_{leucine}$ undergoes mainly a transition at 800K, $E_{glycine}$ shows similar peak ratios at both 300K and 800K, and $E_{threonine}$ has a major transition at 300K. This suggests the following order of stability: *leucine* > *glycine* > *threonine*.

From the results for both global and motif-based observables, we can describe the structural and energetic changes taking place during the dimerization process of glycophorin A. At 800K, the two helices come into contact and interact with a significant inter-helix energy. Structurally, the majority of dimers found at this temperature feature some characteristics of the native structure. Indeed, the motif *leucine*, and to a lesser extent, the motif *glycine*, already exhibit significant contributions to the overall ground state energy. This observation is not surprising considering that the GxxxG motif is known to promote dimerization of many membrane proteins [21,25]. Besides, leucine L75 and isoleucine I76 form a "hook" in the native structure, which stabilizes the dimer interface via hydrophobic interactions and close contact packing between branched residues, (see Fig. 6, right). Another transition, at about 300K, corresponds to the convergence towards the native structure. The average RMSD falls below 2Å and the motif *glycine*, and especially the motif *threonine*, undergo a transition towards the native energies. Stabilization of the dimer is affected via the formation of inter-helical hydrogen bonding.

Our findings on the two-stage dimerization process of glycophorin A agree very well with the hypothesis proposed by Schneider [26] who suggested decomposing the oligomerization into two stages. First, the contact between helices is promoted by a detailed fit between the helical surfaces, leading to close packing and van der Waals interactions. In a second stage, stabilization of the preformed dimer is obtained by electrostatic interactions, i.e. hydrogen bonding, or binding of a cofactor. We found indeed a first dimerization step governed by dispersive interactions (motif *leucine*) and close packing (motif *glycine*), while the second transition involved the formation of hydrogen bonds within the motif *threonine*.

3. Conclusions

Wang-Landau sampling has proven to be effective for studying both folding of lattice HP model proteins and dimerization of glycophorin A. The same basic techniques that are used for models in statistical physics work well, and in both cases a two-step process leads to the low temperature native state. However, it is interesting to see the different level of accuracy needed by the two models to obtain significant results. The HP model, simple in appearance, requires accurate estimation of the density of states to obtain sufficient details about the folding transition. To the contrary, information on the dimerization of glycophorin A can be well tackled by using a rough estimate of the density of states followed by multicanonical sampling of structural and energetical observables.

Acknowledgements

We wish to thank S.-H. Tsai and D. Seaton for valuable discussions.

References

1. F. Wang and D. P. Landau, Phys. Rev. Lett. **86**, 2050 (2001); Phys. Rev. E **64**, 056101 (2001); Comput. Phys. Commun. **147**, 674 (2002).
2. D. P. Landau and K. Binder, *A Guide to Monte Carlo Simulations in Statistical Physics* (Cambridge University Press, Cambridge, UK, 2005), 2nd ed.
3. T. Wüst and D. P. Landau, Phys. Rev. Lett. **102**, 178101 (2009).
4. T. Wüst and D. P. Landau, Comput. Phys. Commun. **179**, 124 (2008).
5. N. Lesh, M. Mitzenmacher, and S. Whitesides, in *RECOMB* (2003), p. 188.
6. J. M. Deutsch, J. Chem. Phys. **106**, 8849 (1997).
7. K. A. Dill, Biochemistry **24**, 1501 (1985); K. F. Lau and K. A. Dill, Macromolecules **22**, 3986 (1989).
8. J. Zhang, S. C. Kou, and J. S. Liu, J. Chem. Phys. **126**, 225101 (2007).
9. Y. Iba, G. Chikenji, and M. Kikuchi, J. Phys. Soc. Jpn. **67**, 3327 (1998); Phys. Rev. Lett. **83**, 1886 (1999).
10. H. Frauenkron *et al.*, Phys. Rev. Lett. **80**, 3149 (1998); U. Bastolla *et al.*, Proteins **32**, 52 (1998); H.-P. Hsu *et al.*, Phys. Rev. E **68**, 021113 (2003).
11. S. C. Kou, J. Oh, and W. H. Wong, J. Chem. Phys. **124**, 244903 (2006).
12. A. D. Sokal, in *Monte Carlo and Molecular Dynamics Simulations in Polymer Science*, edited by K. Binder (Oxford University Press, New York, Oxford, 1995), p. 47.
13. R. Fraser and J. I. Glasgow, in *ICANNGA (1)* (2007), p. 758.
14. M. Bachmann and W. Janke, Phys. Rev. Lett. **91**, 208105 (2003); J. Chem. Phys. **120**, 6779 (2004).
15. K. R. MacKenzie, J. H. Prestegard, and D. M. Engelman, Science **276**, 131 (1997).

16. C. Gervais, T. Wüst, D. P. Landau, and Y. Xu, J. Chem. Phys. **130**, 215106 (2009).
17. E. Neria, S. Fischer, and M. Karplus, J. Chem. Phys. **105**, 1902 (1996).
18. Z. Chen and Y. Xu, Proteins **62**, 539 (2006).
19. Z. Chen and Y. Xu, J. Bioinform. Comput. Biol. **4**, 317 (2006).
20. J. U. Bowie, Curr. Opin. Struct. Biol. **11**, 397 (2001).
21. B. Brosig and D. Langosch, Protein Sci. **7**, 1052 (1998).
22. G. Kleiger, R. Grothe, P. Mallick, and D. Eisenberg, Biochem. **41**, 5990 (2002).
23. S. O. Smith, M. Eilers, D. Song, E. Crocker, W. Ying, M. Groesbeek, G. Metz, M.Ziliox, and S. Aimoto, Biophys. J. **82**, 2476 (2002).
24. J. M. Cuthbertson, P. J. Bond, and M. S. P. Sansom, Biochem. **45**, 14298 (2006).
25. W. P. Russ and D. M. Engelman, J. Mol. Biol. **296**, 911 (2000).
26. D. Schneider, FEBS Lett. **577**, 5 (2004).

STOCHASTIC MATRICES AS A TOOL FOR BIOLOGICAL EVOLUTION MODELS

R. KERNER
Laboratoire LPTMC
Université Paris-VI - CNRS UMR 7600,
Tour 24, Boite 121, 4 Place Jussieu, 75005 Paris, France
E-mail: `rk@ccr.jussieu.fr`

R. ALDROVANDI
Instituto de Física Teórica - São Paulo State University - UNESP
Rua Dr. Bento Teobaldo Ferraz, 271 - Bl. II
01140-070 Barra Funda - São Paulo - SP - Brazil
E-mail: `raldrovandi@pq.cnpq.br`

A simple model is set forth in which discrete changes in the statistical distribution of various features are treated by means of stochastic matrices, which transform one probability distribution into another one. A few examples of application of this method, like phyllotaxis and heredity of blood groups in humans, are presented and discussed.

1. Introduction

When we are dealing with a great number of similar events or objects, we turn naturally to statistical analysis and probability distributions, because it is not only impossible, but also useless to keep the track of all characteristics of separate single items in the swarm of data. The precise knowledge of the exact numbers of items found in a given state at a given time is often out of reach experimentally. Life in general is an off-equilibrium state, and its most important feature is constant variation, change and evolution. In particular, this concerns statistical distributions of different types of cells in an organism, or different types of organisms in a population of a given species, or a distribution of different species in an ecosystem. In all these cases the statistical distributions vary in time, slowly or quite rapidly according to the case, and sometimes they can reach a state of relative equilibrium which may last very long. Mathematical models usually describe the continuous time evolution by means of differential equations. However, in many cases a discrete model is more appropriate, e.g. when analyzing the variations occurring from one generation of organisms to another, or other processes that can be regarded as a series of discrete steps. If a large number of steps is analyzed, then we can always proceed to the continuous limit and recover the corresponding differential equation model. In this article we propose several models using stochastic matrices as generators of time development of probability distributions in living organisms. The first section, in which we define and discuss elementary properties of stochastic matrices, is followed by an exposition of a few applications to biological systems: growth of two-dimensional patterns known as *phyllotaxis*, growth of icosahedral viral capsids, and evolution due to the direct hereditary transmission of genetic features, like e.g. blood groups in humans.

2. Stochastic Matrices and their Properties

Let us suppose that we know the statistical repartition of a few essential parameters among a huge number of items. The simplest example is given by a dychotomic variable which can take on only one of two values; as an example, we can consider long and short segments agglomerating together to form a chain, or black and white balls, etc. When a dynamical process is observed, like agglomeration in clusters, or transformation of one part of the items into another, with or without conservation of the total number of entities, the statistics change with time, and different behaviors can be observed, stable, quasi-stable or unstable. One of the best ways to introduce the stochastic matrix approach and its essential properties is the analysis of linear growth, either by agglomeration or by internal multiplication, of simple chains composed of two types of units (molecules, cells...), which we shall represent by long and short segments. The analysis of an elementary agglomeration process can be formulated using linear mappings represented by matrices acting on the set of probabilities. Consider a linear growth process in which a very long chain is gradually assembled by sticking at its end two types of segments, long (L) and short (S), which may represent two different types of molecules, one by one, yielding the following growth process as seen from the right side of the chain:

$$SLL \to SLLS \to SLLSL \to SLLSLS \to SLLSLSL \to ... \qquad (1)$$

Here, for simplicity, we have only added new segments to the right side, but a more general situation can be envisaged with growth by agglomeration on both sides; as we shall see, there will be no difference from the statistical point of view. Of course, such a growth process can be either totally or partly ordered, i.e. more or less strict rules can be imposed upon the consequent agglomeration steps. One can exclude the pairing between two similar segments, thus allowing only alternating chains $LSLSLS...$; one can exclude only the SS pairs admitting the LL pairs to be formed, as in the Fibonacci sequence given by a simple *inflation rule* $L \to LS, \quad S \to L$, leading to the following growth scheme, here starting from an L segment; but it would be the same if we started with an S segment, just shifted a bit:

$$L \to SL \to LSL \to SLLSL \to LSLSLLSL \to SLLSLLSLSLLSL... \qquad (2)$$

We can represent this process by an action of the so-called *inflation matrix* on a column representing a couple L and S:

$$\begin{pmatrix} 1 & 1 \\ 1 & 0 \end{pmatrix} \begin{pmatrix} L \\ S \end{pmatrix} = \begin{pmatrix} L+S \\ L \end{pmatrix}. \qquad (3)$$

It is well known that if we repeat the action of this matrix on any initial segment, the resulting chain will present the statistics of L and S segments tending to the limit value

$$\left[\frac{N_L}{N_S}\right]_\infty = \tau = \frac{1+\sqrt{5}}{2},$$

known also under the name of the "golden number". One can easily check this assertion by inspecting the numbers N_L and N_S resulting from the subsequent steps of inflation (here starting from a single L segment):

$$(1,0);\ (2,1);\ (3,2);\ (5,3);\ (8,5);\ (13,8);\ ...$$

However, similar chains can be obtained by agglomeration of segments from surrounding medium containing a great number of such building blocks. In the simplest case, let us assume that the probability of agglomeration depends only on the contact interaction between the last segment of the chain and the new one that is about to stick to it, and remain fixed there as the new extreme element, to which another buliding block, S or L as it may be, shall stick in turn. Obviously, the probability that the newly attached segment will be short or long depends also on the relative concentration (denoted here by c) of, say, S-segments among all segments available in the surrounding medium: $c = N_S/(N_S + N_L)$, and in principle can vary with time. Here we shall suppose that the surrounding medium is so huge, that the concentration c can be considered as being constant. Let us suppose that we can observe the growth process of a huge amount of chains, so that a reliable statistics of S and L segments on their free extremities can be established. In principle, one could follow all agglomeration processes observing the growth of all chains one by one; but the same statistics should be obtained if we pick up all the chains containing n segments, then all the chains containing $n+1$ segments, then all the chains containing $n+2$ segments, etc., and get the statistics of short and long building blocks at their extremities; by the way, there is no reason to suppose any dissymetry between two extremities, so that we could look at one of the extremities of each chain. The coincidence of statistics made on a huge number of items and the statistics obtained as function of time while collecting the data on the chains in the process of growth is called the *ergodicity property* of a given statistical ensemble. We can also suppose that the process of agglomeration is progressing in time in such a manner, that the average chain length, $< n >$ is a steadily increasing function of time, so that $\frac{d<n>}{dt} > 0$. In order to get the result as close as possible to the Fibonacci series, we must exclude the pairs of short neighbors, i.e. the SS pairs, leaving three possibilities of agglomeration: $L + L \rightarrow LL$, $L + S \rightarrow LS$ and $S + L \rightarrow SL$. Let us denote the probabilities of finding an S or an L at the end of a chain of average size n by p_S and p_L, respectively; these probabilities must be normalized to 1, so that $p_S + p_L = 1$. They can be arranged in a column as follows:

$$\begin{pmatrix} p_L \\ p_S \end{pmatrix}$$

After the agglomeration of one more segment to all these chains, we obtain the chains by one unit longer, of the size $n+1$. Depending on the concentration c in the reservoir, the purveyor from which the new segments keep arriving, and on the possible "chemical affinities", or energy barriers characterizing the three admissible ways of assembling the segments, the resulting statistics will be modified. The simplest assumption is that the new probabilities are given by linear combinations of the previous ones, which can be described as a certain matrix action on the probability column:

$$\begin{pmatrix} M_{LL} & 1 \\ M_{SL} & 0 \end{pmatrix} \begin{pmatrix} p_L \\ p_S \end{pmatrix} = \begin{pmatrix} p_L{}' \\ p_S{}' \end{pmatrix}. \tag{4}$$

We have set the matrix element M_{LS} equal to one, and the matrix element $M_{SS} = 0$ because only a long segment (L) can be attached to a short one if it happens to be at the extremity of a chain, i.e. the transition probability $S \rightarrow L$ is 1, while the probability of an $S \rightarrow S$ transition is zero. Obviously, we must have $p_L{}' + p_S{}' = 1$, which means that also

$M_{LL}\, p_L + p_S + M_{SL}\, p_L = 1$; recalling that $p_S = 1 - p_L$, we get the condition that $M_{LL} + M_{SL} = 1$. This is a particular example of the so-called *stochastic matrix*, which has the following properties:

i) all its entries are non-negative real numbers less or equal to 1;

ii) the sum of entries in each column is equal to 1.

In our case, this means that $M_{SL} = 1 - M_{LL}$, with $0 \leq M_{LL} \leq 1$. The characteristic equation

$$det \begin{pmatrix} M_{LL} - \lambda & 1 \\ M_{SL} & -\lambda \end{pmatrix} = \lambda^2 - M_{LL}\, \lambda - M_{SL} = 0 \tag{5}$$

has two solutions,

$$\lambda_1 = 1, \quad \lambda_2 = -M_{SL} = -1 + M_{LL}. \tag{6}$$

This is the common feature of stochastic matrices, independent of the rank: there must be at least one eigenvalue equal to 1, other eigenvalues being always of modulus less than one. For larger stochastic matrices, there can be many unitary eigenvalues, but only in a *reducible* case when such a matrix splits into independent blocks. This means that after many consecutive actions of the stochastic matrix on an arbitrary initial probability distribution, the limit will be the eigenvector corresponding to the eigenvalue 1. We easily get the only independent equation:

$$M_{LL}\, p_L^\infty + p_S^\infty = p_L^\infty, \tag{7}$$

which yields the normalized probabilities of the limit distribution

$$p_L^\infty = \frac{1}{1 + M_{SL}}, \qquad p_S^\infty = \frac{M_{SL}}{1 + M_{SL}}. \tag{8}$$

If the sticking probabilities depended only on the relative concentration of long and short segments in the surrounding medium, without any other dependence on energy barriers and the like, then the matrix elements would be as follows:

$$M_{LL} = (1 - c), \quad M_{SL} = c. \tag{9}$$

The resulting limit values for the probabilities will become then

$$p_L = \frac{1}{1 + c}, \quad p_S = \frac{c}{1 + c}. \tag{10}$$

It is easy to see that the final distribution of probabilities in the very long chains (in the limit, the probability of finding a given type of segment at the extremity will be the same as in the entire chain) is not equal to the surrounding distribution: inside the chains, the rate of S-segments is always less than c, the equality being attained only when $c = 0$, when only pure L-chains are obtained from a 100% of L segments. This could have been expected from the beginning, because we have prohibited the $S - S$ pairs, so that there will be always an excess of L type in the chains as compared with the surrounding medium. This phenomenon can be interpreted as the simplest example of *self-organization* when the agglomerates "choose" which building blocks they pick up more eagerly, and in which combinations. Another interesting limit is obtained when $M_{LL} = 0$, i.e. when not only the SS-pairs are prohibited, but also the LL-pairs as well. The corresponding transformation matrix becomes

$$\begin{pmatrix} 0 & 1 \\ 1 & 0 \end{pmatrix}$$

with the eigenvalues 1 and -1; this matrix is an *idempotent*, i.e. its square is equal to the identity matrix. The resulting chain is a perfect one-dimensional crystal:

$$LSLSLSLSLSLSLS..., \tag{11}$$

alternating short and long segments. This result is independent of the surrounding relative concentrations – an example of self-organization.

Now we can generalize the simple example presented above, and give a more rigorous description of stochastic matrices.

An $N \times N$ matrix M is a stochastic matrix if its entries are real numbers satisfying the two conditions

$$0 \leq M_{\alpha\beta} \leq 1 \ , \quad \forall \ \alpha, \beta = 1, 2, \dots N \quad \text{and} \quad \sum_{\alpha=1}^{N} M_{\alpha\beta} = 1 \ . \tag{12}$$

As an immediate consequence, we have

$$\sum_{\alpha=1}^{N} M^2{}_{\alpha\beta} = \sum_{\alpha=1}^{N} \sum_{\gamma=1}^{N} M_{\alpha\gamma} M_{\gamma\beta} = 1 \ . \tag{13}$$

It follows that M^2, as well as all the successive powers M^m, are also stochastic matrices. Suppose there is some physical system whose states are indicated by the discrete labels $\alpha, \beta, \gamma \dots$ The entry $M_{\alpha\beta}$ is then the probability of transition from state β to state α and the power m is a discrete "time" parameter. We shall call a column-vector v a probability distribution when each entry satisfies $0 \leq v_\alpha \leq 1$ and further $\sum_\alpha v_\alpha = 1$. Each column in a stochastic matrix is a probability distribution by itself. In this formalism, the complete characterization of the state of the system is given by a similar distribution, whose meaning is that the system has probability v_α of being in state α, probability v_β of being in state β, etc.

If $P_n(\alpha)$ is the probability of the system being in state α at time n, its evolution is governed by the equation

$$P_{n+1}(\alpha) = \sum_{\beta=1}^{N} M_{\alpha\beta} P_n(\beta). \tag{14}$$

Condition (12) implies that probability is conserved:

$$\sum_{\alpha=1}^{N} P_{n+1}(\alpha) = \sum_{\beta=1}^{N} P_n(\beta). \tag{15}$$

The probability distribution $P(a)$ is an equilibrium distribution if it is an eigenvector of M with eigenvalue 1:

$$P(\alpha) = \sum_{\beta=1}^{N} M_{\alpha\beta} P(\beta), \tag{16}$$

which defines a stationary solution of (14). The spectrum of M is, of course, of primary interest. The fundamental result comes from the Cayley-Hamilton theorem. If the characteristic equation is written as

$$\Delta(\lambda) = \prod_{j=0}^{N} (\lambda - \lambda_j) = \sum_{j=0}^{N} c_j \lambda^j = 0, \tag{17}$$

then $\Delta(M) = 0$ and $\sum_{m=0}^{N} c_m {M^m}_{\alpha\beta} = 0$. Taking $\sum_\alpha$ and using (13), we find $\sum_{m=0}^{N} c_m = 0$, which says that 1 is an eigenvalue of M. Thus, a stochastic matrix has always at least one eigenvector of eigenvalue 1, that is, at least one equilibrium distribution.

3. An exercise in phyllotaxis

Stochastic matrices can be also used in the analysis of groxth of two and three-dimensional patterns. The simple model of linear growth exposed in the previous section can be easily generalized. As a matter of fact, stochastic matrices can be also used in the analysis of two– and three–dimensional growth, in particular, the formation of various patterns in the vegetal world, often known under the name of *phyllotaxis.* And also in animals, in particular in primitive species which develop quite interesting patterns on their surfaces. It is known since Kepler[4] that, when deformable elastic spheres are posed on a flat surface and pressed together, the resulting pattern is very close to the regular hexagonal tiling of the plane, with three hexagons sharing a common vertex. Such patterns are quite common in nature; however, most of the time the shapes are not flat, but curved; this is achieved by inserting a certain amount of pentagons which create angular deficit resulting in positive curvature. This is particularly visible in icosahedral viral capsids, but can be also found on another scale in multicellular organisms. The assembling of polygonal structures, imitating growth of living tissue, can be also encoded in stochastic matrices. Let us consider the simplest case in two dimensions, with a three-coordinate network consisting exclusively of 5-, 6- and 7-sided equilateral polygons. As we know, on the Euclidean plane tri-coordinate and purely hexagonal lattice displays perfect translational symmetry, and can be extended to infinity. In contrast, pentagons and heptagons found in such a lattice represent local *defects*; it is therefore logical to assume that their presence in otherwise hexagonal lattice creates local stress equivalent to certain energy cost. Whatever its value, we know that in order to be able to produce an infinite tiling of a plane, the total number of pentagons must be equal to the total number of heptagons. Figure 1 shows three examples of two-dimensional tissues formed exclusively by pentagonal and heptagonal cells, of which two realize some crystalline symmetries, while the third is a random structure.

The statistical factors that we shall consider when evaluating probabilities of various configurations depend on the way in which the polygons (rings) are formed. If the tiling is produced with already prefabricated equilateral (but deformable) polygons by progressive sticking their edges, then one has to take into account the fact that there are 5 different ways of asembling a pentagon sharing a common edge with another polygonal cell, 6 ways of assembling a hexagon, and 7 ways of assembling a heptagon. The growth of the network, seen as the progressive addition of new polygons to the cells already in place, can be described by a stochastic matrix obtained from the analysis of the possible elementary agglomeration steps. If we denote the probabilities of finding an n-sided cell among the three types of newly produced ones by p_5, p_6 and p_7, with $p_5 + p_6 + p_7 = 1$, then the probabilities of nine possible elementary agglomeration steps can be evaluated by means of the following factors: sticking a new pentagon to another pentagon occurs with probability proportional to $5\, p_5\, P_5^{(0)}\, e^{-\alpha_{55}}$; sticking a

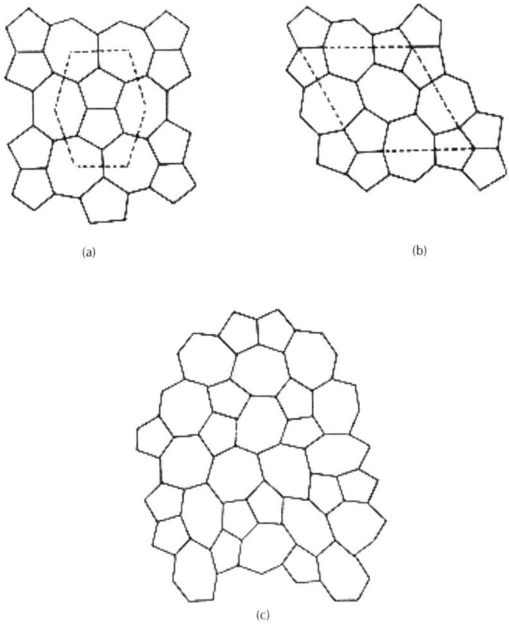

Figure 1. One random (c) and two regular (a, b) tilings with $P_6 = 0$.

pentagon to a hexagon should be proportional to $5\,p_5\,P_6^{(0)}\,e^{-\alpha_{56}}$, agglomerating a hexagon to a pentagon should occur with probability proportional to $6\,p_6\,P_5^{(0)}\,e^{-\alpha_{65}}$, and so on. The exponential factors correspond to the affinities coming from chemical and energetical barriers that may promote or inhibit certain couplings as compared with other possible issues. We shall also suppose that these factors are symmetric, i.e. that $\alpha_{56} = \alpha_{65}$, $\alpha_{67} = \alpha_{76}$ and $\alpha_{57} = \alpha_{75}$, thus leaving us with six free parameters. Then all possible transitions resulting from adjunction of new polygonal cells can be encoded in the following stochastic 3×3 matrix whose action on the initial probabilities $P_5^{(0)}$, $P_6^{(0)}$ and $P_7^{(0)}$ describes the new probability distribution resulting - on the average - from the adjunction of new cells:

$$\begin{pmatrix} 5p_5\,e^{-\alpha_{55}}/Q_5 & 5p_5\,e^{-\alpha_{56}}/Q_6 & 5p_5\,e^{-\alpha_{57}}/Q_7 \\ 6p_6\,e^{-\alpha_{65}}/Q_5 & 6p_6\,e^{-\alpha_{66}}/Q_6 & 6p_6\,e^{-\alpha_{67}}/Q_7 \\ 7p_7\,e^{-\alpha_{75}}/Q_5 & 7p_7\,e^{-\alpha_{76}}/Q_6 & 7p_7\,e^{-\alpha_{77}}/Q_7 \end{pmatrix} \begin{pmatrix} P_5^{(0)} \\ P_6^{(0)} \\ P_7^{(0)} \end{pmatrix} = \begin{pmatrix} P_5' \\ P_6' \\ P_7' \end{pmatrix} \tag{18}$$

with the normalization factors Q_5, Q_6 and Q_7 given by the following sums:

$$Q_5 = 5p_5\,e^{-\alpha_{55}} + 6p_6\,e^{-\alpha_{65}} + p_7\,e^{-\alpha_{75}},$$

$$Q_6 = 5p_5\,e^{-\alpha_{56}} + 6p_6\,e^{-\alpha_{66}} + 7p_7\,e^{-\alpha_{76}},$$

$$Q_7 = 5p_5\, e^{-\alpha_{57}} + 6p_6\, e^{-\alpha_{67}} + 7p_7\, e^{-\alpha_{77}}.$$

In statistical physics, the exponentials would represent the Boltzmann factors depending on the ambient temperature, like $e^{-\frac{E}{k_B T}}$; here, however, the variations of absolute temperature T are a very minor factor, so we shall omit the explicit dependence on T, tacitly assuming that T is constant. In order to simplify our model, let us also assume that the potential barriers are proportional to the departure from flatness associated with a given coupling; we shall suppose that no barrier exists between the most natural association of two hexagons, a single potential should be associated with the couple formed by a hexagon with a pentagon or a heptagon, a double barrier with the couple of two pentagons or two heptagons, and again no barrier for the couple pentagon-heptagon, because their curvatures, respectively positive and negative, should compensate each other.

So, from now on, we shall use the simplified notations

$$e^{-\alpha_{56}} = e^{-\alpha_{67}} = e^{-\alpha}, \quad e^{-\alpha_{55}} = e^{-\alpha_{77}} = e^{-2\alpha}, \quad e^{-\alpha_{66}} = e^{-\alpha_{57}} = e^{0} = 1.$$

The action of the stochastic matrix (18) on the probability column can be represented symbolically as $\mathbf{p}' = M\,\mathbf{p}$; this, in turn, can be interpreted as a differential equation as follows. Supposing that the elementary agglomeration step takes the time Δt on the average, the variation of the probability distribution associated with one single agglomeration step (or the single creation of a new cell) will be $\Delta\mathbf{p} = \mathbf{p}' - \mathbf{p}$. In the continuous limit, we can write then

$$\frac{\delta \mathbf{p}}{\Delta t} \simeq \frac{d\mathbf{p}}{dt} = \mathbf{p}' - \mathbf{p} = \mathbf{p} - M\,\mathbf{p}. \tag{19}$$

A stationary régime occurs when the all over probability distribution remains stable and invariant: one can then write, symbolically, $M\,\mathbf{p}_\infty = \mathbf{p}_\infty$, with the tacit assumption that the probabilities $P_i^{(O)}$ in the agglomerates are the same as the probability distribution of the newly created or attached entities, p_i $(i, j = 5, 6, 7)$. But this is nothing else but the equation for the eigenvector corresponding to the eigenvalue 1. It can be also interpreted as a set of diffferential equations, and the stationary solutions are obtained when all the derivatives are equal to zero. Moreover, the resulting algebraic equations being homogeneous, $(\mathbf{I} - M)\,\mathbf{p} = \mathbf{0}$, the matrix $(\mathbf{I} - M)$ has to be singular in order to admit non-trivial eigenvectors. This is indeed the case, because the entries in the probability column are not independent, satisfying $\sum p_i = 1$. In our example, this means that we can choose only two out of three variables (say, p_5 and p_6) as being independent, setting everywhere $p_7 = 1 - p_5 - p_6$ and keeping only two equations out of three. They can be written as follows:

$$\frac{dp_5}{dt} = 5p_5^2 e^{-2\alpha} Q_6 Q_7 + 6p_5 p_6 e^{-\alpha} Q_5 Q_7 + 7p_5 p_7 Q_5 Q_6 - p_5 Q_5 Q_6 Q_7 = 0,$$

$$\frac{dp_6}{dt} = 5p_5 p_6 e^{\alpha} Q_6 Q_7 + 6p_6^2 Q_5 Q_7 + 7p_6 p_7 e^{-2\alpha} Q_5 Q_6 - p_6 Q_5 Q_6 Q_7 = 0\,. \tag{20}$$

The third equation, with dp_7/dt, is linearly dependent of these two: the variable p_7 should be replaced everywhere by $1 - p_5 - p_6$, leaving two equations for two variables, p_5 and p_6. In a real agglomeration process, at least at its initial stage, one can observe a mixture of single polygons and freshly created doublets; later on also triplets, quadruplets of agglomerated polygons are created but, at the beginning, only single polygons ("singlets") and edge-sharing pairs ("doublets") dominate. The assembling of bigger clusters can be also represented by an action of an stochastic matrix, only a larger one because it has to take into account more congurations. For example, there are six doublets made of three kinds of polygons considered above, 55, 56, 57, 66, 67 and 77, and even more triplets. Real problems are, however, often reduced to a fewer number of congurations, many of them being prohibited for energetic or geometric reasons. Although it looks quite complicated (and we shall not display explicit forms), it is quite easy to perform the analysis of phase trajectories of this differential system, displayed in Fig. 2. The probabilities p_5 and p_6 are confined to a triangle because they

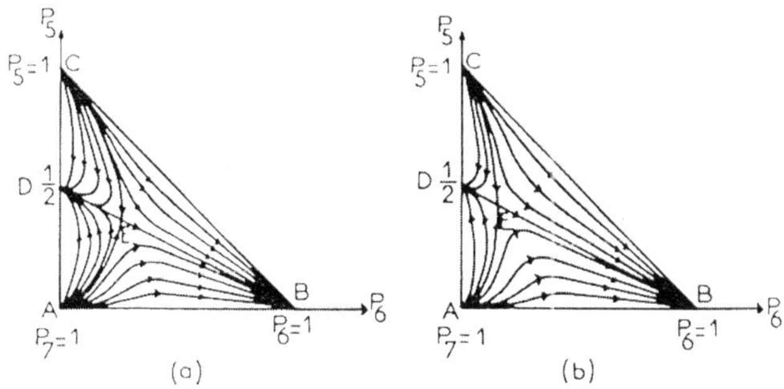

Figure 2. Two phase potraits of probability trajectories describing the agglomeration process of 5-, 6- and 7-sided polygons.

have to remain positive but smaller than 1, as well as $p_7 = 1 - p_5 - p_6$. First of all, we find out the positions of *singular points*, corresponding to constant solutions at which the two derivatives vanish simultaneously. There are five such solutions, three of which are found at the vertices of the simplex of probabilities:

$$A:\ p_7 = 1,\quad p_5 = 0,\ (p_6 = 0);\quad B:\ p_6 = 1\quad (p_5 = p_7 = 0);$$

$$C:\ p_5 = 1,\quad p_7 = 0,\quad (p_6 = 0); \tag{21}$$

two other solutionss are:

$$D:\quad p_5 = p_7 = \frac{1}{2}\quad (p_6 = 0), \tag{22}$$

and the fifth lies inside the triangle:

$$E:\quad p_5 = \frac{1}{3 - e^{-\alpha}},\quad p_6 = \frac{1 - e^{-\alpha}}{3 - e^{-\alpha}},\quad p_7 = 1 - p_5 - p_6 = \frac{1}{3 - e^{-\alpha}}. \tag{23}$$

It is easy to check (by linearization of the system (20) that A and C are *repulsive* singular points, B and D are *attractive* singular points, while E is a *saddle point.* There is another saddle point at infinity, a fact coming also from Euler's formula for a sphere, but for the probabilities it has, of course, no physical meaning. The points A,B,C,D keep their position steady independently of the value of parameter α, whereas the position of the saddle point E does depend on the value of that parameter, falling towards the attractive point D when $\alpha \to 0$. Typical phase trajectories are displayed in Fig. 2. It is worthwhile to note that the two attractive points and the saddle point are found on the line $p_5 = p_7$, which satisfies the flatness constraint for three-coordinate lattices. The separatrix curve AEC divides the simplex into two regions: on the right, the system is driven towards crystallization ($P_6 = 1$), whereas if the initial conditions happen to be on the left, the system will prefer another attractive singular point corresponding to an amorphous mixture of sole pentagons and heptagons. Besides, in a more realistic model taking into account second-order effects, the system can remain infinitely long in the vicinity of the meta-stable saddle point E. In most of the examples of two-dimensional growth the resulting structure is not flat; more often curvature is not only a welcomed, but also a needed feature. In this case the condition $p_5 = p_7$ should be postponed; and one can do without heptagons at all, like what can be observed in capsid viruse shells[6,7]. A better insight in the growth and agglomeration process can be obtained if one looks at the evolution of *pairs* of polygona, as represented in Fig. 3.

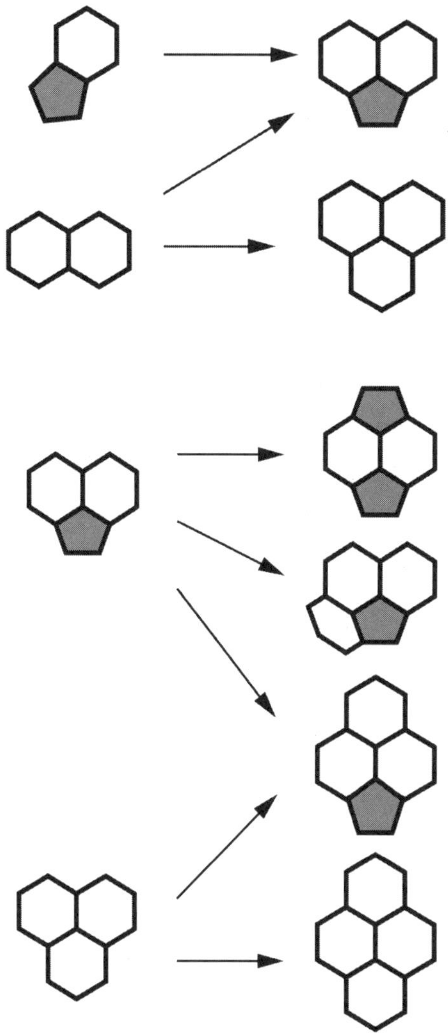

Figure 3. Formation of a new polygon in a (6,5)-type and in a (6,6)-type cavities.

Here again, it is easy to follow the creation of new pairs when a new polygon is added in a cavity between two polygons already in place. If only 5 and 6-sided cells can be formed, then only three kinds of pairs are at hand: 55, 56 and 66. Assuming that the creation of a 55 pair gives rise to an excessive stress due to the strong local curvature it creates, we shall consider only the last two kinds of pairs, 56 and 66. Suppose that the pentagonal and hexagonal cells (capsomers) are produced in the medium with relative probabilities p_5 and $p_6 = 1 - p_5$.. Let the distribution of pairs on the rims of the already formed agglomerates be P_{56} and P_{66}. Then it is easy to count the probabilities of obtaining two new pairs from a given one (we show here only the allowed agglomerations, excluding the creation of 55-pairs):

$$(56) + 6 \to (1 - p_5) \times [(56) + (66)]; \ \ (66) + 5 \to p_5 \times [(56) + (56)];$$

$$(66) + 6 \to (1 - p_5) \times [(66) + (66)]. \tag{24}$$

Now it is easy to count the few possibilities and to form the corresponding matrix, which after normalization yields the following evolution equation:

$$\begin{pmatrix} P'_{56} \\ P'_{66} \end{pmatrix} = \begin{pmatrix} \frac{1}{2} & p_5 \\ \frac{1}{2} & (1 - p_5) \end{pmatrix} \begin{pmatrix} P_{56} \\ P_{66} \end{pmatrix}. \tag{25}$$

Its eigenvector corresponding to the eigenvalue 1 satisfies $P_{56} = 2\alpha P_{66}$. An interesting case occurs when the formation of triplets 666 is prohibited by a chemical or steric potential barrier. Then the eigenvector does not depend on the surrounding distribution of polygons, and is equal to $P_{56} = 2/3$, $P_{66} = 1/3$. This is exactly the proportion of doublets realized in smallest icosahedral viral capsids, C_{60} fullerene molecules or soccer balls. The method of stochastic matrix can be applied to the analysis of other interesting objects, including the DNA chains.

4. Evolution of Blood Groups

A very interesting application of the Stochastic Matrix approach is concerned with blood groups. Although only a toy model, it provides insight into the nontrivial mechanisms of transmission of a given blood group from one generation to another. Before we expose that model, let us show a simpler version of the stochastic matrix approach to genetics, namely, a model for the heredity transmission in yellow (Y) and green (g) peas observed by Georg Mendel[8] in the ninetieth century. If account is taken of all possible combinations, they are encapsulated in the "chemical reactions"

$$\begin{gathered} YY + YY \to 4\,YY, \\ YY + Yg \to 2\,YY + 2\,Yg, \\ YY + gg \to 4\,Yg, \\ Yg + Yg \to YY + 2\,Yg + gg, \end{gathered}$$

$$Yg + gg \to 2\,Yg + 2\,gg,$$
$$gg + gg \to 4\,gg.$$

It is useful to represent the above rules by a matrix with positive integers acting on a column representing the three available "states", i.e., starting from the top, YY, Yg and gg. By simple counting we see that from a YY pair one can obtain, by different couplings, six new YY pairs, six Yg pairs and no gg pairs, etc. Doing the same counting for the other cases, we arrive at the matrix

$$\begin{pmatrix} 6 & 3 & 0 \\ 6 & 6 & 6 \\ 0 & 3 & 6 \end{pmatrix} . \tag{26}$$

From this "raw" matrix describing the multiplicities of the various issues, we can produce the corresponding stochastic matrix acting on the probabilities represented by columns, by normalizing each column to 1:

$$\begin{pmatrix} 1/2 & 1/4 & 0 \\ 1/2 & 1/2 & 1/2 \\ 0 & 1/4 & 1/2 \end{pmatrix} . \tag{27}$$

The unique eigenvector of this matrix corresponding to the eigenvalue 1 represents the asymptotic probability distribution no matter what was the initial sample composition, provided it was large enough to include all possible alleles. Here it is, denoted by $\mathbf{p}^{\infty}$:

$$\mathbf{p}^{\infty} = \begin{pmatrix} p^{\infty}_{YY} \\ p^{\infty}_{Yg} \\ p^{\infty}_{gg} \end{pmatrix} = \begin{pmatrix} 1/4 \\ 1/2 \\ 1/4 \end{pmatrix} \tag{28}$$

It is an easy task to examine examples of the statistical evolution of the three genetic types in a population of peas, starting from different initial distributions. They all converge quite rapidly to the stable final distribution, known also as the Hardy-Weinberg equilibrium[2], (see figures 4 and 5).

One may ask the question whether all combinations of parent peas are occurring with the same frequency, or the same statistical weight. Were it so, the stochastic matrix describing the process would be as in (27). But if certain combinations were privileged and other enhanced, one should acknowledge it by extra factors which we represent by means of exponentials (so that they are always positive) looking like the usual Boltzmann factors in statistical physics; here, however, they should not depend on the temperature, but rather on the set of parameters that influence the formation

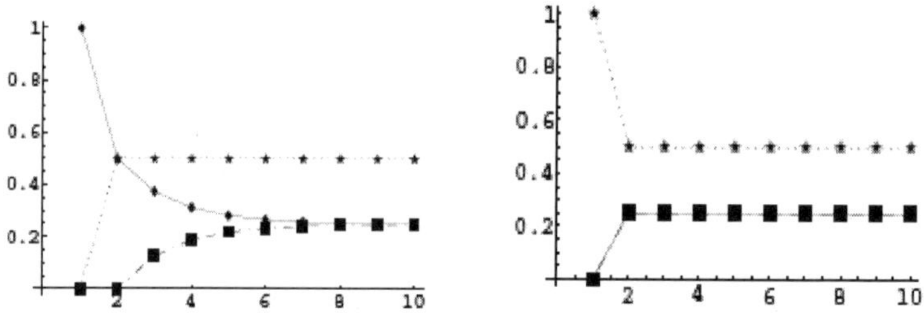

Figure 4. The evolution of YY, Yg and the gg population of peas starting from a pure YY state and from a hybrid Yg state.

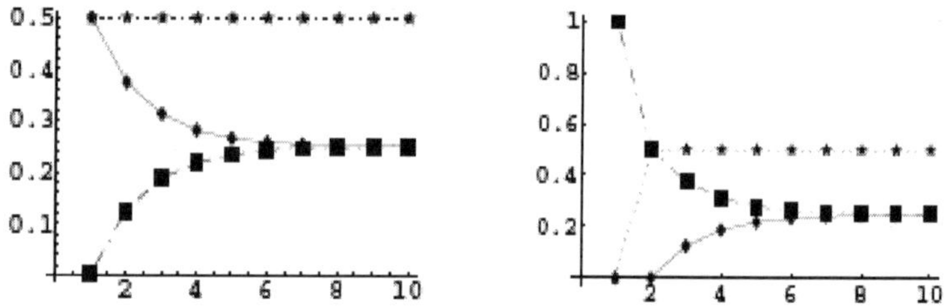

Figure 5. The evolution of YY, Yg and the gg population of peas starting from a yellow $\frac{YY+Yg}{2}$ state and from a pure gg state.

of couples, or the better or poorer survival of their posterity. In such a case, the "chemical reactions" above become

$$\begin{gathered}
YY + YY \to 4\,YY\,e^{-\mu_{YY}}, \\
YY + Yg \to 2\,YY\,e^{-\mu_{YY}} + 2\,Yg\,e^{-\mu_{Yg}}, \\
YY + gg \to 4\,Yg\,e^{-\mu_{Yg}}, \\
Yg + Yg \to YY\,e^{-\mu_{YY}} + 2\,Yg\,e^{-\mu_{Yg}} + gg\,e^{-\mu_{gg}}, \\
Yg + gg \to 2\,Yg\,e^{-\mu_{Yg}} + 2\,gg\,e^{-\mu_{gg}}, \\
gg + gg \to 4\,gg\,e^{-\mu_{gg}}
\end{gathered}$$

which yield, after column normalization, the stochastic matrix

$$\begin{pmatrix} 4e^{-\mu_{YY}} + 2e^{-\mu_{Yg}} & 2e^{-\mu_{Yg}} + e^{-\mu_{gg}} & 0 \\ 4e^{-\mu_{Yg}} + 2e^{-\mu_{YY}} & 2e^{-\mu_{YY}} + 2e^{-\mu_{gg}} + 2e^{-\mu_{Yg}} & 2e^{-\mu_{gg}} + 4e^{-\mu_{Yg}} \\ 0 & 2e^{-\mu_{Yg}} + e^{-\mu_{YY}} & 4e^{-\mu_{gg}} + 2e^{-\mu_{Yg}} \end{pmatrix}. \tag{29}$$

After the normalization, only the ratios of the exponentials actually remain relevant; we can introduce the following abbreviated notations:

$$A = e^{\mu_{Yg} - \mu_{YY}}, \quad B = e^{\mu_{Yg} - \mu_{gg}},$$

and the stochastic matrix with normalized columns becomes

$$\begin{pmatrix} \frac{1+2A}{3+3A} & \frac{2+B}{6+3A+3B} & 0 \\ \frac{2+A}{3+3A} & \frac{2+2A+2B}{6+3A+3B} & \frac{2+B}{3+3B} \\ 0 & \frac{2+A}{6+3A+3B} & \frac{1+2B}{3+3B} \end{pmatrix}. \tag{30}$$

It is easy to see that this matrix reduces to that obtained with totally random pairings (27) if we set $A = 1$ and $B = 1$. This can be regarded as a proof that the peas do not choose partners for fecondation, which occurs in a totally random manner. Moreover, there is apparently no other bias present, like e.g. a lesser or higher resistance to the external conditions of one particular genotype. In what follows we shall see that this is not always the case with humans. Since their discovery, the repartition of different blood groups in human populations has been subject of intense studies. We have a vast amount of knowledge concerning most of the various human groups constituting the actual population of our planet[9]. It was established that *four phenotypes* (inherited with specific genes) are observed, known as O, A, B and AB types. Type O is recessive with respect both to A and B, which means that a person with blood group O has two O genes, and may be noted OO, whereas a person with blood of group A can be either AO or AA; the same with the blood group B, which may come from the combination of genes OB or BB. The corresponding "chemical reactions" are:

$$\begin{gathered} OO + OO \to 4\,OO, \\ OO + AO \to 2\,OO + 2\,AO, \\ OO + AA \to 4\,AO, \\ OO + BO \to 2\,OO + 2\,BO, \end{gathered}$$

$$OO + BB \to 4\,BO,$$
$$OO + AB \to 2\,AO + 2\,BO,$$

$$AO + AO \to AA + 2\,AO + OO,$$
$$AO + AA \to 2\,AA + 2\,AO,$$
$$AO + BO \to AB + OO + BO + AO,$$
$$AO + BB \to 2\,AB + 2\,BO,$$
$$AO + AB \to AA + BO + AB + AO,$$

$$AA + AA \to 4\,AA,$$
$$AA + BO \to 2\,AB + 2\,AO,$$
$$AA + BB \to 4\,AB,$$
$$AA + AB \to 2\,AA + 2\,AB,$$

$$BO + BO \to BB + 2\,BO + OO,$$
$$BO + BB \to 2\,BB + 2\,BO,$$
$$BO + AB \to AB + BO + BB + AO,$$

$$BB + BB \to 4\,BB,$$
$$BB + AB \to 2\,BB + 2\,AB,$$

$$AB + AB \to AA + BB + 2\,AB$$

As in the case of Mendelian peas, the outcomes of all possible pairings can be displayed as a matrix:

$$\begin{pmatrix} 8 & 4 & 0 & 4 & 0 & 0 \\ 8 & 8 & 8 & 4 & 0 & 4 \\ 0 & 4 & 8 & 0 & 0 & 4 \\ 8 & 4 & 0 & 8 & 8 & 4 \\ 0 & 0 & 0 & 4 & 8 & 4 \\ 0 & 4 & 8 & 4 & 8 & 8 \end{pmatrix} \tag{31}$$

And, to interpret the results statistically, as probabilities, all columns of the matrix should be normalized to 1. This leads to the following stochastic matrix, which rules the variations of probability distributions after each generation change:

$$\begin{pmatrix} 1/3 & 1/6 & 0 & 1/6 & 0 & 0 \\ 1/3 & 1/3 & 1/3 & 1/6 & 0 & 1/6 \\ 0 & 1/6 & 1/3 & 0 & 0 & 1/6 \\ 1/3 & 1/6 & 0 & 1/3 & 1/3 & 1/6 \\ 0 & 0 & 0 & 1/6 & 1/3 & 1/6 \\ 0 & 1/6 & 1/3 & 1/6 & 1/3 & 1/3 \end{pmatrix} . \tag{32}$$

The asymptotic probability distribution is given by the eigenvector of this matrix corresponding to the unique eigenvalue 1:

$$\mathbf{p}^\infty = \begin{pmatrix} p^\infty_{OO} \\ p^\infty_{OA} \\ p^\infty_{OB} \\ p^\infty_{AA} \\ p^\infty_{AB} \\ p^\infty_{BB} \end{pmatrix} = \begin{pmatrix} 1/9 \\ 2/9 \\ 1/9 \\ 2/9 \\ 1/9 \\ 2/9 \end{pmatrix} . \tag{33}$$

This result is in obvious contradiction with the known data concerning the blood group distribution in various human populations, as can be seen in the following table[10]:

Regions, Peoples	Group A %	Group B %	Group O %
American natives	1.7%	0.3%	98%
Australian natives	22%	2%	76%
Africa	18%	13%	69%
Europe average	27%	8%	65%
Basques	23%	2%	75%
Italian	20%	7%	73%
English	25%	8%	67%
East Asia	20%	19%	61%

In order to explain the discrepancy between the probabilities of blood group distribution resulting from totally random matings and the real situation we must admit that - contrary to what happens with peas - there are some extra factors that play in favor of certain pairings of genes, and disfavor other ones. An example would come from a higher resistance of

O-type people to syphilis, which would account for its very large occurrence among american natives. The mathematical expression of such facts can be included into extra factors modifying the probabilities of couplings, which results in the following rules:

$$\begin{gathered}
OO + OO \to 4\,e^{-\mu_{OO}}\,OO, \\
OO + AO \to 2\,e^{-\mu_{OO}}\,OO + 2\,e^{-\mu_{AO}}\,AO, \\
OO + AA \to 4\,e^{-\mu_{AO}}\,AO, \\
OO + BO \to 2\,e^{-\mu_{OO}}\,OO + 2\,e^{-\mu_{BO}}\,BO, \\
OO + BB \to 4\,e^{-\mu_{BO}}\,BO, \\
OO + AB \to 2\,e^{-\mu_{AO}}\,AO + 2\,e^{-\mu_{BO}}\,BO,
\end{gathered}$$

$$\begin{gathered}
AO + AO \to e^{-\mu_{AA}}\,AA + 2\,e^{-\mu_{AO}}\,AO + e^{-\mu_{OO}}\,OO, \\
AO + AA \to 2\,e^{-\mu_{AA}}\,AA + 2\,e^{-\mu_{AO}}\,AO, \\
AO + BO \to e^{-\mu_{AB}}\,AB + e^{-\mu_{OO}}\,OO + e^{-\mu_{BO}}\,BO + e^{-\mu_{AO}}\,AO, \\
AO + BB \to 2\,e^{-\mu_{AB}}\,AB + 2\,e^{-\mu_{BO}}\,BO, \\
AO + AB \to AA + BO + AB + AO,
\end{gathered}$$

$$\begin{gathered}
AA + AA \to 4\,e^{-\mu_{AA}}\,AA, \\
AA + BO \to 2\,e^{-\mu_{AB}}\,AB + 2\,e^{-\mu_{AO}}\,AO, \\
AA + BB \to 4\,e^{-\mu_{AB}}\,AB, \\
AA + AB \to 2\,e^{-\mu_{AA}}\,AA + 2\,e^{-\mu_{AB}}\,AB,
\end{gathered}$$

$$\begin{gathered}
BO + BO \to e^{-\mu_{BB}}\,BB + 2\,e^{-\mu_{BO}}\,BO + e^{-\mu_{OO}}\,OO, \\
BO + BB \to 2\,e^{-\mu_{BB}}\,BB + 2\,e^{-\mu_{BO}}\,BO, \\
BO + AB \to e^{-\mu_{AB}}\,AB + e^{-\mu_{BO}}\,BO + e^{-\mu_{BB}}\,BB + e^{-\mu_{AO}}\,AO,
\end{gathered}$$

$$\begin{gathered}
BB + BB \to 4\,e^{-\mu_{BB}}\,BB, \\
BB + AB \to 2\,e^{-\mu_{BB}}\,BB + 2\,e^{-\mu_{AB}}\,AB,
\end{gathered}$$

$$AB + AB \to e^{-\mu_{AA}}\,AA + e^{-\mu_{BB}}\,BB + 2\,e^{-\mu_{AB}}\,AB$$

These "reactions" lead to the following unrenormalized matrix describing the issues of all couplings:

$$\begin{pmatrix}
8\,e^{-\mu_{OO}} & 4\,e^{-\mu_{OO}} & 0 & 4\,e^{-\mu_{OO}} & 0 & 0 \\
8\,e^{-\mu_{AO}} & 8\,e^{-\mu_{AO}} & 8\,e^{-\mu_{AO}} & 4\,e^{-\mu_{AO}} & 0 & 4\,e^{-\mu_{AO}} \\
0 & 4\,e^{-\mu_{AA}} & 8\,e^{-\mu_{AA}} & 0 & 0 & 4\,e^{-\mu_{AA}} \\
8\,e^{-\mu_{BO}} & 4\,e^{-\mu_{BO}} & 0 & 8\,e^{-\mu_{BO}} & 8\,e^{-\mu_{BO}} & 4\,e^{-\mu_{BO}} \\
0 & 0 & 0 & 4\,e^{-\mu_{BB}} & 8\,e^{-\mu_{BB}} & 4\,e^{-\mu_{BB}} \\
0 & 4\,e^{-\mu_{AB}} & 8\,e^{-\mu_{AB}} & 4\,e^{-\mu_{AB}} & 8\,e^{-\mu_{AB}} & 8\,e^{-\mu_{AB}}
\end{pmatrix} \tag{34}$$

The exponents play the role of "chemical potentials" characteristic for each type of coupling. After normalization of all columns to 1, all entries become homogeneous fractions containing the exponentials. In order to simplify the notations, we can multiply all the entries by, say, $e^{\mu_{OO}}$ before the normalization; then, after the matrix is normalized, introduce the following parameterization:

$$\alpha = e^{\mu_{OO}-\mu_{AO}}, \quad \alpha' = e^{\mu_{OO}-\mu_{AA}}, \quad \beta = e^{\mu_{OO}-\mu_{BO}},$$

$$\beta' = e^{\mu_{OO}-\mu_{BB}} \quad \gamma = e^{\mu_{OO}-\mu_{AB}}.$$

Then the unrenormalized matrix can be written as

$$\begin{pmatrix} 1 & 1 & 0 & 1 & 0 & 0 \\ \alpha & 2\alpha & \alpha & \alpha & 0 & \alpha \\ 0 & \alpha' & \alpha' & 0 & 0 & \alpha' \\ \beta & \beta & 0 & 2\beta & \beta & \beta \\ 0 & 0 & 0 & \beta' & \beta' & \beta' \\ 0 & \gamma & \gamma & \gamma & \gamma & 2\gamma \end{pmatrix} \tag{35}$$

We shall not write down the normalized stochastic matrix; it will be enough to show the stable eigenvector corresponding to the eigenvalue 1:

$$\mathbf{p}^{\infty} = \begin{pmatrix} p^{\infty}_{OO} \\ p^{\infty}_{OA} \\ p^{\infty}_{OB} \\ p^{\infty}_{AA} \\ p^{\infty}_{AB} \\ p^{\infty}_{BB} \end{pmatrix} = \frac{1}{\Sigma} \begin{pmatrix} 1+\alpha+\beta \\ \alpha\,(1+2\alpha+\alpha'+\beta+\gamma) \\ \alpha'\,(\alpha+\alpha'+\gamma) \\ \beta\,(1+\alpha+2\beta+\beta'+\gamma) \\ \beta'\,(\beta+\beta'+\gamma) \\ \gamma\,(\alpha+\alpha'+\beta+\beta'+2\gamma) \end{pmatrix}, \tag{36}$$

where the overall normalization factor Σ is given by the following sum:

$$\Sigma = 1+2\alpha\,(1+\alpha+\alpha'+\beta+\gamma)+2\beta\,(1+\beta+\beta'+\gamma)+2\gamma\,(\alpha'+\beta'+\gamma)+\alpha'^2+(\beta')^2.$$

At this stage it is difficult to decide which properties of particular gene combinations are responsible for a given distribution of blood groups; what we can get from the parametrization of "affinities" are certain sets of parameters leading to the observed distributions. Luckily enough, one may observe that the number of unknown parameters (five) is equal to the number of independent equations to be satisfied (because of the six components of the probability column, normalized to 1, only five are independent). It does not mean, nevertheless, that each distribution corresponds to a unique set of values of "affinity parameters" $(\alpha, \alpha', \beta, \beta', \gamma)$, because the equations are highly non-linear.

Here are a few examples found numerically:

• close to the european average:

$$\alpha = \beta = 0.10,\ \alpha' = 0.48,\ \beta' = 0.17,\ \gamma = 0.02 \rightarrow \begin{pmatrix} 0.637 \\ 0.095 \\ 0.152 \\ 0.079 \\ 0.026 \\ 0.009 \end{pmatrix}$$

• close to american natives:

$$\alpha = 0.07,\ \beta = \gamma = 0.001,\ \alpha' = 0.08,\ \beta' = 0.04 \rightarrow \begin{pmatrix} 0.980 \\ 0.010 \\ 0.007 \\ 0.001 \\ 0.001 \\ 0.000 \end{pmatrix}$$

We see that smaller values of the parameter α produce an excess of the OO type, which is reasonable, because they lower the "chemical barrier" for the creation of this characteristic. Also, equal parameters $\alpha = \alpha' = \beta = \beta' = \gamma$ imply $AO = BO$, $BB = AA = AB/2$ whatever their common value. When all the parameters are set to 1, we come back to the pure random case (32). In practice, the types AO and AA can be collapsed into a single group, and the same with the types BO and BB, reducing the problem to four probabilities only, corresponding to the observed blood groups, the phenotypes O, A, B and AB, and to three parameters, α, β and γ. The results of this simplified model are essentially the same.

5. Conclusion

We have presented a few examples illustrating how the stochastic matrix method can be used to modell and analyse statistical distributions of various biological features. This method can be used in many situations where statistical correlations bring important information about the system and its evolution. The "system" can have various meanings, from a viral capsid assembling via agglomeration of capsomers, or an organism considered as an assembly of various species of cells, to an ecosystem comprising various coexisting species. The model can be substantially amended in a situaton when the total number of items considered is not conserved. If this is the case, like e.g. in a living organism subjected to an iflux of pathogens from the exterior and loosing a substantial amount of its healthy cells, then the evolution equations are no more homogeneous, but have to include the external influences on the right-hand side.

The generalized models of this type will be the subject of a forthcoming paper.

References

1. R. Aldrovandi, *Special Matrices in Mathematical Physics*, World Scientific, Singapore, 2001.
2. D. Hartl, A.G. Clark, *Principles of population genetics*, Simauer Ass. Inc. Publishers, Sunderland, Mass. (1985).
3. R. Kerner, D.M. dos Santos, *Phys. Rev. B*, **37**, pp. 3881-4000, (1988).
4. J. Kepler, *De nive sexangula*, Teubneri ed., Leipzig, 1611.
5. R. Kerner, *Models of Agglomeration and Glass Transition*, Imperial College Press, London-New-York-Singapore, (2007).
6. R. Kerner, *Journal of Theoretical Medicine*, **6** (2), p.95-97 (2005).
7. R. Kerner, *Computational and Mathematical Methods in Medicine*, **9** (3 and 4), p. 175-181, (2008).
8. Gregor Mendel, *Experiments in Plant-Hybridization*, in James Newman (ed.), *The World of Mathematics*, Dover, Mineola, New York, 2000, vol.2, ps. 937-49. Also in the web address www.mendelweb.org/Mendel.html.
9. L. L. Cavalli-Sforza and W. F. Bodmer, *The genetics of human populations*, Dover, Mineola, New York, 1999.
10. L. L. Cavalli-Sforza, *Genes, Peoples and Languages*, Penguin Books, London, 2001. The given table is taken from that at page 15.

FRACTAL AND NONLINEAR ANALYSIS OF BIOCHEMICAL TIME SERIES

H. PUEBLA

Departamento de Energa, Universidad Autonoma Metropolitana, Azcapotzalco
Av. San Pablo No. 180, Reynosa-Tamaulipas
Azcapotzalco, 02200, D.F. Mexico
E-mail hpuebla@correo.azc.uam.mx

E. HERNANDEZ-MARTINEZ AND J. ALVAREZ-RAMIREZ

Departamento de IPH, Universidad Autonoma Metropolitana, Iztapalapa
Apartado Postal 55-534
Iztapalapa, 09340, D.F. Mexico
E-mail jjar@correo.azc.uam.mx

In recent years, it has been used time-series analysis to characterize complex biochemical oscillations. The aim of this work is to show that fractal analysis has the ability to provide further information on biochemical mechanisms leading to complex biochemical oscillations. Analysis of simulated and experimental time-series data for two systems, intracellular calcium and circadian oscillations, is made in order to illustrate the relation between fractal parameters and biochemical mechanisms leading to complex oscillations. For completeness, standard non-linear analysis, such as delayed phase-plane reconstruction and lyapunov exponents, is also performed.

1. Introduction

Living organisms are thermodynamically open systems, that is, they are in a state of permanent flux, continuously exchanging energy and matter with their environment. Moreover, they are characterized by a complex organization, which results from a vast network of molecular interactions involving a high degree of nonlinearity. Under appropriate conditions, the combination of these two features, openness and nonlinearity, enables complex systems to exhibit properties that are self-organizing [9,5]. In biochemical systems, such properties may express themselves through the spontaneous formation, from random molecular interactions, of long-range correlated, macroscopic dynamical patterns in space and time the process of self-organization.

Time series signals derived from living organisms are extraordinarily complex, as they reflect ongoing processes involved, and can be used to diagnose incipient pathological conditions [20,23]. Many of these time series contain hidden long-range correlations that can provide interesting and useful information on the structure and evolution of theses dynamical system. A deep insight on the fundamental mechanisms of biochemical systems can be achieved by the identification of stochastic and deterministic components of the time-series and the presence of multiple time-scales as biochemical systems are ruled by mechanisms operating across multiple temporal scales [15,23].

Many approaches have been developed in the last years to analyze complex time-series, including, for example, Fourier spectra, chaotic dynamics, scaling properties, multifractal properties and multivariate autoregressive methods [14,19,22,17]. The idea behind the time series analysis in biochemical systems is that this class of analysis can extract fundamental information on the nature of the biochemical mechanisms leading to oscillation. In this way, a suitable analysis of biochemical time-series can extract useful parameters to identify pathological states.

The aim of this work is to show that fractal analysis has the ability to provide further information on biochemical mechanisms leading to complex biochemical oscillations. Analysis of time-series data for two systems, intracellular calcium and circadian oscillations, is made in order to illustrate the relation between fractal parameters and biochemical mechanisms leading to complex oscillations. This is done by performing three spectral analysis: detrended fluctuation analysis, the Jeanish coefficient and the Hurst scaling exponent. The Hurst analysis is a method designed for the detection of multifractal characteristics in stochastic time series [11,14]. The detrended fluctuation analysis (DFA) is a simple and efficient scaling method commonly used for detecting long-range correlations in non-stationary fractal sequences [17,1]. The fractal technique developed by Jaenisch et al is obtained by calculating a coefficient, similar to the Hurst coefficient [16]. The Jaenisch coefficient indicates transition to chaos by fluctuation between anti-persistent values and the anti-persistent transition point. To the best of our knowledge these techniques have been not yet applied to intracellular calcium and circadian time series. It is found that, in general, complex biochemical trajectories can display persitence and anti-persistence behavior and a multiscale temporal behavior. For completeness, standard non-linear analysis, such as delayed phase-plane reconstruction and Lyapunov exponents, is also performed.

This work is organized as follows: In Section 2, the case studies and corresponding time series are presented. In Section 3 we briefly describe the main features of spectral methods that we have applied. In section 4 the application of the fractal and standard analysis are presented. Section 4 also presents the discussion of our results. Finally, some concluding remarks are given in Section 5.

2. Complex Time-Series from Biochemical Oscillations

Biochemical oscillations are observed in all types of organisms from the simplest to the most complex. Biochemical oscillations are present because they confer positive functional advantages to the organism. The advantages fall into five general categories: temporal organization, spatial organization, prediction of repetitive events, efficiency and precision of control [9,3,23]. Examples include the oscillation dynamics of the mRNA concentration, insulin-secreting cells of the pancreas, circadian cycles, and oscillations and waves in the concentration of free intracellular calcium Ca^{2+} [9,20]. In this section we present the case studies that we are addressing: intracellular calcium oscillations and the circadian cycle. Simulated (model- based) and real data sets were used.

2.1. *Intracellular Calcium oscillations*

Over the last 15 years oscillations in intracellular Ca^{2+} have become a major example of oscillatory behavior at the cellular level. Ca^{2+} oscillations trigger different cellular functions, including muscle contraction, heart beat, cell death, brain processing and store information[3]. These oscillations are observed in a large variety of cell types, with periods ranging from seconds to minutes. The mechanisms for calcium oscillations have been mainly modeled as deterministic processes. However, it has been recognized recently that several different aspects of calcium signaling in cells definitely require stochastic treatment. We consider the model of [10], which exhibits a diversity of calcium responses, notably steady states, spiking and bursting oscillations, multirhythmic and chaotic regimes. The model contains three variables, namely the concentrations of free Ca^{2+} in the cytosol (x_1) and in the internal pool (x_2), and the IP3 concentration (x_3).

$$\frac{dx_1}{dt} = V_{in} - V_2 + v_3 + k_f x_2 - kx_1, \tag{1}$$

$$\frac{dx_2}{dt} = V_2 - V_3 - k_f x_2, \tag{2}$$

$$\frac{dx_3}{dt} = \beta V_4 - V_5 - \sigma x_3 \tag{3}$$

$$V_{in} = V_0 + V_1\beta, V_2 = V_{M2}\frac{x_1^2}{K_2^2+x_1^2}$$
$$V_3 = V_{M3}\frac{x_1^m}{K_1^m+x_1^m}\frac{x_2^2}{K_2^2+x_2^2}\frac{x_3^4}{K_3^4+x_3^4}$$
$$V_5 = V_{M5}\frac{x_3^p}{K_5^p+x_3^p}\frac{x_1^n}{K_d^n+x_1^n}$$

With appropriate parameter values the model can display complex Ca^{2+} oscillations, including bursting, chaos and quasiperiodicity. Two sets of parameter values corresponding to bursting and chaos behavior were used to generate the time-series shown in Figure 1.

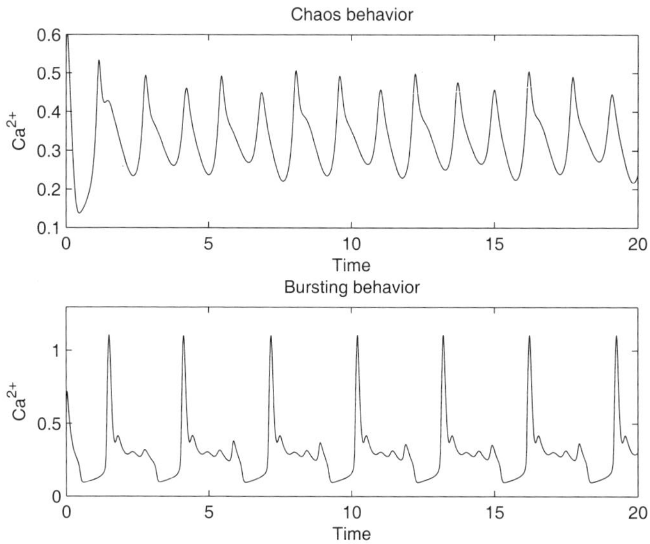

Figure 1. Simulated time-series of circaian oscillations.

Bursting intracellular Ca^2+ oscillations take the form of abrupt spikes, sometimes preceded by a gradual increase in cytosolic Ca^2+. Chaotic intracellular Ca^2+ oscillations shows an apparent erratic random behavior, which however, was obtained from a deterministic model.

2.2. *Circadian oscillations*

The biological functions of most living organisms are organized along an approximate 24-h time cycle or circadian rhythm. Circadian rhythms, are endogenous because they can occur in constant environmental conditions, e.g. constant darkness. These endogenous rhythms govern daily events like sleep, activity, hormonal secretion, cellular proliferation and metabolism [8,9]. Circadian rhythm can also be entrained by external forcing of modified light-darkness cycles or phase-shifted when exposed to light pulses. Concerning the modeling of this phenomenon, it has to be stressed that the mechanism can be considerably different for the different living beings in which it has been studied, ranging from unicellular organisms to mammalians, going through fungi and flies [9]. Such models generally take the form of a system of coupled ordinary differential equations. The time series we analyzed corresponds to telemeter frequency temperature recordings made at 5 minute intervals over 3 months on a nocturnal mammal, Peromyscus bardii (prairie deer mouse) [2]. Higher counts indicating higher temperature. The animal was given 20 days of 12 hours light and 12 hours dark as an adjustment period, followed by 10 days of constant dark, 12 hours light and 26 days constant dark [2]. The analyzed data are from days 31 to 86. Figure 2 shows the data from days 31 to 86. The time series apparent exhibit persistent oscillations with a cycle period of approximately 20-24 hrs.

3. Time Series Analysis Methods

In the last few years, time-series analysis have been applied to molecular and cell-biology in order to understand the complex dynamics of the cell. Its applications range from DNA sequences [17,22], to heart rate dynamics [6], neuron spiking [4], among others [12,13,20]. In this section we briefly describe the time-series techniques that we have applied to the case studies.

3.1. *Fractal Hurst method*

Multifractality is a useful tool for explaining many of the patterns seen in nature. In particular, multifractal analysis allow us to analyze the mixing state of fractal dimensions. Multifractal analysis is a technique first introduced by Mandelbrot[14] in the context of turbulence, and then studied as a mathematical tool. We have applied the Hurst method, which is widely used for fractal analysis of time series, to compute the generalized Hurst

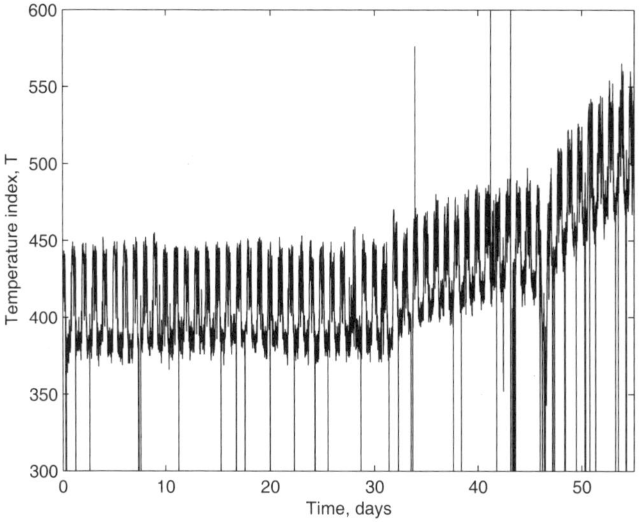

Figure 2. Experimental time-series of circaian oscillations.

exponent, called H_q, for the biochemical time-series shown in Figures 1 and 2. The Hurst method involves calculating the qth-order height-height correlation $F_q(\tau)$, where τ is a time scale, as follows [14,11]:

$$F_q(\tau) =< |x(t) - x(t+\tau)|^q >_t \tag{4}$$

where only non-zero terms are considered in the average, and angular brackets denote an average quantity. If $x(t)$ is scaling, then a power law behavior is expected as

$$F_q(\tau) \propto \tau^{qH_q} \tag{5}$$

where H_q is the generalized qth-order Hurst exponent. The aim in the use of generalized height-height correlations $F_q(\tau), q \gg 1$, is to magnify large fluctuations of $x(t)$. In this way, large fluctuations are the main contribution to $F_q(\tau)$, so that their effects in the dynamical behavior of $x(t)$ can be explored. Some comments regarding the Hurst exponents H_q are in order[14]:

(i) If H_q varies with q, a nontrivial multiaffine spectrum is obtained. Basically, multifractal analysis allow us to analyze the mixing state of fractal dimensions as displayed in the complex nature of the time series $x(t)$.

(ii) The relation between the fractal dimension D_f and the Hurst exponent H_2 ($q = 2$) can be expressed as $D_f = 2 - H_2$. So, by finding H_2, we can estimate the fractal dimension of the time series. Notice that $-(2H_2 + 1) = -2$ corresponds to an uncorrelated noise with $H_2 = 0.5$.

(iii) It has been shown that the correlation function $C(\tau)$ of future values, $x(\tau)$, with past values, $x(-\tau)$, is given by $C(\tau) \propto 2(2^{2H_q-1} - 1)$. A value of H_q=0.5 results from uncorrelated time series and corresponds to a purely random walk or Brownian motion (BM). In this case, $x(t)$ is characterized by white noise, which means that future predictions of the time series is impossible. But for $H_q < 0.5$, one has that $C(\tau) \neq 0$ independent of the time horizon . This indicates infinitely long correlations and leads to scale-invariance associated with positive long-range correlations (persistence) for $H_q > 0.5$ (i.e., in increasing trend in the past implies an increasing trend in the future) and to a scale-invariance associated with negative long-range correlations (anti-persistence) for $H_q < 0.5$ (i.e., an increasing trend in the past implies a decreasing trend in the future).

3.2. *DFA*

The detrended fluctuation analysis (DFA) developed by Peng[17] et al. is a simple and efficient scaling method commonly used for detecting long-range correlations in non-stationary fractal sequences. The basic DFA algorithm can be described as follows[17]. In a first step, the time series is integrated to obtain a type of random-walk profile. Then, the resulting integrated sequence is divided into non-overlapping boxes of equal size n. In each box the local trend is estimated by m-degree polynomial fitting. In a subsequent step, the root-mean square fluctuation (denoted by $F_m(n)$) of the difference between the integrated sequence and the local polynomial fits is calculated for each box n. Finally, assuming that such fluctuations meet a power-law with respect to the box size, a scaling factor a_m is computed as the slope of the following fluctuation plot:

$$F \stackrel{Def}{=} log(n) \; versus \; log(F_m(n)) \tag{6}$$

Normally, a linear fit is adequate to obtain the correlation properties. The following comments are in order:

(i) In the case of having only short-range correlations (or not correlations at all) the detrended walk profile displays properties of a standard random walk (e.g., white noise) with $a_m = 0.5$. On the other hand, if $a_m < 0.5$ the correlations in the signal are anti-persistent, and if $a_m > 0.5$ the correlations in the signal are persistent.

(ii) The values $a_m = 1.0$ and $a_m = 1.5$ correspond to $1/f$-noise and Brownian motion, respectively. A value $a_m > 1.5$ corresponds to long-range correlations that are not necessarily related to stochastic processes. Indeed, $a_m > 1.5$ can be reflecting deterministic correlations.

(iii) The value of $a_m = 0$ corresponds to a pure anti-persistent process represented by a sequence with alternating values. In fact, a pure anti-persistent sequence oscillates between two values a and b, so that an increment is to be followed by a decrement, and vice versa. The time scale crossover at which $a_m \to 0$ can be taken as an estimate of the characteristic time scale or dominant period (i.e., the inverse of the dominant frequency in Hz) of the harmonic components.

3.3. *Jeanisch method*

The fractal technique developed by Jaenisch[16] is obtained by calculating a coefficient, similar to the Hurst coefficient. This is an entropic fractal statistic sensitive to micro-trends (resolution of fine structure) in the data, and it does not require calculating the accumulated departure from the mean of the data. It uses the actual number of points in a data set and weighs the range of the actual data in a recursive method. The method of depends only upon the range R, number of points N, and standard deviation σ of the data. The range, standard deviation and number of points are combined together into a single number, which can be used to interpret the data in terms of macro- or micro-trends. In general, different combinations of the quantities R, N and σ may be considered. Musielak[16] et al. defines the Jaenisch coefficient as follows,

$$J = \frac{log(R\sigma N)}{log(1/N)} \tag{7}$$

the fractal dimension D_J is determined by using the following relationships:

$$\begin{array}{lll} D = \frac{1}{1-J} + \varepsilon & 0 < J \leq 0.5 & (anti-persistent) \\ D = 2 + \varepsilon & J = 0.5 & (non-persistent) \\ D = \frac{1}{J} + \varepsilon & 0.5 < J \leq 1.0 & (persistent) \end{array}$$

where ε is the embedding space, which is fixed. The Jaenisch coefficient indicates transition to chaos by fluctuation between anti-persistent values and the anti-persistent transition point. The Jaenisch fractal technique has the advantage of being robust, requiring only the range, the standard deviation, and the number of points of a data set.

4. Results and Discussion

In this section numerical results obtained from the application of the time series analysis methods presented in the above section to the case studies are shown. Moreover, for completeness, standard results such as delayed phase plane reconstruction and Lyapunov exponents are presented.

4.1. *Intracellular calcium oscillations*

4.1.1. *Phase plane and Lyapunov exponents*

Reconstructed phase-planes from the time series corresponding to bursting and chaotic dynamics of intracellular calcium are shown in Fig. 3. Such phase-planes provide a means to shown the topological characteristics of the corresponding attractors obtained from the time series. Maximum Lyapunov exponents obtained from the time series are 0.9143 and 1.0831 for bursting and chaotic behavior respectively, such that the complex chaotic behavior is corroborated.

4.1.2. *Fractal analysis*

Fractal dimensions for the intracellular calcium time series are 1.347 and 1.38, which are computed with the Jeanish method. Figures 4 and 5 shows the data computed from the Hurst analysis for bursting and chaotic behavior for x_1. It can be seen in Figure 4 that there is at least two time scales involved in the dynamics obtained from the intracellular calcium time series. Moreover, from Figure 4 it is apparent that the curves corresponding to versus $(R/S)_{\tau 2}$ are well described by straight lines, from where the Hurst exponent H_2 is estimated. For the time series with bursting behavior, the

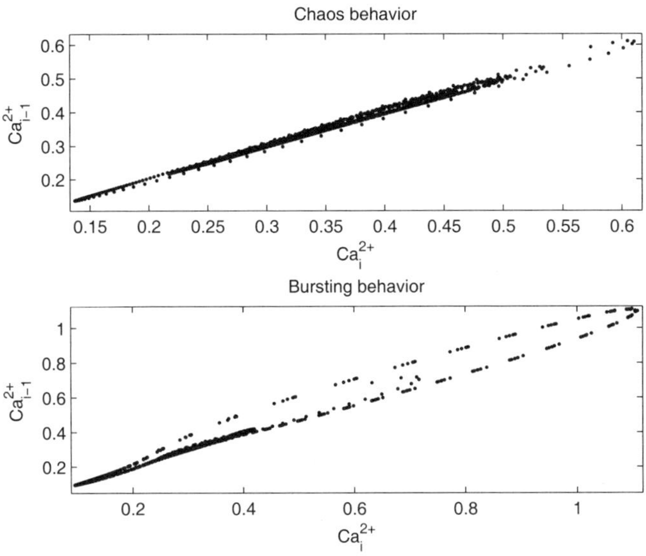

Figure 3. Phase-plane of intracellular calcium oscillations.

computed values for H_2 are 0.8546 and 0.1283 and for the time series with chaotic behavior the values of H_2 are 0.9608 and 0.3387. Then, in the scaling domains $0 \leq \tau_b \leq 0.105$ and $0 \leq \tau_c \leq 0.0.085$ time series shows persistence (i.e. in an increasing trend in the past implies an increasing trend in the future), and for $\tau_b > 0.105$ and $\tau_c > 0.085$ time series shows anti-persitence (i.e. an increasing trend in the past implies a decreasing trend in the future), for the bursting and chaotic behavior respectively. Figure 5 shows the dependence of H_q. For the time series with bursting behavior H_q depends nonlinearly on q, such that the time series displays mixing of Hurst exponents, and consequently multiaffinity nature. For the time series with chaotic behavior H_q depends linearly on q, and consequently monofractal nature. To find an explanation to the multi-scale behavior, one should look at the temporal structures presented in complex biochemical oscillations. Indeed, the underlying dynamics of some biochemical systems spans a great range of frequencies. For instance, the central nervous system can display a wide spectrum of spatially synchronized, rhythmic oscillatory patterns of activity with frequencies in the range from 0.5Hz, 20Hz, to 30-80 Hz (rhythm) and even higher up to 200Hz.

Figure 6 shows the corresponding data from the DFA for bursting and

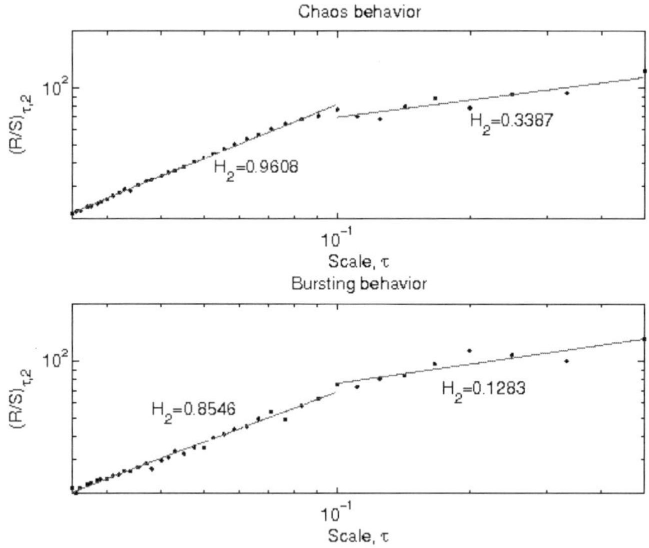

Figure 4. R/S stadistic with two time scales for intracellular calcium oscillations.

chaotic dynamics. As can be seen, Figures 6 and 4 have a very similar shape and corroborates the values that indicates long-range correlations (persistence) for $0 \leq \tau_b \leq 0.10$ and negative long-range correlations (anti-persistence) for $\tau_c > 0.10$. In particular, for time scales $\tau_b < 0.1$ and $\tau_c < 0.085$, $a_m(\tau) \sim 1.0$ for both time series, which implies that the time correlations beyond such time scale can here be indeed interpreted from an stochastic standpoint. For $\tau_b > 0.1$ and $\tau_c > 0.085$, $a_m(\tau) \to 0$, the dynamics of the time series is dominated by harmonic components.

4.2. *Circadian oscillations*

4.2.1. *Phase plane and Lyapunov exponents*

For time series from circadian oscillations Figure 7 shows the reconstructed phase-plane. As can be seen from Figure 6 the phase-plane displays a simple cycle limit corrupted with noise. The maximum Lyapunov exponent obtained in this case is positive (0.1652) such that complex chaotic oscillatory behavior is displayed.

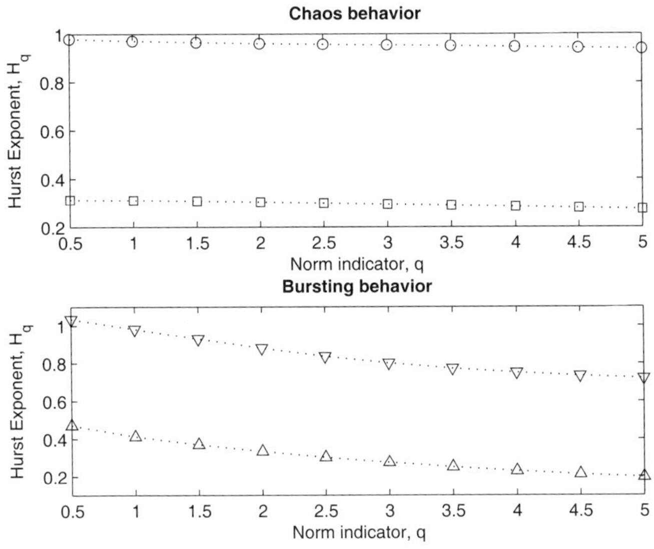

Figure 5. Hurst exponent as function of q.

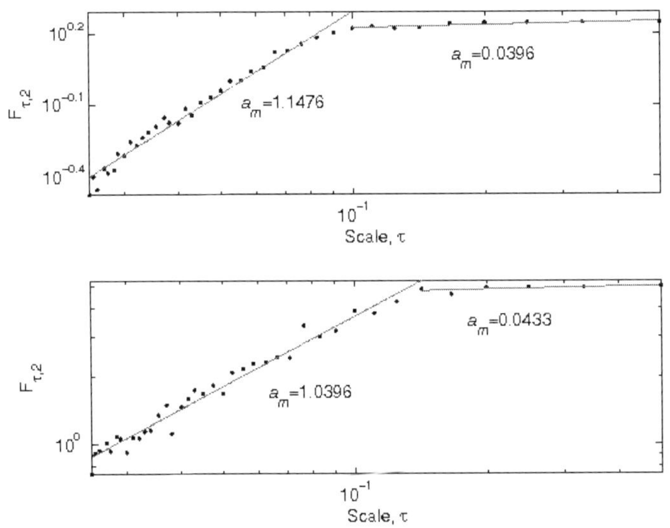

Figure 6. Log-log plots of the fructuation function $F_{\tau,2}$ as function of time scale τ.

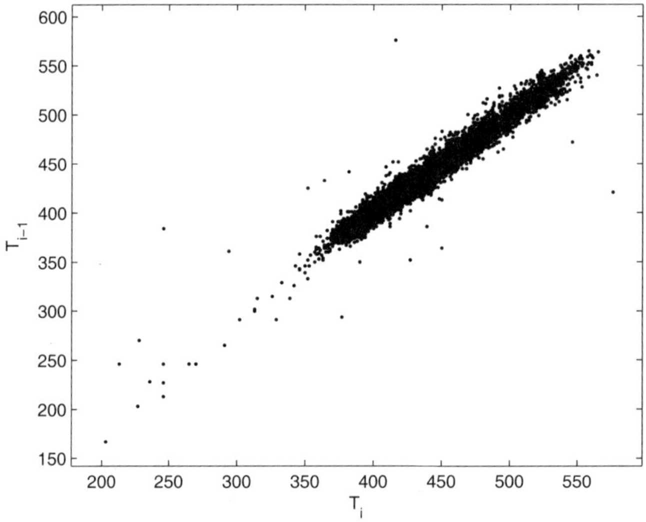

Figure 7. Phase-plane of circadian oscillations.

4.2.2. *Fractal analysis*

Figure 8 shows the Hurst analysis and DFA results for the time series. As can be seen, the circadian time series is dominated by two time scales. Contrary to the intracellular calcium oscillations, in this case a persistent process is obtained for large time scales (> 0.1). For small time scales, it can be observed an anti-persisten process (the sequence oscillates between two values a and b, so that an increment is to be followed by a decrement, and vice versa, as in typical cycle limit) near of uncorrelated time series ($a_m(\tau), H_q \sim 0.5$).

5. Conclusions

The deterministic and stochastic components of living processes turns out to be crucial for understanding how biochemical oscillations can display an apparent random noise-like behavior or be dominated by harmonic oscillations for large time scales. In this work we have applied methods of nonlinear time series analysis in order to determine their stochastic/deterministic nature of simulated and real time-series from two benchmark examples from biochemical oscillations: intracellular calcium and circadian oscilla-

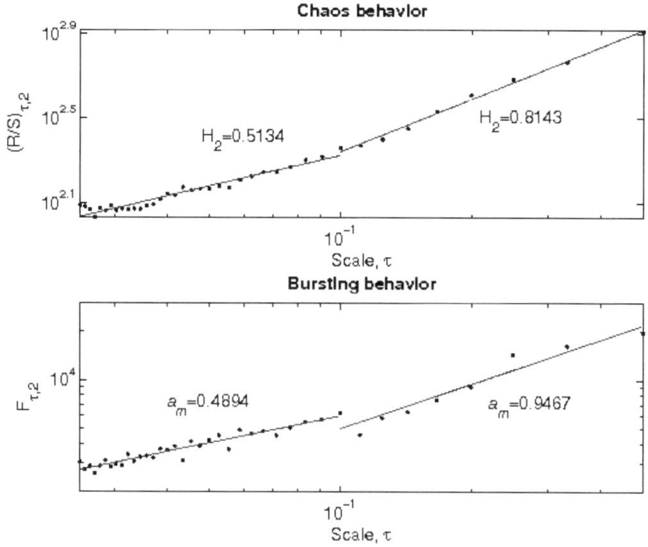

Figure 8. Scaling analysis methods ($R/S_{\tau,2}$ and $F_{\tau,2}$) for time series circadian oscillations.

tions. Three spectral methods were applied: multifractal hurst analysis, DFA, and the jeanish method. Our results show that the Hurts exponent, DFA and the Jeanish coefficient are useful methods for characterizing the "randomness" of complex biochemical oscillations at different time scales. For the intracellular calcium simulated time series the system shows persistence and antipersistence for some time scales. For the circadian oscillations for small scales the time series is uncorrelated, which however, deserves a further analysis, as noise in original time series can change the results obtained from our analysis.

Acknowledgments

This research was supported under project PROMEP "Analisis dinamico, control y sincronizacion de ritmos biologicos"

References

1. Alvarez-Ramirez, J., Rodriguez, E., Echeverria, J.C., Puebla, H. *Chaos Solitons Fractals* **36**, 1157-1169 (2008).
2. Andrews, D.F., Herzberg, A.M. *Springer*, New York. Data Set 48.2 (1985).

3. Berridge, M.J., Bootman, M.D., Lipp, P. *Nature* **395**, 645-648 (1998).
4. Blesic, S., Milosevic, D., Stratimirovic, D., Ljubisavljevic, M. *Physica A* **268**, 275-282 (1999).
5. Blumenfeld, L.A., Tikhonov, A.N. *Springer Verlag*, New York (1994).
6. Echeverra, J.C., Aguilar, S.D., Ortiz, M.R., Alvarez-Ramirez, J., Gonzlez-Camarena, R. *Physiol. Measurement* **27**, N19-N25 (2006).
7. Feder, J. *Plenum Press*, New York (1998).
8. Fu, L.N., Lee, C.C. *Nature rev.* **3**, 350-361 (2003).
9. Goldbeter, A. *Cambridge University Press*, Cambridge, UK (1996).
10. Houart, G., Dupont, G., Goldbeter, A. *Bull. Math Biol.* **61**, 507 (1999).
11. Hurst, H.E. *Trans. Amer. Soc. Civil Eng.* **116**, 770 (1950).
12. Letellier, C. *Acta Biotheor.* **50**, 1-13 (2002).
13. Maqueta, J., Letelliera, C., Aguirre, L.A. *J. Theor. Biology* **228**, 421-430 (2004).
14. Mandelbrot, B.B. *Springer-Verlag*, New York (1997).
15. Mikhailov, A.S., Hess, B. *J. Theor. Biol.* bf 176, 185-192 (1995).
16. Musielak, D., Musielak, Z., y Kennamer K. *Fractals* **13**, 19-31 (2004).
17. Peng, C-K, Buldyrev, S.V., Havlin, S., Simons, M., Stanley, H.E., Goldberger, A.L. *Phys Rev E* **49**, 1685 (1994).
18. Puebla, H. *J. Biol. Sys.* **13**, 173 (2005).
19. Schuster, H.G. *Deterministic Chaos: An Introduction*, Weinheim, VCH (1988).
20. Shelhamer, M. *World Scientific Publishing*, (2007).
21. Small, M. *World Scientific Publishing*, (2005).
22. Stanley, H.E., Amaral, L.A.N., Goldberger, A.L., Havlin, S., Ivanov, P.Ch., Peng, C-K. *Physica A* **270**, 309-324 (1999).
23. West, B.J. *World Scientific Publishing*, (1990).

CONTROL AND SYNCHRONIZATION OF HODGKIN-HUXLEY NEURONS

H. PUEBLA, R. AGUILAR-LOPEZ, E. RAMIREZ-CASTELAN

Departamento de Energa, Universidad Autonoma Metropolitana, Azcapotzalco
Av. San Pablo No. 180, Reynosa-Tamaulipas
Azcapotzalco, 02200, D.F. Mexico
E-mail hpuebla@correo.azc.uam.mx

E. HERNANDEZ-MARTINEZ AND J. ALVAREZ-RAMIREZ

Departamento de IPH, Universidad Autonoma Metropolitana, Iztapalapa
Apartado Postal 55-534
Iztapalapa, 09340, D.F. Mexico
E-mail cbi206380874@xanum.uam.mx

Control and synchronization of HH neurons is a central topic in understanding the rhythmicity of living organisms in neuroscience. By using a feedback control approach, we provide a method for both control and synchronization of single and coupled HH neurons. Our main purpose is to elucidate how the synchronization between coupled HH neurons is achieved by a plausible external stimuli. Numerical simulations shown the effectiveness of our control approach.

1. Introduction

The Hodgkin-Huxley (HH) neurons are usually used as realistic models of neuronal systems, for studying neuronal synchronization [7,14]. The HH model is a quantitative model that describes how the action potentials in neurons are initiated and propagated. It's generally realized that the control and synchronization of neuron's activities is important for memory, calculation, motion control and diseases such as epilepsy [4,8,11]. Moreover, it plays an important role in the realization of associative memory, image segmentation and binding [11,14,24]. Indeed, in the past decade it has been shown that synchronized activity and temporal correlation are fundamental tools for encoding and exchanging information for neuronal information processing in the brain [24,15].

Understanding both the processes that influence the synchronization of individual neurons and the functional role of synchronized activity of coupled neurons are important goals in neurobiology [11,14,17,21]. Indeed, clarifying the mechanisms behind HH neurons synchrony could ultimately provide critical information to exploit the synchronized behavior in living organisms. For instance, the application of the knowledge of dynamical systems in biology and medicine is giving rise to new therapeutic approaches, such as the treatment of Parkinson's disease by means of neuronal desynchronization [26,27].

In this work, we study from a feedback control approach the regulation and synchronization of both single and random coupled HH neurons via an external applied current. To this end, different control and synchronization methods can be applied [1,16,5,6,2]. However, since most control designs are based on the mathematical model of the system to be controlled, the presence of disturbances, dynamic uncertainties, and nonlinearities pose great challenges. We introduce a control approach that has two nice features for biological applications [3]: (i) robustness against model uncertainties, and (ii) simplicity in the design. Numerical simulations results indicate good regulation and tracking performance of the closed-loop system. Moreover, our study can be useful to find a possible explanation on the generation mechanism of synchronization between HH neurons.

This work is organized as follows: In Section 2, a generic simple model of HH neurons and its coupling are presented. In Section 3 we introduce our control approach for control and synchronization of single and coupled HH neurons. Numerical simulations in Sections 4 shows the control and synchornization capabilities of our control approach. Finally, some concluding remarks are given in Section 5.

2. Mathematical models of simple and coupled HH neurons

The HH model describes how action potentials in neurons are initiated and propagated. It is a set of nonlinear ordinary differential equations that approximates the electrical characteristics of excitable cells [13]. From its inception, there have been continuing investigations of the HH model to bring it more in accord with modern accurate experimental measurements [22,7]. However, the conceptual and mathematical structure of these equations remains with little modification. Here we consider as the basic single HH neuron the following four-state HH model [13,14]:

$$C_m \frac{dx_1}{dt} = u - g_k x_2^4 (x_1 - x_{1k}) - g_{Na} x_3 x_4 (x_1 - x_{1Na}) - g_l (x_1 - x_{1l}) = u + f_1(x_i) \tag{1}$$

$$\frac{dx_2}{dt} = \alpha_n(x_1)(1 - x_2) - \beta_n(x_1) x_2 = f_2(x_i), \tag{2}$$

$$\frac{dx_3}{dt} = \alpha_m(x_1)(1 - x_3) - \beta_m(x_1) x_3 = f_3(x_i), \tag{3}$$

$$\frac{dx_4}{dt} = \alpha_h(x_1)(1 - x_4) - \beta_h(x_1) x_4 = f_4(x_i), \tag{4}$$

where x_1, x_2, x_3 and x_4 represent the membrane potential, the activation of the potassium flow current, the activation and inactivation of the sodium flow current, respectively. C_m is the membrane capacitance, g_k, g_{Na} and g_l are the maximum ionic and leak conductances, while x_{1k}, x_{1Na} and x_{1l} stand for the ionic and leak reversal potentials [13,14]. The external stimulus current can be modelled by the term u, usually a tonic or periodic forcing. The explicit form of the functions $\alpha_i(x_1)$ and $\beta_i(x1)$ $(i = n, m, h)$, which describes the transition rates between open and closed states of the channels in Eqs. (1)-(5), are given by[14],

$$\begin{aligned} \alpha_n(x_1) &= 0.01(10 - x_1)/(exp[(10 - x_1)/10] - 1) \\ \beta_n(x_1) &= 0.125 exp(-x_1/80) \\ \alpha_m(x_1) &= 0.1(25 - x_1)/(exp[(25 - x_1)/10] - 1) \\ \beta_m(x_1) &= 4 exp(-x_1/18) \\ \alpha_h(x_1) &= 0.07 exp(-x_1/20) \\ \beta_h(x_1) &= 1/(exp[(30 - x_1)/10] + 1) \end{aligned} \tag{5}$$

The HH model (1)-(5) is able to exhibit a plethora of dynamical behavior, from regular to chaotic one, and especially a mixing of spiking and bursting [14,7]. The main interesting collective phenomena in ensembles of neurons is synchronization [14,11,23]. We consider that there are N subsystems in a lattice $x_{i,j}, i = 1, .., 4$ and $j = 1, 2, .., N$. In the absence of coupling, the dynamics of $x_{i,j}$ is given by the single HH neuron (1)-(5). That is, the dynamics of $x_{i,j}$ satisfies,

$$\begin{aligned} \dot{x}_{1,j} &= u_j + f_{1,j}(x_i), j = 1, ..., N \\ \dot{x}_{k,j} &= f_{k,j}(x_i), k = 2, 3, 4 \end{aligned} \tag{6}$$

The N subsystems are coupled as follows,

$$\begin{aligned} \dot{x}_{1,j} &= C(x_{1,j}) + u_j + f_{1,j}(x_i), j = 1, ..., N \\ \dot{x}_{k,j} &= f_{k,j}(x_i), k = 2, 3, 4 \end{aligned} \tag{7}$$

where $C(x_{1,j})$ is a coupling function. We consider non-local random coupling, where $C(x_{1,j}) = \sigma A x_{1,j}$, with

$$A_{lm} = \begin{pmatrix} 0 \text{ if } r_{lm} < r_{min} \\ \\ 1 \text{ if } r_{lm} \geq r_{min} \end{pmatrix} \tag{8}$$

where $r_{lm} \in (0,1)$ is a uniformly distributed random number and the threshold $r_{min} \in (0,1)$. The elements A_{lm} of the matrix A are either 0 or 1 and are assigned in a random way. This coupling structure resembles that of neural networks.

3. Robust Feedback Control Design

In this section, it is presented a robust feedback control approach for both control and synchronization of the dynamic behavior of single and coupled oscillators described by (1)-(5) and (6)-(8).

3.1. *Control problem*

In the HH neurons the control problem is related to explore new means of the regulation of response characteristics of neural circuits. Possible applications are blocking of nerve cell firing and control of neural oscillation [9], as well as neuromodulation [14,25]. Some papers have been addressed the control problem in HH neurons [9,29,18,28]. In particualr, Frohlich and Jezernik[10] introduces a state feedback controller based on model predictive control for action potential generation, suppression of oscillations and blockage of action potential transmission via manipulation of the injected current. Following the same ideas, in this paper the control problem is stated as either the suppression or tracking with respect to a desired signal y_{ref} via the manipulation of the external input I.

3.2. *Synchronization problem*

One of the most challenging in neurobiological systems is an explanation of synchronization of coupled neurons. The synchronization problem consists of making two or more systems oscillate in a synchronized way. In this case we require that the signals are identical, at least asymptotically when $t \to \infty$. From the synchronization point of view, several papers have been published about this topic, they can be classified into two general groups. The first one is related with a natural coupling (self-synchronization) [19,20,23]. The second one is related with an artificial coupling forced via explicit feedback control [6,2]. The synchronization problem addressed in this paper can be stated as follows. Let $y^j_{ref}(t) = (y^1_{ref}(t), ..., y^n_{ref}(t))$ be a desired dynamic behavior for the membrane potential, $x_{1,j}$. The control problem correspond to a synchronization problem with respect to a single synchronization signal $y_{ref}(t)$ via manipulation of the external input I. The control and synchronization problems description are completed by the following assumptions:

A1. The measurement of the variable to be controled or synchronized y, is available for control or synchronization design purposes.

A2. Nonlinear functions $f_1(x_i)$ and $f_{1,j}(x_i)$ are uncertain, and there are available rough estimates of these terms.

A3.The coupling function $C(x_{1,i})$ is not available for synchronization design purposes.

The following comments are in order:

(i) $A1$ is a reasonable assumption since the measurement of the membrane potential is standard. Even in the absence of such measurements, a state estimator can be designed.

(ii) $A2$ considers that functions $f_1(x_i)$ and $f_{1,j}(x_i)$ can contain uncertain parameters, or in the worst case the whole terms are unknown. Indeed, the parameters (diffusion coefficient, kinetics, etc.) in the HH model have some degree of uncertainties, as these parameter values commonly are estimated from experimental data, which contain errors due to both the estimation procedure adopted to fit data and the experimental errors of the data themselves.

(iii) The use of I as the manipulable variable is a realistic since it has a significant effect on the dynamics of HH neurons. Indeed, it has been reported that in HH neurons a small pulse of applied current applied produces a small positive perturbation of the membrane potential (depolarization), which results in a small net current that drives V back to rest (repolarization) [13,14]. However, an intermediate size pulse of current produces a perturbation that is amplified significantly because membrane conductances depend on V [13,14]. Such a non-linear amplification causes V to deviate considerably from V_{rest}. On the other hand, several experimental studies have shown that the synchronization of coupled neurons depends on external stimulus properties [25].

3.3. *Control design*

Let $e = y - y_{ref}$ be the regulation error, and define the modeling error function η as,

$$\eta = [-g_k x_2^4(x_1 - x_{1k}) - g_{Na} x_3 x_4 (x_1 - x_{1Na}) - g_l(x_1 - x_{1l})]/C_m = f_1(x) \quad (9)$$

notice that the complete function $f_1(x)$ is considered unknown, as the worst-case design. System (1)-(5) can be written as,

$$\dot{e} = \eta + C_m^{-1} u - \dot{y}_{ref} \quad (10)$$

where $\dot{y}_{ref}$ is the first derivative of y_{ref}. In the MEC approach the idea is to use a high-gain reduced-order observer to compute an estimate η, of the real uncertain term $\hat{\eta}$, based on the available measurements [3]. After some direct algebraic manipulations the reduced order observer can be written as,

$$\begin{aligned} \dot{w} &= -C_m^{-1} u + y_{ref} - \hat{\eta} \\ \hat{\eta} &= \omega_e^{-1}(w + e) \end{aligned} \quad (11)$$

where ω_e is an observer design parameter. Replacing in model (10) the estimate uncertain term $\hat{\eta}$, instead the real one η, an inverse dynamics control law is given as follows,

$$u = -C_m(\hat{\eta} - \dot{y}_{ref} + \omega_c^{-1} e) \quad (12)$$

where ω_c is a control design parameter, so that the controlled system is given by,

$$\dot{e}(t) = -\omega_c e(t) \quad (13)$$

Since $\omega_c > 0$, the controlled system is asymptotically stable about the zero tracking error, *i.e.* $e(t) \to 0$ asymptotically.

3.4. *Synchronization design*

As discussed above, the synchronization problem design is framed as a feedback control problem. In this case, by assumptions A2 and A3, Eqs. (6)-(8) can be written as,

$$\frac{dy_j}{dt} = -C_{mj}^{-1} u_j + \eta_j \qquad (14)$$
$$\eta_j = C_i(x_{1,j}) + f_{1,j}(x_i)$$

where η_j are the modeling error functions of the N lattice subsystems. In this way, similar to the control design based for the single HH neuron presented above, the modeling error functions are estimated with uncertainty estimators, which after some algebraic manipulations can be written as,

$$\dot{w}_j = -C_{m,j}^{-1} u_j + y_{ref,j} - \hat{\eta}_j \qquad (15)$$
$$\hat{\eta}_j = \omega_{e,j}^{-1}(w_j + e_j)$$

where $e_j(t) = y^j(t) - y^j_{ref}(t)$ is the synchronization error. The inverse dynamics feedback control function is given by,

$$u_j = -C_{m,j}(\hat{\eta}_j - \dot{y}^j_{ref} + \omega_{c,j}^{-1} e_j) \qquad (16)$$

Notice that the resulting feedback control design for synchronization (14)-(15) purposes depends only on measured signals $[y^j, \dot{y}^j_{ref}]$ and estimated values of parameter $C_{m,j}$, and do not relies on a good mathematical model of system. The tuning of parameters ω_c and ω_e, can be set in two steps: (i) determine a value of ω_c up to a point where a satisfactory nominal response is attained, and (ii) the estimation time constant ω_e, which determines the smoothness of the modeling error and the velocity of the time-derivative estimation respectively, can be chosen as $\omega_e > 2.0\omega_c$.

4. Numerical Simulations

In this section, simulation results are presented for the control of a single HH neuron and for synchronization of five random coupled HH neurons. We consider lower and upper limits for the minimum and maximum amplitude of the control inputs as $u_{min} = 0$, and $u_{max} = 200.0$. Our simulation results indicate good regulation and tracking performance of the closed-loop system. Although a rigorous robustness analysis is beyond the scope of this study, numerical simulations will show that our feedback control approach is able to control and synchronize HH neurons despite significant parameter uncertainties and external disturbances. Moreover, for both control and synchronization objectives, the rough estimate of the term C_m has a parameter mismatch between 5-15 % of the nominal parameter used in simulations.

4.1. *Control of single HH neuron*

Control tasks include the regulation to a constant reference, *i.e.* suppression of oscillatory behavior, and tracking of a sinusoidal signal, *i.e.* enforcing to a new oscillatory behavior. In both cases the control action is activated at $t = 50$ time units. We have set the control design parameters ω_c and ω_e as 2.0 and 5.0 respectively.

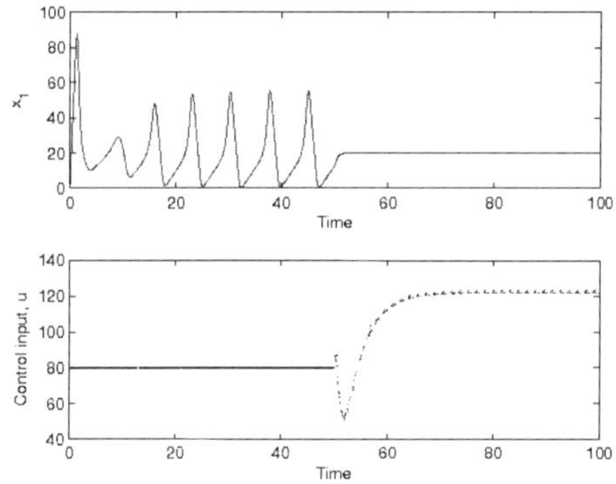

Figure 1. Regulation of HH neuron dynamic to a constant reference.

Figure 1 shows simulation results for regulation to a constant reference $y_{ref} = 20.0$. It can be seen that we can successfully perform the regulation of the oscillatory behavior via a step external input.

For the tracking task, let the desired controlled behavior be a sinusoidal signal $y_{ref} = 20 + 10sin(t)$. Control design parameters are the same of the regulation case. Simulation results are shown in Figure 2. It can be seen from Figure 2 that the control inputs is a periodic influx of u. Numerical results are in accordance with experimental and theoretical studies. On the one hand, some theoretical works have shown that when the amplitude of

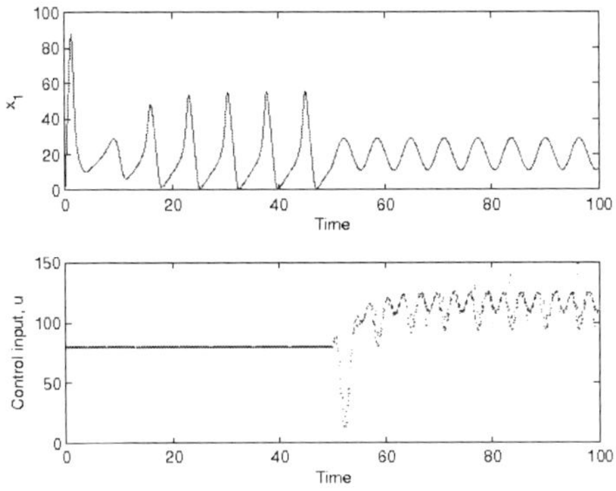

Figure 2. Tracking of HH neuron dynamic to a sinusoidal reference.

the applied current is small, the cell is quiescent, an when the amplitude of the applied current is large, the cell fires repetitive spikes. When the amplitude of the applied current is changed, the cell undergoes a transition from quiescence to repetitive spiking [11,13,14]. For a step current of sufficient amplitude, the HH neuron responds with a train of action potentials, which corresponds to a stable limit cycle.

4.2. *Synchronization of coupled HH neurons*

Synchronization objective is the tracking of a sinusoidal reference. The control methodology for synchronization purposes is implemented in an array of 5 random coupled HH neurons and random initial conditions and the parameters of each HH neuron are all equal to the nominal parameter values. The control is connected at t=100 and $y^j_{ref}(t) = 20 + 10sin(0.5t)$, $i = 1, ..., N$. Figures 3 and 4 presents the behavior of the controlled array for $\omega_{ci} = 1.5$ and $\omega_e = 6.0$, $i = 1, ..., N$. It can be seen that, after a short transient, the array of HH neurons synchronizes about the desired periodical dynamical behavior. Figure 4 shows that by using a pattern of the applied current we can force the HH neuron periodicity. The perturbation depends on the current state of the HH neuron which receives the external impulse. Depending on the spatio-temporal distribution of the input currents that depolarize the membrane voltage, firing threshold (sufficient membrane voltage depolarization) can be reached and an action potential (AP) triggered [13,14].

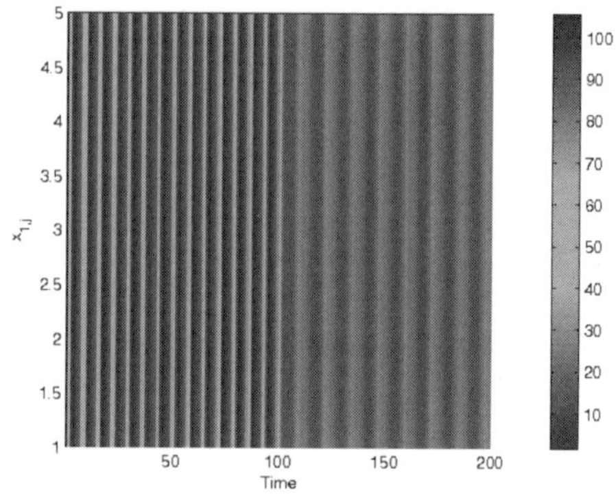

Figure 3. Synchronization of 5 random coupled HH neurons.

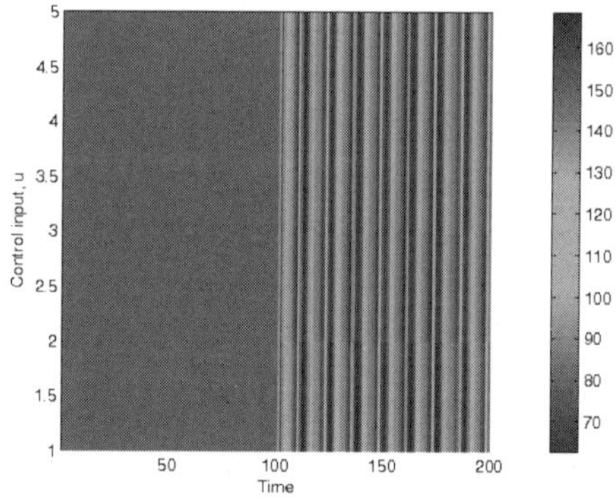

Figure 4. Control input for synchronization of 5 random coupled HH neurons.

5. Conclusions

In this paper, a feedback control approach to control and synchronize single and random coupled HH neurons was presented. Spatio-temporal patterns in HH neurons are modeled via the random coupling of single HH neurons. By using a simple robust control approach, we provide a simple method to achieve synchronization of HH neurons via the modulation of the external applied current. Although we have used both a particular neuron model and random coupling in this article, the control approach can be extended to a wide class of systems describing neural systems. We hope that our control approach can be used to study the effect of electrical stimulation of nerve cell, which has a range of clinical applications.

Acknowledgments

This research was supported under project PROMEP "Analisis dinamico, control y sincronizacion de ritmos biologicos"

References

1. Afraimovich, V. S., Chow, S. N., Hale, J. K. *Physica D*, **103**, 442 (1997).
2. Aguilar-Lopez, R., Martinez-Guerra, R. *Chaos Solitons Fractals*, **37**, 539-546 (2008).
3. Alvarez-Ramirez, J. *Int. J. Robust Nonlin. Control*, **9**, 361-371 (1999).
4. Bar-Gad, I., Bergman, H. *Curr. Opin. Neurobiol.* **11**, 689, (2001).
5. Boccaletti, S., Kurths, J., Osipov, G., Valladares, D.L., Zhou, C.S. *Phys. Rep.* **36** 1, (2001).
6. Cornejo-Perez, O., Femat, R. *Chaos Solitons Fractals* **25**, 43-53 (2005).
7. Fall, C.P., Marland, E.S., Wagner, J.M., Tyson, J.J. *Springer-Verlag*, New York (2002).
8. P Fries. *Trends Cogn. Sci.*, **9**, 474-480, (2005).
9. Frohlich, F., Jezernik, S. *Cont. Eng. Practice* **13**, 1195-1206 (2005).
10. Frohlich, F., Jezernik, S. *J. Comput. Neurosci.* **17**, 165-178 (2004).
11. C M Gray. *J. Comput. Neurosci.*, **1**, 11-38, (1994).
12. Goldbeter, A. *Cambridge University Press*, Cambridge, UK (1996).
13. Hodgkin, A.L., Huxley, A.F. *J. Physiology* **177**, 500-544 (1952).
14. Izhikevich, E. *The MIT Press* (2005).
15. G. Laurent, *Trends Neurosci.* **19**, 489, (1996).
16. Mirollo, R. E., Strogatz, S. H. *SIAM J. Appl. Math.*, **50**, 1645 (1990).
17. E. Niebur, S.S. Hsiao, K.O. Johnson, *Curr. Opin. Neurobiol.* **12**, 190, (2002).
18. Parmananda, P., Mena, C.H., Baier, G. *Phys. Rev. E* **66**,47202 (2002).
19. Pasemann, F. *Physica D* **128**, 236-249 (1999).
20. A.S. Pikovsky, J. Kurths, M.G. Rosenblum, *Europhys. Lett.* **34**, 165, (1996).
21. R. Ritz, T.J. Sejnowski, *Curr. Opin. Neurobiol.* **7**, 536, (1997).
22. Sangrey, T.D., Friesen, W.O., Levy, W.B. *J. Neurophysiol.* **91**, 2541-2550 (2004).
23. Shuai, J-W., Durand, D.M. *Phys. Lett. A* **264**, 289-297 (1999).
24. Singer, W., Gray, C.M. *Annu. Rev. Neurosci.* **18**, 555-586 (1995).
25. J M Schoffelen, R Oostenveld, and P Fries. *Science*, **308** 111-113, (2005).
26. P.A. Tass, *Europhys. Lett.* **53**, 15-21, (2001).
27. P.A. Tass, *Europhys. Lett.* **57**, 164-170, (2002).
28. Wang, W., Wang, Y., Wang, Z.D., Wang, W. *Phys. Rev. E* **57**, 2527 (1998).
29. Yu, Y., Wang, W., Wang, J., Liu, F. *Phys. Rev. E* **63**, 21907 (2001).

A CORRELATION BETWEEN ATOM SITES AND AMIDE PLANES IN PROTEIN STRUCTURES

R.P. MONDAINI

Federal University of Rio de Janeiro

Centre of Technology, 21945-672 P.O. Box 68511, Rio de Janeiro, RJ, Brazil

The consideration of generalized Fermat problems for modelling the localization of atom sites in biomolecules is a successful approach for studying their structures. In the present note, the generic methods of this kind of modelling are introduced and an application is made to protein structures.

1. Introduction

Let us suppose that a molecular cluster can be formed by the interaction of an atom site A with n atom sites $A_1, A_2, \cdots, A_n$. The potential energy of the cluster can be considered to be written as

$$U = \sum_p \sum_{j=1}^{n} Q_{j(p)} \left\| \vec{r}_{A_j} - \vec{r}_A \right\|^p \tag{1}$$

where $\| \cdots \|$ stands for the Euclidean norm and $Q_{j(p)}$ is a characteristic object of the interaction, v.g. :$Q_{j(-1)} = q_A q_{A_j}$ for Coulombian interaction, q_A, q_{A_j} being the charges of A and A_j atom sites; $Q_{j(-6)}$, $Q_{j(-12)}$, corresponding to Lenard-Jones interaction and so on.

The necessary conditions for a minimum of the potential energy can be given by

$$0 = \frac{\partial U}{\partial x^s_{A_j}} = \sum_p \sum_{j=1}^{n} p Q_{j(p)} \left\| \vec{r}_{A_j} - \vec{r}_A \right\|^{p-1} \frac{(x^s_{A_j} - x^s_A)}{\left\| \vec{r}_{A_j} - \vec{r}_A \right\|} \tag{2}$$

After multiplying the equation above by the unit vector $\hat{\mathrm{i}}_s$ of the s-coordinate axis and summing up in s, we get

$$0=\sum_{s=1}^{3}\hat{\mathrm{i}}_s\frac{\partial U}{\partial x^s_{A_j}}=\sum_{s=1}^{3}\sum_{p}\sum_{j=1}^{n}pQ_{j(p)}\left\|\vec{r}_{A_j}-\vec{r}_A\right\|^{p-1}\hat{\mathrm{i}}_s\frac{(x^s_{A_j}-x^s_A)}{\left\|\vec{r}_{A_j}-\vec{r}_A\right\|} \tag{3}$$

or

$$0=\sum_{p}\sum_{j=1}^{n}pQ_{j(p)}\left\|\vec{r}_{A_j}-\vec{r}_A\right\|^{p-1}\hat{\mathrm{r}}_{A,A_j} \tag{4}$$

where

$$\hat{\mathrm{r}}_{A,A_j}=\frac{\vec{r}_{A_j}-\vec{r}_A}{\left\|\vec{r}_{A_j}-\vec{r}_A\right\|} \tag{5}$$

For any given p-value, we assume that

$$Q_{j(p)}\left\|\vec{r}_{A_j}-\vec{r}_A\right\|^{p-1}=Q_{k(p)}\left\|\vec{r}_{A_j}-\vec{r}_A\right\|^{p-1}=F_{(p)},\forall j,k \tag{6}$$

This means that the interaction forces between the atom sites A and A_j are assumed to be the same for $\forall j,k$.

We then have

$$\sum_{p}pF_{(p)}\sum_{j=1}^{n}\hat{\mathrm{r}}_{A,A_j}=0;\Rightarrow\sum_{j=1}^{n}\hat{\mathrm{r}}_{A,A_j}=0 \tag{7}$$

2. A Generalized Fermat Problem

Let us take again the molecular cluster of the 1st section. We wish now to know the coordinates of the A atom site such that the sum of the Euclidean distances to the other atom sites A_j is a minimum. We then look for a minimum of the convex function

$$L=\sum_{j=1}^{n}\left\|\vec{r}_{A_j}-\vec{r}_A\right\| \tag{8}$$

We have

$$0=\frac{\partial L}{\partial x^s_{A_j}}=\sum_{j=1}^{n}\frac{x^s_{A_j}-x^s_A}{\left\|\vec{r}_{A_j}-\vec{r}_A\right\|} \tag{9}$$

and

$$0 = \sum_{s=1}^{3} \hat{\mathrm{i}}_s \frac{\partial L}{\partial x^s_{A_j}} = \sum_{j=1}^{n} \hat{\mathrm{r}}_{A,A_j} \tag{10}$$

We then see that this geometric problem is the same problem of minimizing the potential energy of the clusters, eqs. (1-7), if the assumption of identical interaction forces between nearest neighbours is taken into account, eq. (6). Eqs. (8) and (11) lead to the system of n equations and C^2_n unknowns:

$$\sum_{j=1}^{n} \hat{\mathrm{r}}_{A,A_j} . \hat{\mathrm{r}}_{A,A_k} = 0, \; k = 1, 2, \cdots, n \tag{11}$$

For $n \geq 2$, there are solutions of C^2_n equal angles, given by

$$\hat{\mathrm{r}}_{A,A_j} . \hat{\mathrm{r}}_{A,A_k} = -\frac{1}{n-1}(1 - n\delta_{jk}) \tag{12}$$

For $n = 2, 3$, this kind of solution is unique and for $n \geq 4$, these are special solutions since for the generic case we have to introduce $C^2_n - n = n(n-3)/2$ additional geometric conditions to solve the problem. The cases $n = 3, 4$, correspond to the centre of an equilateral triangle and the centre of a regular tetrahedron for the A atom site, respectively.

3. Evenly Spaced Atom Sites

Let us consider a sequence of atom sites in a biomolecular structure. We wish to know the 3-dimensional curve through a sequence of evenly spaced atom sites. These will be characterized by their position vectors[1] as

$$\vec{r}_j = (\rho(j\omega)\cos(j\omega), \rho(j\omega)\sin(j\omega), z(j\omega)) \tag{13}$$

The conditions for evenly spaced points will be given by

$$\|\vec{r}_{j+1} - \vec{r}_j\| = \|\vec{r}_j - \vec{r}_{j-1}\| \tag{14}$$

or

$$(\rho((j+1)\omega) - \rho((j-1)\omega))(\rho((j+1)\omega) + \rho((j-1)\omega)$$

$$-2\rho(j\omega)\cos\omega) + (z((j+1)\omega) - z((j-1)\omega))(z((j+1)\omega)$$

$$+z((j-1)\omega) - 2z(j\omega)) = 0 \tag{15}$$

From eq.(15), we can have four cases:

$$\rho(j\omega) = r(\omega); z(j\omega) = jh(\omega) \tag{16}$$

$$\rho(j\omega) = r(\omega); z(j\omega) = h(\omega) \tag{17}$$

$$\rho(j\omega) = r(\omega)\cos(j\omega); z(j\omega) = h(\omega) \tag{18}$$

$$\rho(j\omega) = r(\omega)\cos(j\omega); z(j\omega) = jh(\omega) \tag{19}$$

We will choose the case given by eq.(16) and we write

$$\vec{r}_j(\omega) = (r(\omega)\cos(j\omega), r(\omega)\sin(j\omega), jh(\omega)) \tag{20}$$

For a solution of the form given by eq.(12), we have

$$(\vec{r}_{j+1} - \vec{r}_j).(\vec{r}_{j-1} - \vec{r}_j) = -\frac{1}{n-1}\left\|\vec{r}_{j+1} - \vec{r}_j\right\|.\left\|\vec{r}_{j-1} - \vec{r}_j\right\| \tag{21}$$

From eqs.(20),(21), we get a relation between the functions for evenly spaced consecutive points or

$$h^2(\omega) = \frac{2}{n-2}r^2(\omega)(1-\cos\omega)(1-(n-1)\cos\omega) \tag{22}$$

Some remarks are now in order:

(1) Eqs.(17)-(19) lead only to trivial solutions.

(2) It should be noted that for evenly spaced consecutive points given by eq.(20), the only admissible n-values[2] are $n = 3, 4$, which is in agreement with the observed structure of biomolecules[3].

(3) The equation $z(j\omega) = jh(\omega)$ has no sense for a function of only one variable except if $z(\omega) \equiv h(\omega)$ which also means that $h(\omega)$ is a linear function. A fortiori, since we now have a homogeneous function of one variable, we have from Euler's theorem,

$$\omega \frac{\partial h}{\partial \omega} = h \Rightarrow h(\omega) = \alpha\omega \tag{23}$$

And $\vec{r}(\omega)$ is a right circular helix. These remarks should be taken as a confirmation of the insight for the existence of helix curves in the modeling of biomolecules and in particular of the alpha-helices observed on the 3-dimensional structure of proteins.

4. The Modelling Process of Biomolecules

We now consider two sequences of evenly spaced atom sites,

$$\vec{r}_k(\omega) = (r(\omega)\cos(k\omega), r(\omega)\sin(k\omega), kh(\omega)) \tag{24}$$

$$\vec{s}_k(\omega) = (s(\omega)\cos(k\omega), s(\omega)\sin(k\omega), kh_s(\omega)) \tag{25}$$

where $r(\omega) \geq s(\omega)$. We introduce the non-dimensional quantities

$$H \equiv h/s; H_s \equiv h_s/s; R \equiv r/s; z \equiv \cos\omega \tag{26}$$

And we define the unit vectors

$$\hat{\mathrm{s}}_{l.m} = \frac{\vec{s}_m - \vec{s}_l}{\|\vec{s}_m - \vec{s}\|} \quad , \quad \hat{\mathrm{r}}_{l.m} = \frac{\vec{r}_l - \vec{s}_m}{\|\vec{r}_l - \vec{s}_m\|} \tag{27}$$

The following internal products of some of these unit vectors should be useful in the present modelling process:

$$\hat{\mathrm{s}}_{k+1,k+2}.\hat{\mathrm{r}}_{k+1,k+1} = \frac{(R-1)(z-1)+(k+1)H_s(H-H_s)}{A_{k+1}C} \tag{28}$$

$$\hat{\mathrm{s}}_{k-1,k-2}.\hat{\mathrm{r}}_{k-1,k-1} = \frac{(R-1)(z-1)-(k-1)H_s(H-H_s)}{A_{k-1}C} \tag{29}$$

$$\hat{\mathrm{r}}_{k,k+1}.\hat{\mathrm{r}}_{k+1,k+1} = \frac{(R-1)(Rz-1)+(k+1)(H-H_s)[k(H-H_s)-H_s]}{A_{k+1}B_{k+1}} \tag{30}$$

$$\hat{\mathrm{r}}_{k,k-1}.\hat{\mathrm{r}}_{k-1,k-1} = \frac{(R-1)(Rz-1)+(k-1)(H-H_s)[k(H-H_s)+H_s]}{A_{k-1}B_{k-1}} \tag{31}$$

$$\hat{\mathrm{s}}_{k+1,k+2}.\hat{\mathrm{r}}_{k,k+1} = \frac{(z-1)(2Rz+R-1)+H_s[k(H-H_s)-H_s]}{B_{k+1}C} \tag{32}$$

$$\hat{\mathrm{s}}_{k-1,k-2}.\hat{\mathrm{r}}_{k,k-1} = \frac{(z-1)(2Rz+R-1)-H_s[k(H-H_s)+H_s]}{B_{k-1}C} \tag{33}$$

where

$$A_{k\pm 1} \equiv [(R-1)^2+(k\pm 1)^2(H-H_s)^2]^{1/2} \tag{34}$$

$$B_{k\pm 1} \equiv [(R-1)^2+2R(1-z)+(k(H-H_s)mH_s)^2]^{1/2} \tag{35}$$

$$C \equiv [2(1-z)+H_s^2]^{1/2} \tag{36}$$

In order to proceed with the modeling, we should note that the internal products above cannot change if we make $k+1 \rightarrow k-1, k+2 \rightarrow k-2$. Eqs.(28)-(33) will then lead to

$$H_s = H \tag{37}$$

and

$$\hat{\mathrm{s}}_{k+1,k+2}.\hat{\mathrm{r}}_{k+1,k+2} = \hat{\mathrm{s}}_{k-1,k-2}.\hat{\mathrm{r}}_{k-1,k-2} = \cos\Gamma \tag{38}$$

$$\hat{\mathrm{r}}_{k,k+1}.\hat{\mathrm{r}}_{k+1,k+1} = \hat{\mathrm{r}}_{k,k-1}.\hat{\mathrm{r}}_{k-1,k-1} = \cos\Delta \tag{39}$$

$$\hat{\mathrm{s}}_{k+1,k+2}.\hat{\mathrm{r}}_{k,k+1} = \hat{\mathrm{s}}_{k-1,k-2}.\hat{\mathrm{r}}_{k,k-1} = \cos(\Gamma+\Delta) \tag{40}$$

5. An Application to the Modelling of Protein Structure

We now use these results to model the structure of a protein molecule. The first calculation scheme to be used is given in Figure 1.

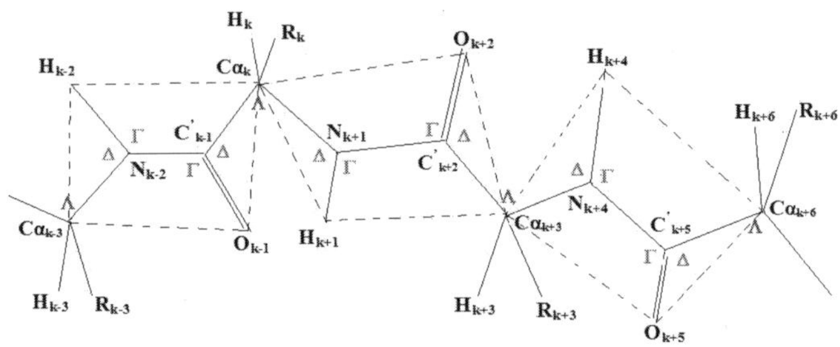

Figure 1. Scheme for modeling the protein structure with Amide planes. Plane bond angles Γ, Δ, Λ are defined in the text.

We then have a geometric modelling of the protein structure and we consider Steiner points to be the representatives of nitrogen (N) and carbonyl (C') atom sites. There are Fermat problems with $n = 4$ edges to determine the alpha-carbon atom sites. We now assume that the molecular structure of the protein is obtained from a first order perturbation of the values

$$\cos\Gamma_0 = -\frac{1}{2}; \tag{41}$$

$$\cos\Delta_0 = -\frac{1}{2}; \tag{42}$$

$$\cos(\Gamma_0 + \Delta_0) = -\frac{1}{2}; \tag{43}$$

$$\cos\Lambda_0 = -\frac{1}{3} \tag{44}$$

according to eq.(23).

A sequence of the form given by eq.(24) is then assumed through the alpha carbon (C_α), hydrogen (H) and oxygen (O) atom sites ($\vec{r}_{C_{\alpha k}}, \vec{r}_{H_{k+1}}, \vec{r}_{O_{k-1}}$), for example. Another sequence (eq.(25)) is also assumed through nitrogen

(N) and carbonyl (C') atom sites ($\vec{s}_{N_{k+1}}, \vec{s}_{C'_{k+2}}$). The part of the problem corresponding to first order perturbation of Steiner atom sites is given by

$$\frac{z-1}{[H^2+2(1-z)]^{1/2}} = \cos\Gamma = -\frac{1}{2}+\gamma \tag{45}$$

$$\frac{Rz-1}{[H^2+(R-1)^2+2R(1-z)]^{1/2}} = \cos\Delta = -\frac{1}{2}+\delta \tag{46}$$

$$\frac{(z-1)(2Rz+R-1)-H^2}{[H^2+2(1-z)]^{1/2}[H^2+(R-1)^2+2R(1-z)]^{1/2}} = \cos(\Gamma+\Delta)$$

$$\approx -\frac{1}{2}-(\gamma+\delta) \tag{47}$$

where we are working with first order perturbations $O(\gamma^2)\approx 0$, $O(\delta^2)\approx 0$, $O(\gamma\delta)\approx 0$. This is already taken into account for writing the right hand side of eq.(18). There is of course a solution for $\gamma=0$, $\delta=0$ and we should have

$$H^2 = 2(1-z)(1-2z) \tag{48}$$

and this is eq. (22) for $n=3$. Eqs. (45)-(47) are identically satisfied. In the generic case, we can write

$$H^2 = (1-z)[1-N-(1+(2-N)z)R] \tag{49}$$

where

$$N \approx -2\frac{1+2(\gamma+\delta)}{1-2(\gamma+\delta)} \tag{50}$$

From eqs. (45),(49), we also get

$$R = \frac{3-N-4(1-z)(1-2\gamma)^{-2}}{1+(2-N)z} \tag{51}$$

From eqs.(45), (50), (51), we can write,

$$4(1-2(\gamma+\delta))^2(1-z)^2(1+4z-2(\gamma+\delta))^{-2}(az^2+bz+c)\approx 0 \tag{52}$$

where

$$a = 16(1-2\gamma)^{-2}[(1-2\gamma)^{-2}(1-2\delta)^{-2} - (1-2(\gamma+\delta))^{-2}] \tag{53}$$

$$b = 8[(1-2\gamma)^{-2}(1-2\delta)^{-2} - (1-2(\gamma+\delta))^{-2}] \tag{54}$$

$$c = (1-2\gamma)^{-2} + (1-2\delta)^{-2} - 2(1-2(\gamma+\delta))^{-1} - 4[(1-2\gamma)^{-2}$$

$$- (1-2(\gamma+\delta))^{-1}]^2 \tag{55}$$

and we have $a \approx 0$, $b \approx 0$, $c \approx 0$ for first order perturbations.

The second part of the problem of first order perturbation of atom sites corresponds to the perturbation of alpha carbon geometry and we have at zeroth order regular tetrahedra whose vertices are the atom sites ($\vec{s}_{C'_{k-1}}$, $\vec{s}_{N_{k+1}}$, $\vec{r}_{H_k}$, $\vec{r}_{R_k}$).

This solution $\vec{r}_{C_{\alpha k}}$ is obtained via a Fermat problem with $n = 4$, eq.(12), or

$$(-\vec{r}_{k,k-1}) + \frac{\vec{r}_{H_k} - \vec{r}_k}{\|\vec{r}_{H_k} - \vec{r}_k\|} + \frac{\vec{r}_{R_k} - \vec{r}_k}{\|\vec{r}_{R_k} - \vec{r}_k\|} + (-\vec{r}_{k,k+1}) = 0 \tag{56}$$

Eq. (44) corresponds to the zeroth order solution of the problem given into eq. (56) above. We look for a first order perturbation solution of the form

$$(-\vec{r}_{k,k-1}).(-\vec{r}_{k,k+1}) = \cos\Lambda = -\frac{1}{3} + \lambda \tag{57}$$

and we have $O(\lambda^2) \approx 0$, $O(\lambda\gamma) \approx 0$, $O(\lambda\delta) \approx 0$.

There are some works in the literature about the perturbations of this angle due to the change of residue R on dipeptides[4]. Our aim is to relate these perturbations and those on Steiner atom sites to the existence of Amide planes. We will need the linearized forms of eqs. (49)-(51) in all subsequent calculations. We can write,

$$N \approx -2[1 + 4(\gamma + \delta)] \tag{58}$$

$$R \approx 1 - 8(1 + 4z)^{-1}(1 - z)(\gamma - \delta) \tag{59}$$

$$H^2 \approx 2(1 - z)(1 - 2z) + 16(1 - z)^2\gamma \tag{60}$$

After substituting eqs. (24)-(27) into eq. (57), we get the equation for small perturbations of alpha carbon geometry, or

$$H^2 - 2z^2 + \frac{2}{3} + \lambda - R^2(\frac{4}{3} - \lambda) + Rz(\frac{7}{3} - \lambda) = 0 \tag{61}$$

From eqs. (58)-(60), we get

$$(1 + 4z)(z^2 - 2Az + A^2 + B^2)(z - z_R) \approx 0 \tag{62}$$

where A and B are the real and imaginary parts of complex roots are given by

$$A \equiv -\tfrac{1}{72}(\xi + 9\alpha)^{1/3} + 18\zeta(\xi + 9\alpha)^{-1/3} + \tfrac{19}{36} + \tfrac{1}{6}\lambda + \tfrac{10}{9}\delta \tag{63}$$

$$B \equiv 3^{1/2}[\tfrac{1}{72}(\xi + 9\alpha)^{1/3} + 18\zeta(\xi + 9\alpha)^{-1/3}] \tag{64}$$

The non-trivial real root of the alpha carbon geometry equation can be written as

$$z_R = \frac{1}{36}(\xi + 9\alpha)^{1/3} - 36\zeta(\xi + 9\alpha)^{-1/3} + \frac{19}{36} + \frac{1}{6}\lambda\frac{10}{9}\delta \tag{65}$$

where is small auxiliary variable in the sense that

$$\frac{\alpha^2}{\xi^2} << 1 \tag{66}$$

We have for the linear parts of α^2, ξ, ζ:

$$\alpha^2 \approx -2.4367.10^5 - 1.4999.10^6\gamma + 4.2113.10^6\delta + 2.9857.10^5\lambda \tag{67}$$

$$\xi \approx -406 - 51.948\gamma + 31.548\delta - 10.917\lambda \tag{68}$$

$$\zeta \approx -\frac{271}{1296} - \frac{5}{18}\gamma + \frac{179}{162}\delta + \frac{25}{216}\lambda \tag{69}$$

After using eqs. (67)-(69) into eq. (65) and keeping only the linear terms on γ, δ, λ and the auxiliary variable α, we get

$$z_R \approx 1.7501 + 35.937\gamma - 25.265\delta + 6.8724\lambda - 0.0059921\alpha \tag{70}$$

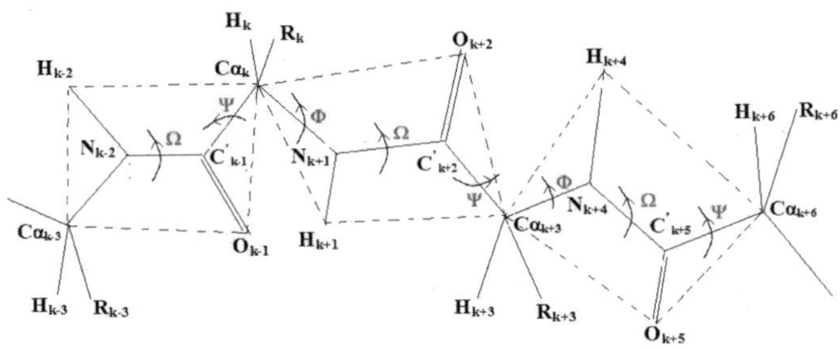

Figure 2. Scheme for modeling the protein structure with Amide planes. Dihedral angles Ω, Ψ, Φ are defined in the text.

In the modeling which we are proposing, we have to take into account the Amide plane structure of the protein molecule. This is done by introducing the small perturbations of dihedral angles (Fig. 2). The hypothesis of planarity of the Amide group on peptides has been checked by many authors[5,6,7]. After intensive statistical analysis of some dipeptide structures, the literature has registered the mean values of all relevant bond and dihedral angles for eight different residues. In this work a more direct approach is used which is based on linear programming techniques of Optimization theory[8]. From eqs. (27) and Fig. 2 above, the dihedral angles[9] can be defined by

$$\widehat{\Omega}_{k,k+1,k+2}.\widehat{\Omega}_{k+1,k+2,k+3} = \cos\Omega = -1 + \varepsilon, \varepsilon \geq 0 \tag{71}$$

$$\widehat{\Psi}_{k-2,k-1,k}\cdot\widehat{\Psi}_{k-1,k,k+1} = \cos\Psi \tag{72}$$

$$\widehat{\Phi}_{k-1,k,k+1}\cdot\widehat{\Phi}_{k,k+1,k+2} = \cos\Phi \tag{73}$$

where the unit vectors above are obtained from the following cross products:

$$\begin{aligned}\vec{\Omega}_{k,k+1,k+2} &= (-\vec{r}_{k,k+1}) \times \vec{s}_{k+1,k+2},\\ \vec{\Omega}_{k+1,k+2,k+3} &= \vec{s}_{k+1,k+2} \times \vec{r}_{k+3,k+2}\end{aligned} \tag{74}$$

$$\begin{aligned}\vec{\Psi}_{k-2,k-1,k} &= \vec{s}_{k-2,k-1} \times \vec{r}_{k,k-1},\\ \vec{\Psi}_{k-1,k,k+1} &= \vec{r}_{k,k-1} \times (-\vec{r}_{k,k+1})\end{aligned} \tag{75}$$

$$\begin{aligned}\vec{\Phi}_{k-1,k,k+1} &= \vec{r}_{k,k-1} \times (-\vec{r}_{k,k+1}),\\ \vec{\Phi}_{k,k+1,k+2} &= (-\vec{r}_{k,k+1}) \times \vec{s}_{k+1,k+2}\end{aligned} \tag{76}$$

In eq. (71), we have allowed for a small perturbation of the dihedral angle Ω to correspond to Amide planes through a minimization process of a linear programming approach. The perturbation $\varepsilon \geq 0$ will be the objective function of the subsequent analysis. Eqs.(72), (73) are written here for completeness and they are useful for verifying some conclusions obtained[9] from Ramachandran's plot $\Psi \times \Phi$ in terms of allowed perturbations.

From eqs. (24)-(27), we can write eq. (71) in the following form:

$$H^2[3z + 1 - (2z + 5)\varepsilon + (-2z^2 + 3 - (2z^2 + 2z - 1)\varepsilon)R$$

$$+(4z^3 - 3z + 1 - \varepsilon)R^2] + (1 - z^2)[1 + (1 - 2z)R]^2(2 - \varepsilon) = 0 \tag{77}$$

After substituting the linear approximations of the functions H and R, eqs. (59), (60) and discarding all the non linear terms, we get the sixth degree polynomial,

$$(128 - 512\gamma - 512\delta)z^6 + (-256 + 640\gamma + 1152\delta - 32\varepsilon)z^5$$

$$+(8 + 832\gamma - 192\delta + 112\varepsilon)z^4 + (284 - 1504\gamma - 1120\delta - 194\varepsilon)z^3$$

$$+(-22 - 1120\gamma + 584\delta - 8\varepsilon)z^2 + (-58 + 528\gamma + 64\delta + 40\varepsilon)z$$

$$-9 + 176\gamma - 36\delta = 0 \quad (78)$$

We connect this result with the perturbation of alpha-carbon geometry by substituting the linearized form of the real root $z = z_R$, eq. (70) into eq. (78). We can write

$$8.9427.10^2 + 1.0989.10^5\gamma - 4.7680.10^4\delta + 2.2314.10^4\lambda$$

$$-17.676\alpha - 4.6907.10^2\varepsilon \approx 0 \quad (79)$$

In order to derive the objective function, we eliminate α between eqs. (79) and (67). After solving the resulting equation for ε, we get,

$$\varepsilon = 91.701 + 7.9289.10^2\gamma - 1.6700.10^3\delta - 63.618\lambda \quad (80)$$

The constraints of this linear programming problem can be obtained from

$$\alpha^2 \geq 0 \quad , \quad z_R^2 \leq 1 \quad , \quad H^2 \geq 0 \quad , \quad \varepsilon \geq 0 \quad (81)$$

The systematic derivation of these constraints in linearized form will be done in the following section. We notice that α is an auxiliary variable and it should be eliminated from the final formulation of the problem. The objective function is also a constraint and it should be also eliminated from the set of variables.

6. A Summary of the Optimization Problem and the example of a Feasible Solution

The Optimization process starts from the linearized equations:

$$\alpha^2 = L + N\gamma + M\delta + P\lambda \tag{82}$$

$$z_R = l + n\gamma + m\delta + p\lambda + q\alpha \tag{83}$$

$$W + X\gamma + Y\delta + T\lambda + U\alpha + V\varepsilon \approx 0 \tag{84}$$

These equations are the expression for the auxiliary variable α in terms of the fundamental variables of this process, the real root obtained from the alpha-carbon geometry and the equation obtained by substituting this real root into the equation for perturbation of the Amide planes, eq. (77). We eliminate α from eqs. (83), (85) and we solve for ε, the objective function

$$\varepsilon = \frac{1}{2VW}[LU^2 - W^2 + (NU^2 - 2WX)\gamma + (MU^2 - 2WY)\delta$$

$$+(PU^2 - 2WT)\lambda] \tag{85}$$

The 1st constraint is given by $\alpha^2 \geq 0$, or

$$L + N\gamma + M\delta + P\lambda \geq 0 \tag{86}$$

The 2nd constraint (z_R^2) is obtained from eqs. (86), (84) and (85). We have,

$$-1 + (l - \frac{qW}{U})[l - \frac{qLU}{W} + (2n - \frac{qNU}{W})\gamma + (2m - \frac{qMU}{W})\delta$$

$$+(2p - \frac{qPU}{W})\lambda] \leq 0 \tag{87}$$

The 3rd constraint is now obtained by substituting eq. (83) into eq. (60) and eliminating from eq. (82):

$$C_{H^2} = (1-2l)^2 - 4q^2L - 4[(1-2l)(n-2) + q^2N]\gamma$$

$$4[(1-2l)m + q^2M]\delta - 4[(1-2l)p + q^2P]\lambda \geq 0 \tag{88}$$

The 4th constraint is $\varepsilon \geq 0$. The values of the parameters (L, N, M, P), (l, n, m, p, q) and (W, X, Y, T, U, V) are obtained from the numerical steps of the process described above and given in eqs. (67), (70) and (79), respectively. We can now solve the linear programming problem

$$min \quad \varepsilon \tag{89}$$

such that

$$\alpha^2 \geq 0 \quad , \quad C_{z_R} \leq 0 \quad , \quad C_{H^2} \geq 0 \quad , \quad \varepsilon \geq 0 \tag{90}$$

and we get

$$\Omega \approx 180^\circ \quad , \quad \Gamma \approx 117.48^\circ \quad , \quad \Delta \approx 115.27^\circ \quad , \quad \Lambda \approx 109.47^\circ \tag{91}$$

7. Concluding Remarks

The results of the last section are in very good agreement with the results of the scientific literature[6] which have been obtained by a statistical analysis. The existence of Amide planes is then supported by a linear programming calculation confirming Pauling's insight. We notice that this approach is able to unveil the behaviour of the protein structure under small perturbation of bond angles and the angle corresponding to alpha-carbon geometry The consideration of evenly spaced atom sites is the fundamental assumption of the present modeling. It is related to the structure of the protein since it precludes atom sites with more than $n = 3, 4$ links to their nearest neighbours. It is also related to other Pauling's insight as alpha helices are concerned. A systematic analysis by using the technique of Ramachandran's plot will be useful for understanding the protein structure in terms of their feasible small perturbations. Work along this line is now in progress.

References

1. R.P .Mondaini - "Modelling Biomolecular Structure with Geodesic curves through Ordered Sets of Atom Sites", *International Journal of Bioinformatics Research and Applications* IJBRA 5(2) (2009) 197-206.
2. R.P. Mondaini - "The Steiner Tree Problem and its Application to the Modelling of Biomolecular Structures in Mathematical Modelling of Biosystems", *Applied Optimization Series*, Vol. 102 (2008) 199-219, Springer Verlag Berlin-Heidelberg.

3. D. Voet, J.D. Voet - "Biochemistry", J. Wiley, Inc. (2004), 3/Edition
4. C.Dale Keefe, J.K. Pearson - "*Ab Initio* investigations of Dipeptide Structures", *J. Mol. Structure* (Theochem) 679 (2004) 65-72.
5. L. Pauling, R.B. Corey, H.R. Branson - "The Structure of Proteins: two Hydrogen-bonded helical configurations of the Polypeptide Chain", *Proc. Natl. Acad. Sci.*, USA, Vol.37, No. 729, (1951) 205-211.
6. A.S.Edison - "Linus Pauling and the Planar Peptide Bond", *Nature Structural Biology* 8(3) (2001) 201-202.
7. M.S. Bazaraa, J.J. Jarvis, H.D. Sherali - "Linear Program and Network Flows", J. Wiley, Inc. (1990), 2nd edition.
8. A. Neumaier - "Molecular Modelling of Proteins and Mathematical Prediction of Protein Structures", *SIAM Rev.* 39 (1997), 407-460.
9. C. Branden, J. Tooze - "Introduction to Protein Structure", Garland Publ. Inc., New York, 1991.

MATHEMATICAL MODELLING OF SUSTAINABLE DEVELOPMENT: AN APPLICATION TO THE CASE OF THE RAIN-FOREST OF MADAGASCAR

C. BERNARD

Cemagref de Clermont-Ferrand,
LISC (Laboratory of engineering for complex systems),
24, avenue des Landais - BP 50 085 - 63 172 Aubière Cedex 1, France
E-mail: `claire.bernard@cemagref.fr`

Protecting the rain forest in Madagascar is essential to conserve biodiversity on the island. However, regulations must take into account the development of the local population. The aim of this paper is to determine if there are viable solutions which lead to the conservation of the forest and the development of the inhabitants of the forest corridor of Fianarantsoa. We will also determine what decisions must be taken in accordance with the two aspects of the problem, by elaborating a model and studying it with the mathematical tools of the viability theory. This theory deals with controlled dynamical systems and consists in finding controls which keep the evolutions governed by these systems in a defined constraint set. The set which gathers all the situations for which such controls exist is called the viability kernel. The computation of the viability kernel with a condition on equity between the different generations excludes a large part of situations. However, adding monetary transfers enables to increase the size of the viability kernel.

1. Introduction

Sustainable development has become an essential concept these last years. The most widely acknowledged definition appeared in 1987 at the World Commission on Environment and Development in the Brundtland report: a sustainable development "implies meeting the needs of the present without compromising the ability of future generations to meet their own needs". From this conceptual definition, many operational criteria have been proposed to distinguish, build or describe developments that should be considered as sustainable on different case study. However, there is no consensus. We do not fulfill such a top down approach, but we are interested in the common characteristics of sustainable management problems. In the case of the reduction of CO_2 emissions, of the preservation of the forest area, or in the management of energy resources, the aim is always to satisfy economic, ecological and social constraints for a long horizon of time.

We suggest here an application of the sustainable management by the case of the rain-forest of Madagascar. Currently, the malagasy forest, in particular the forest corridor of Fianarantsoa, in the east of the country is threatened by human activities. In fact, because of a high demography rate the local population clears the forest and converts it in rice growing to ensure its subsistence[7]. However, protecting the rain forest is essential to conserve biodiversity. Thus we have two major stakes: an ecological one, preserving the forest area to conserve the biodiversity and an economic one, the economic development of the local population. To ensure a sustainable management, these two constraints must be satisfied now and for ever or for a long period, given that hypotheses on which models are built are not indefinitely valid. How to conciliate protection of the forest and economic development of the local population? Promoting outside-working, establishing a birth policy or introducing monetary transfer could be some solutions. Then, we have the characteristics of a problem of sustainable management: determining the policy actions which allow the satisfaction of the economic and ecological constraints in a long range plan.

In order to study this problem for a long period, it is necessary to have a dynamical vision. Thus, we elaborate a stylised dynamical model. The aim is to study the mechanism which have an important role in the forest corridor. And to determine what are their consequences on the viability of the system. Then, to evaluate the pertinence of previously cited solutions and their consequences on the economic development and the conservation of the forest, we have to consider actions which may govern the possible evolutions of this dynamical system. So we introduce control variables. Finally, we are going to study a dynamical and controlled system submitted to constraints. The framework of viability theory[2] is well-adapted to that kind of study. This theory deals with controlled dynamical systems and consists in finding controls which keep the evolutions governed by these systems in an economic and ecological constraint set. The set which gathers all the situations for which such controls exist is called the viability kernel. Viability theory has already been applied to the study of sustainable management in Refs. [1, 3, 4, 5].

In this paper, we study the dynamics of the forest area, the population and the capital of the population. And we evaluate the pertinence of measures such as promoting outside-working, monetary transfer or birth policy. We introduce the following control variables: the proportion of outside workers, the monetary transfer, the population growth and the effort of development of new area. Our aim is to determine evolutions starting from an initial state with a given population, forest area and capital, which satisfy the ecological and economic constraints for a long time, and also to determine what control variables are necessary to remain in a viable situation.

The paper proceeds as follows. Section 2 describes the context in the forest corridor of Fianarantsoa and introduces a model of sustainable development in Madagascar. Then, in section 3, viability theory is developed while in section 4 results are presented.

2. The model

2.1. *Context in the forest corridor of Fianarantsoa*

The rain forest corridor in the region of Fianarantsoa is located in the east of Madagascar, it connects the national park of Ranomafana in the north to the Andringitra one in the south. The forest altitude is included between 800 meters and 1300 meters, and this altitude contributes to the high level of endemism which characterizes this area. Moreover, we can find precious wood and medicinal plants there. The forest is also important for the ecological regulation of the region. The east area of the corridor is inhabited by the Tanala. They live on itinerant rice-growing with the technique of slash and burn and on the culture of coffee, banana and sugar cane. In the west, we meet the Betsileo people. They practice an irrigated rice-growing in the shoals, and pluvial culture on the slopes. The Tanala and the Betsileo are connected. The seasons of rice-growing are different on each side of the corridor, so the Tanala work in Betsileo's area during the period of yield and vice-versa. The population of the corridor is rapidly growing. So they have to increase their rice production. We can see a growth of fields to the detriment of the forest.

How to conciliate development of the local population and conservation of the forest? Is it possible and what are the actions which lead to the respect of the two parts of this dilemma?

To answer to these questions, we elaborate a mathematical model which describes the malagasy situation.

2.2. *The Model*

In this part, the dynamical model we elaborate describes the economic behavior of the malagasy inhabitants from the region of the Ranomafana-Andringitra corridor. We recall that we are interested in the tradeoff between biodiversity conservation and population survival. In this basic model, biodiversity conservation is represented by the forest surface variable F, so one variable of the model is the built area S linked with F by $S = F_0 - F$ where F_0 is the forest area of the corridor in 1900. The population is described by the number of individuals P and the global capital K. As we aim at determining management rules, we are interested in the effects of "control" variables on the evolution of the three state variables S, P and K. The more obvious controls to consider are monetary transfer (τ), proportion of outside-workers (V), and decision to build new areas (δ). Moreover, as we can not forecast the future population growth, we also study the model in function of different demography rates (r). Thus, our model is a 3-dimensional model with 4 control variables (τ, δ, v, r). Lastly, to describe the two main objectives of the problem, conservation of the forest and sufficient economic development of the population, we define a set of constraints in the 3-dimensional state space.

2.2.1. *Dynamics of the model*

Surface The area under study is the Ranomafana-Andringitra corridor. A part of this area is forest area F, i.e. "primary" and secondary forest. The other part is considered like a built area S, it means that the area was cleared and irrigated. This area S grows up depending of the new area which are developed. We call δ this effort of development per person and per time unit which is one control variable of the model. We denote $v(t)$ the proportion of outside-workers. So, the variation of the built area is described by :

$$S'(t) = \delta(t)(1 - v(t))P(t)$$

The variation of the forest area can be deduced from the one of the built area.

Population Then, the second state variable which characterizes the system is the number of individuals living in the corridor P. However, we do not know the population growth for the next years. So, we consider that the population growth r fluctuates between two boundaries r_{min} and r_{max}.

$$P'(t) = r(t)P(t)$$

Capital Lastly, the capital K common to the whole population increases depending on incoming from the agriculture, wages of outside-workers and monetary transfers. And it decreases in function of the consumption and the effort of developing new areas.

If $\underline{c}$ is the minimal consumption per person and per time unit and β is the cost to develop one surface unit per year, then the capital decreases by $-\underline{c}P(t) - \beta\delta(t)(1 - v(t))P(t)$, and if ω is their wage per year, so incoming from out-workers is $\omega v(t)P(t)$.

Another origin of the revenue is due to the exploitation of the developed areas. If they are all cultivated, i.e. the proportion of inside-workers is important enough, the farming income is $pxS(t)$ where p is the rice price and x the land productivity per year. However, if there are not enough workers in the corridor, the farming income become $px\gamma(1-v(t))P(t)$ where γ is the maximal surface area one person can cultivate per year.

If τ represents the monetary transfer, the evolution of the capital is described by the following equation :

$$K'(t) = -\underline{c}P(t) - \beta\delta(t)(1 - v(t))P(t) + px\ \min(S(t), \gamma(1 - v(t))P(t))$$

$$+\omega v(t)P(t) + \tau(t)$$

Model The model is a three dimensional model which describes the variation of three state variables : $S(t)$, $P(t)$ and $K(t)$. Moreover, we have four control variables: $d(t)$, $r(t)$, $v(t)$ and $\tau(t)$. We suppose that the values taken by these controls over time are bounded :

- $0 < \delta(t) < \delta_{max}$: the surface area a person can clear per year is bounded;
- $r_{min} < r < r_{max}$: r_{min} and r_{max} values depend on the assumptions made on the birthrate policy fulfilled by the government;
- $0 < v < 1$;
- $0 < \tau < \tau_{max}$ depends on the assumption made on the maximal monetary transfer other countries agree to spend to preserve biodiversity.

We will test different values of these controls in part 3.2 and 3.3 to determine those that allow to conciliate preservation and economic development.

To sum up, we have the following model:

$$\begin{cases} S'(t) = \delta(t)(1 - v(t))P(t) \\ P'(t) = r(t)P(t) \\ K'(t) = -\underline{c}P(t) - \beta\delta(t)(1 - v(t))P(t) + px \ \min(S(t), \gamma(1 - v(t))P(t)) \\ \qquad\qquad +\omega v(t)P(t) + \tau(t) \end{cases} \tag{1}$$

$$\begin{cases} 0 & < \delta(t) < \delta_{max} \\ r_{min} & < r(t) < r_{max} \\ 0 & < v(t) < 1 \\ 0 & < \tau(t) < \tau_{max} \end{cases} \tag{2}$$

2.2.2. *Constraints*

As said in the introduction, implementing sustainable development implies generally several types of stakes. In our study, we face ecological and economic issues.

In this case, one of the objectives is to protect the biodiversity in the rain forest of the corridor. This ecological constraint can be expressed by preserving a minimal fixed area of forest F_{min}.

$$0 \leq S(t) \leq F_0 - F_{min} = S_{max}$$

The other objective is an economic one : to ensure the development of the local population. We force the future incomes *per capita* to be higher than the actual one k_{min} :

$$k_{min} \leq K/P$$

Moreover, in the second part of the study we impose a stronger constraint that the capital per capita nerver decreases :

$$(K/P)' >= 0$$

Lastly, the population must higher than a minimum level P_{min} to ensure the spawning of new generations. And it must be lower than a maximal level P_{max} corresponding to the carrying capacity of the corridor:

$$P_{min} \leq P \leq P_{max}$$

2.2.3. *Variable change to perform the mathematical analysis*

To make easier the mathematical study of the model, we work with the variable $s = S/P$ which is the built area *per capita* and with the variable $k = K/P$ which is the capital *per capita*. Thus the model becomes:

$$\left\{\begin{array}{l} s'(t) = \delta(t)(1-v(t)) - r(t)s(t) \\ P'(t) = r(t)P(t) \\ k'(t) = -\underline{c} - \beta\delta(t)(1-v(t)) + px\ \min(s(t), \gamma(1-v(t))) \\ \qquad\qquad +\omega v(t) + \frac{\tau(t)}{P(t)} - r(t)k(t) \end{array}\right. \tag{3}$$

with the constraints:

$$0 \leq s(t)P(t) \leq F_0 - F_{min} \tag{4}$$

$$k_{min} \leq k \tag{5}$$

$$k' >= 0 \tag{6}$$

$$P_{min} \leq P \leq P_{max} \tag{7}$$

2.2.4. *Parameters*

The aim of our model is not to reproduce the reality nor to predict the future. This model represents the most important dynamics of the forest corridor context and we study what are their consequences. So, defining realistic parameters will be the subject of a future work.

3. Viability Analysis

In this part, we present the main tools of the viability theory. Then, we apply these tools in a first study: to determine the policy actions to conciliate preservation of the forest and economic development of the local population.

3.1. *Translating our sustainable management issue into the viability theory framework*

To model the implementation of a policy of sustainable development of the rain-forest of Madagascar, we have developed a dynamical model with three dimensions and four control variables. We have also defined a constraint set which contains the ecological and economic properties necessary to a sustainable development. The framework of the viability theory is particularly well-adapted to study that kind of mathematical model.

We recall that our model is a dynamical system $\mathcal{S}$ of three equations. These equations represent the dynamics $x'(t) = (s'(t), P'(t), k'(t))$ of each state variable. These dynamics depend on the state variables $x(t) = (s(t), P(t), k(t))$ which belong to the state set X which is

$$X = [0; +\infty[\times[0; +\infty[\times[0; +\infty[\subset \mathbb{R}^3$$

and on the control variables $u(t) = (\delta(t), r(t), v(t), \tau(t))$. These control variables belong to a set of admissible controls U which is

$$U = [0; \delta_{max}] \times [r_{min}; r_{max}] \times [0; 1] \times [0; \tau_{max}] \subset \mathbb{R}^4$$

Different controls can be chosen in this set U. And we call an evolution a solution to the dynamical system depending on a path of chosen controls and starting from an initial state $x \in X$. Thus, for each initial state in X, it exists much as evolutions as combination of controls. Moreover, we denote $\mathcal{K}$ the constraint set, that is the set of all the state variables which satisfy the ecological and economic constraints define by the inequalities (4), (5) and (7):

$$\mathcal{K} = \left\{ (s, P, k) \in K \left| \begin{array}{c} 0 \leq s(t)P(t) \leq F_0 - F_{min} \\ k_{min} \leq k \\ P_{min} \leq P \leq P_{max} \end{array} \right. \right\}$$

We say that an evolution is viable if it always stays in the constraint set.

Is it possible from an initial state to find at least one path of controls such as the dynamics always stay in the constraint set? Or whatever the chosen control the dynamics starting from an initial state in the constraint set always leave the constraint set? The set of all the initial states for which it exists a viable evolution is called the viability kernel $Viab_{\mathcal{S}}(\mathcal{K})$.

In our model, the constraints have been defined in part 2.2.2 by the set $\mathcal{K}$ of states $(s, P, k) \in X$ such as satisfy inequalities (4), (5) and (7), we would like to determine for each initial state $(s_0, P_0, k_0) \in \mathcal{K}$ if it exists controls, $(\delta(.), r(.), v(.), \tau(.))$, that is to say policy actions or measures, such as the evolution starting at (s_0, P_0, k_0) and satisfying the system $\mathcal{S}$ always remains in $\mathcal{K}$. It amounts to determine the built area, the number of inhabitants and the capital necessary to respect the ecological and economic constraints for a very long time. For each triplet (surface/ inhabitants / capital), we test if it exists at least one series of policy actions which enables the respect of the constraints. For example this policy action could consist in converting a lot of new area ($\delta = \delta_{max}$) at the right time, and then in sorely increase

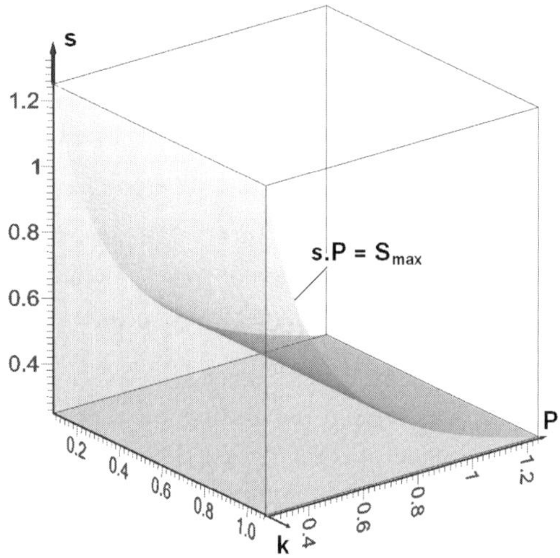

Figure 1. Constraint set $\mathcal{K}$

the outside work ($v = 0,7$), or in having considerable monetary transfer ($\tau = \tau_{max}$) during a finite time.

In the viability theory framework, answering such questions amounts to compute the viability kernel of the dynamics $\mathcal{S}$ submitted to the constraints $\mathcal{K}$.

3.2. *Solving the management issue described by the dynamics S (3) and the constraints (4), (5) and (6)*

The first part of the study consists in determining the viability kernel of the dynamics $\mathcal{S}$ submitted to the constraints $\mathcal{K}$ numerically. We use the algorithm developed by Saint-Pierre[6].

In this first study, there is no any monetary transfer, we consider that $\tau_{max} = 0$. And we have to determine which initial situations in built area, population an capital (s, P, k) are viable without monetary transfer. The figure 2 shows the set of all these initial situations computed numerically, it is the viability kernel. And in this case, the viability kernel is equal to the whole constraint set.It means that for all the situations from the constraint set $\mathcal{K}$, we can find a series of policy actions at least to ensure the viability. Thus, the next step is to determine such a series of policy actions for a given initial situation, *i.e.* to determine what are the controls which keep the system starting from an initial triplet $(s_0, P_0, k_0) \in \mathcal{K}$ into the constraint set for ever.

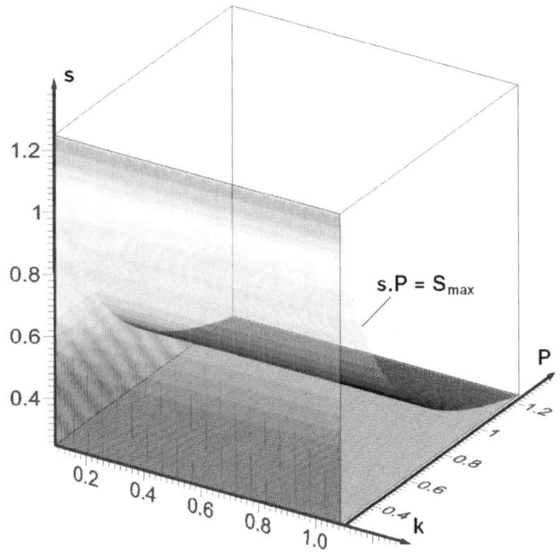

Figure 2. Viability Kernel $Viab_S(\mathcal{K})$

The figure 3 gives us an example of evolutions starting from the same initial situation. We study an initial situation where the built area s is very low, and the population P and the capital k are quite high. From this situation, we have represented three trajectories. They correspond to three possibilities to ensure viability. In the little one, the viability is ensured by a reduction of the population, and consequently by an increase of the capital *per capita*. For the two others evolutions, it is necessary to reduce the capital to maintain the viability and increase the built area. The economic constraint is preserved ($k \geq k_{min}$) but the capital decreases ($k' < 0$).

3.3. *Adding an other constraint*

In the previous study, we have seen that some viable evolutions may experience a decrease of the capital *per capita*. It is not satisfying because in this there is no equity between the different generations. One generation will have a lower capital than its parents, in order to ensure the survey of the next generations. To avoid this situation, we add a new constraint.

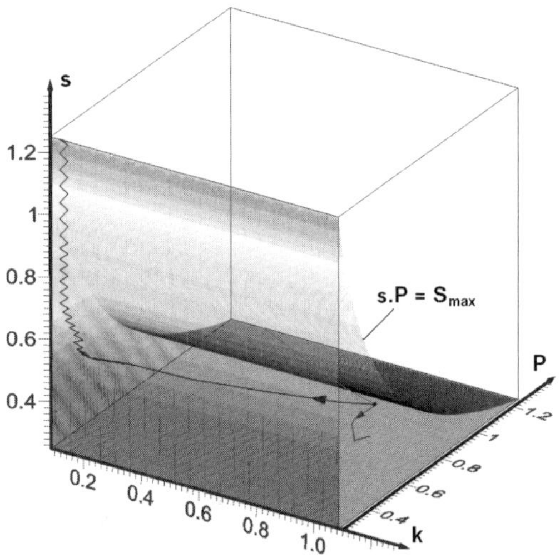

Figure 3. Trajectory in the viability kernel $Viab_{\mathcal{S}}(\mathcal{K})$

The capital *per capita* must increase.

$$k' \geq 0 \tag{8}$$

Like in the previous study, we compute the viability kernel of the dynamics $\mathcal{S}$ submitted to the constraints $\mathcal{K}$ with the additionnal constraint $k' \geq 0$. We always consider that the monetary transfer is equal to zero $\tau_{max} = 0$.

The figure 4 shows the computed viability kernel. It is just a part of the constraint set. The situations with a quite low capital *per capita* k are not viable. It does not exist any evolution which satisfies all the constraints. In particular, the constraint $k' \geq 0$ is not fulfiled for an initial situation of low capital.

So only the starting situations with a high capital are viable. In this case, what are the policy actions which maintain the viability for decades?

The figure 5 shows two viable evolutions. From an initial situation with a high population and capital and a very low built area, we find at least two possibilies to satisfy the ecological and economic constraints. For the longest evolution, the built area must increase and the population decreases to maintain a capital *per capita* constant. It means that the population must invest in clearing the rain forest to built new area and has to decrease. In this case, the evolution reaches the frontier $s = S_{max}$, and

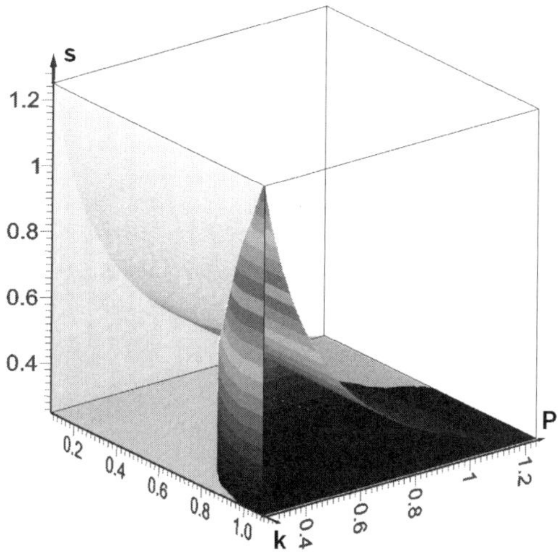

Figure 4. Viability Kernel with the additionnal constraint $k' \geq 0$

we have an equilibrium. For the shortest evolution, we have an equilibrium in a situation where the built area is inferior to the maximal built area S_{max}.

Like the siuations with a low capital are not viable in the second study, we are going to observe the effect of monetary transfer. So in this part of our study, we consider a monetary transfer not equal to zero ($\tau_{max} \neq 0$). We always have the constraint $k' \geq 0$.

We can see the computed viability kernel in the figure 6. This figure displays the constraint set $\mathcal{K}$, the previous viability kernel with no transfer and the one with transfer. When we add monetary transfer, the variation of the capital *per capita* increase and thus the set of viable situations is larger.

4. Conclusion

The first part of our study shows that the whole constraint set is viable on condition that one or many generations are sacrified. In this case, we can ensure the preservation of the forest and the economic development of the local population, however, some generations may have a capital *per capita* lower than its predecessors or its descendants.

The second part of the study takes into account the equity between

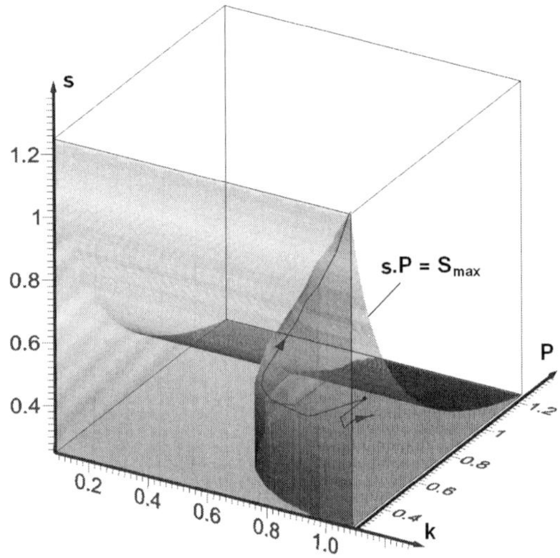

Figure 5. Two exemples of viable evolutions starting from the same situation inside the vaibility kernel with the dynamics $\mathcal{S}$ with th constraints $\mathcal{K}$ and with the additionnal constraint $k' \geq 0$

the different generations and the constraint of a non-decreasing capital per capita is added. We note that only the situations where the capital *per capita* is quite high are viable and show that adding monetary transfers increases the size of the viability kernel.

The next step is to link this work with the present situation of the corridor of Fianarantsoa. Is the current Malagasy situation viable? If it is not viable, which monetary transfer values would make it viable? Are such monetary transfers admissible or would it be more efficient to help the local population by increasing the rice price or improving the rice productivity? Answers will be given by determining realistic parameter values and comparing viability kernels range for different values of such controls as monetary transfers, the rice price or the rice productivity.

Another related interesting issue is to give a price to the conservation by comparing the monetary transfer needed to preserve a given forest area and the monetary transfer needed to preserve one additonal hectare, keeping in mind the necessity of ensuring a sufficient standard of life for inhabitants of the corridor.

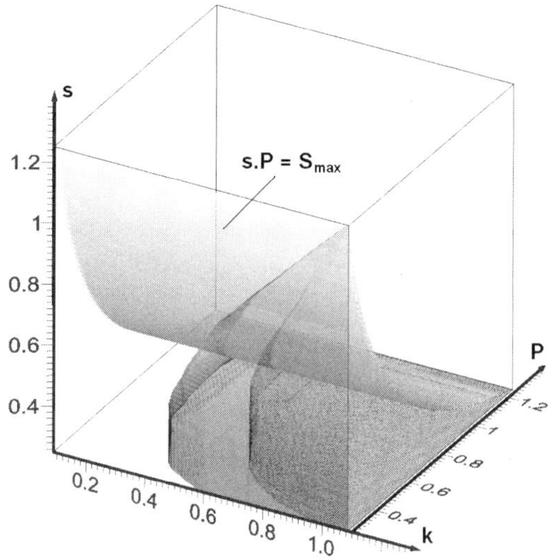

Figure 6. Viability Kernels with the additionnal constraint $k' \geq 0$ and with and without monetary transfer.

Acknowledgments

I thank Sophie Martin and Patrick Saint-Pierre for their useful comments and helpful suggestions during the elaboration of this article.

I thank Patrick Saint-Pierre for the results due to its viability algorithm.

This work is supported by the ANR project Deduction.

References

1. Bernardo T. Saint-Pierre P. Aubin, J.-P. A viability approach to global climate change issues. *Advances in Global Change Research*, 2004.
2. J-P. Aubin. *Viability Theory*. Birkhuser, 1991.
3. S. Martin. The cost of restoration as a way of defining resilience: a viability approach applied to a model of lake eutrophication. *Ecology and Society*, 2004.
4. V. Martinet and L. Doyen. Sustainability of an economy with an exhaustible resource: A viable control approach. *Resource and Energy Economics*, 2007.
5. C. Mullon, P. Cury, and L. Shannon. Viability model of trophic interactions in marine ecosystems. *Natural Resources Modelling*, 2004.
6. P. Saint-Pierre. Approximation of the viability kernel. *Applied Mathematics & Optimisation*, 1994.
7. G. Serpanti and A. Toillier. Dynamiques rurales betsileo lórigine de la deforestation actuelle. In Transitions agraires, dynamiques cologiques et conserva-

tion, Actes du Seminaire GEREM, Antananarivo, 9-10 novembre 2006, pages 5768. Serpanti G., Rasolofohariniro, Carrire S., 2007.

POPULATION DYNAMICS ON COMPLEX FOOD WEBS

L. BEREC

Dept of Theoretical Ecology,
Institute of Entomology, Biology Centre ASCR,
Branišovská 31, 37005 České Budějovice, Czech Republic
E-mail: berec@entu.cas.cz

Food webs are typically complex ecological networks describing who eats whom in a community of species. Presumably the fundamental question of ecology is how species diversity is maintained in nature or, alternatively, why food webs of different complexity are found in different ecosystems. Today, facing negative impacts of anthropogenic activities in virtually every corner of the world, the question of utmost importance for assessing current threats to biodiversity is how fragile or robust are species communities or, alternatively, how do food webs respond to species extinctions and/or to alien species invasions. In the quest for answers to these questions, mathematical models play an inevitable role, since experimenting on or observing (long-term) changes in species communities is risky and often in fact logistically infeasible. While food web complexity shapes and constrains dynamics of the involved populations, these dynamics feed back to shape and constrain food web complexity, so that the two need to be studied simultaneously. Research on food webs is currently among the topical ones and the number of studies that deal with population dynamics on complex food webs via mathematical models steadily increases. Some of these studies explore how food web complexity increases with considering various ecological mechanisms known to stabilize simpler consumer-resource systems. Other studies address the question of how complex food webs arise in the first place, via adding species one by one and looking at whether these new species persist in a growing community or not. The goal of this paper is to show these two types of modelling approaches, discuss their pros and cons, and suggest some avenues for further, systematic research in this exciting and relatively novel branch of ecological science. We also present a specific model of each type. Using the top-down approach, we study effects of simultaneous operation of several stabilizing mechanisms, showing, somewhat surprisingly, that they might suppress one another. Through the bottom-up approach, we examine an extent to which food webs formed via speciation vs. invasion events differ, showing that they differ in size, structure as well as propensity to collapse.

1. Introduction

Presumably the fundamental question of ecology is how species diversity is maintained in nature or, alternatively, why species communities of different complexity are found in different ecosystems as arid zones, tropical forests, coastal waters or open ocean. Despite some 100 years of research, since Eugenius Warming founded ecology as a scientific discipline[28], we are still far from a complete answer. Actually, the route to this holy grail of ecology is paved with simpler and more accessible questions. These include questions such as how are species communities structured[21] or what ecological mechanisms tend to stabilize community dynamics[37,47]. In the light of negative impacts of anthropogenic activities in virtually every corner of the world, two other questions become of utmost importance for assessing current threats to biodiversity: how do species communities respond to species extinctions due to, habitat destruction or overexploitation[24,33], and how do they react to unwanted or deliberate introductions of non-native species[55]. For example, the Indian mongoose *Herpestes auropunctatus*, introduced to many islands as a biological control agent of rats and snakes in agricultural (sugarcane) habitats, has already caused at least seven amphibian and reptile extinctions in Puerto Rico and other islands in the West Indies[a]. In many species communities, such as in some coastal marine ones, these two processes are likely to occur simultaneously[10]. In the quest for answers to these questions, ecologists work with the concept of food webs, complex ecological networks describing, in the simplest case, who eats whom in a species community, and have already developed several mathematical models that attempt to reproduce structures of real food webs[21]. As a matter of fact, food web structure cannot be fully understood without simultaneously studying dynamics of populations that form it; while food web structure shapes and constrains population dynamics, these dynamics feed back to shape and constrain food web structure. Mathematical models of population dynamics are thus often superimposed on such model-generated food webs and community dynamics thus studied[8,57], as experimenting on or observing (long-term) changes in species communities is risky and often in fact logistically infeasible. Speaking of dynamics, we usually mean ecological dynamics. Once the examined timescales are large, however, it becomes necessary to consider evolutionary dynamics as well[20,58]. To help understand how species diversity is maintained in nature, models need to

[a]http://www.columbia.edu/itc/cerc/danoff-burg/invasion_bio/inv_spp_summ/Herpestes_auropunctatus.html

be developed that result in food webs of given complexity; features that underlie such models then need to be studied. For the two applied questions, food webs of given complexity form a starting point to experiment on: either remove or add species, follow population dynamics on the shrunk or augmented food web, and assess how the original food web gets eventually modified. Two modelling approaches are currently available. First, structural food web models are combined with models of population dynamics – the "top-down" approach[8,57]. Second, complex food webs are let to evolve from an initially small number of species by combining population-dynamic models with dynamic processes of speciation and alien species invasion – the "bottom-up" approach[20,58]. None of the published models within any of these approaches is currently entirely satisfactory and the question of how to build dynamically persistent, complex food webs that resemble those found in nature still remains a big challenge. As emphasized by[21], "integration of plausible ecological and evolutionary dynamics and network structure is a grand challenge for ecological modeling, with potential benefits for conservation and management of ecosystems [...], as well as for fundamental scientific understanding of complex adaptive systems." In this paper, we aim to bring near these two types of modelling approaches, discuss their pros and cons, present a specific model of each type, including some novel and interesting results, and suggest some avenues for further, systematic research in this exciting and relatively novel branch of ecological science.

2. Structure of real food webs

Plenty of real food webs have already been collected and published, often related to marine or lake environments [16,45,27,22,2,3] . They differ by complexity and degree of resolution, and are either of the snapshot type, describing the network of consumer-resource (predator-prey, herbivore-plant, parasitoid-host) interactions at a time instant, or cumulative, aggregating acquaintance of such interactions over a longer period. Also, they can be trophic, aggregating species with sufficiently overlapping sets of their resources and consumers, or non-trophic. Unfortunately, there is currently no standard way of how to collect real food webs, an issue that complicates studies of food web dynamics (see below) and that definitely needs some more thoughts in the future. For a long time, food webs have been looked at from the structural point of view, i.e., how large and "wired" they are, without simultaneously considering dynamics of the involved populations; see Ref. [21] for a brilliant and comprehensive review of this topic. Actually, much of the current research on both food web structure and dynamics has been strongly affected by the Robert May's paper published in Nature in 1972[49]. May carried out a local stability analysis of random food webs supplemented by a generic population model and showed that food web stability decreases with increasing food web complexity. More precisely, using species richness (S = the number of species forming the web) and connectance (C = the actual number of consumer-resource links in the web divided by the number of potential links S^2) as the only quantities characterizing food web structure, he showed that randomly assembled food webs would be locally stable if $i\sqrt{SC} < 1$, where i is the average interaction strength in the food web, and unstable if $i\sqrt{SC} > 1$. This implies that for a more wired food web to be locally stable it needs to be composed of a sufficiently small number of species; on the other hand, large, locally stable food webs need to contain only a small number of consumer-resource links (Fig. 1). Although the May's analysis has been repeatedly criticized[42,15], it has become strongly influential for further development of food web theory, in at least two respects. First, it stimulated a fascinating quest for simple models capturing regularities in the structure of real food webs, just in terms of S and C (see below). Second, suggesting that more complex food webs are less stable, it triggered a chase for ecological mechanisms thanks to which species richness might increase with increasing connectance (see Sec. 3.1). Empirical data suggest that while real food webs may contain from units to thousands of species, connectance is constrained to $C \sim 0.03 - 0.3$, with an average about $0.1 - 0.15$[21].

A number of models have been developed that aim to reproduce several central properties characterizing complexity of real food webs, using a minimum number of input parameters[16,59,14,2,56,47]. Following the tradition of May, all these models use just species richness S and connectance C as the input parameters, and apply different rules to distribute CS^2 links between S nodes. Among other things, these structural food web models demonstrate that the structure of real food webs is far from random yet that relatively simple rules can yield quite complex food web structures comparable to those of many real food webs. Although none of these models is perfect, the niche model developed by Ref. [59] has been considered by many quite successful[12,13,23,46], and has thus become a sort of standard for many modelling studies that concern food webs[60,2,56,47]. The niche model is as follows[59]. Given S and C, it assigns a random value drawn uniformly from the interval [0, 1] to each species. This value is called the species' niche value, n_i. Each species consumes all species within a range of niche

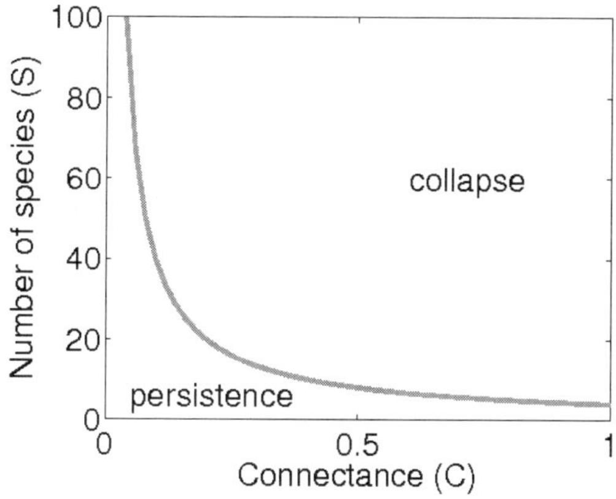

Figure 1. Complexity-stability relationship in randomly assembled food webs due to May (1972). For a more wired food web to be stable it needs to be relatively species-poor. On the other hand, large, stable food webs can only contain a small number of consumer-resource links.

values $[c_i - r_i/2, c_i + r_i/2]$ where the width of this range r_i is randomly assigned using a beta distribution and its centre c_i is drawn uniformly from the interval $[r_i/2, n_i]$ (or $[r_i/2, 1 - r_i/2]$ if $n_i > 1 - r_i/2$). The beta distribution is parameterized so that the actual connectance of generated food webs lies close to the prescribed connectance C[59]. However, not all generated food webs are accepted for further processing. The webs that contain an isolated species (i.e., species with neither incoming nor outgoing link) and webs whose actual connectance differs from the prescribed one by more that 3% are discarded. The niche model allows for the occurrence of cannibalism and loops in the food web structure[59]. Basic structure of food webs generated by the niche model with $S = 30$ and $C = 0.15$ is summarized in Fig. 2. We observe a small proportion of basal species (B, species without resources), majority of intermediate species (I, species with both a resource and a consumer), and a few or no top species (T, species without consumers), a pattern common also to other values of S and C and to other structural food web models. A consequence of this is that any two food webs with identical S and C yet different structure, such as "pyramid" ($B > I > T$) versus "hump-shaped" ($B < I > T$) food webs[10], cannot be generated by the same structural food web model. Last but not least, any structural food web model is only as good as the data on real food webs are. As more resolved data are available and/or any standardized procedure for collecting real food webs adopted, even the niche model might turn inadequate; higher resolution is especially needful to achieve at the basal species level, as basal species appear to be fundamental drivers of food web dynamics (see Sec. 5.1).

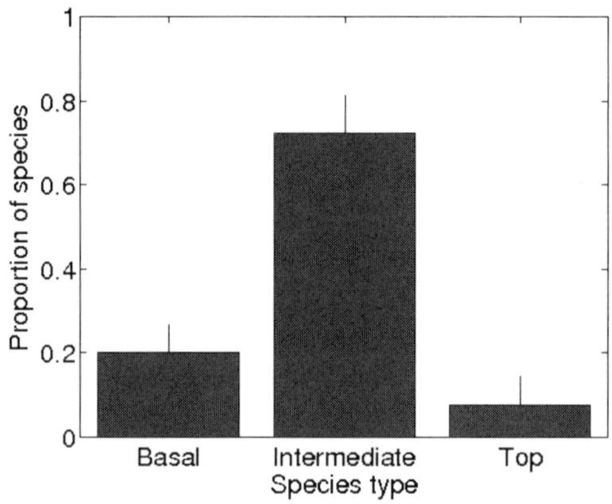

Figure 2. Basic structure of food webs generated by the niche model with $S = 30$ and $C = 0.15$. Basal species have no resource, top species have no consumer, and intermediate species have at least one resource and one consumer. Most species in the web are here intermediate species. Results based on 1000 replicates. Vertical strokes denote respective values of standard deviation.

3. Population dynamics on complex food webs

As we state above, food web structure cannot be fully understood without simultaneously studying dynamics of populations that form it; food web structure shapes and constrains population dynamics and these dynamics in turn feed back to shape and constrain food web structure. Since the pioneering work of Robert May on the relationship between food web complexity and stability [49], researchers have become interested in ecological mechanisms that promote species diversity, i.e., help generate and maintain complex food webs. However, population dynamics has long been devoted to studies of simple food webs composed of two, three or at most four species, sometimes referred to as community modules, including food chains, competition, apparent competition and omnivory[30]. Mathematical analysis of such community modules is relatively straightforward and has demonstrated a significant effect of their structure on dynamics of the involved populations, through both direct and indirect interactions[30,34]. One of the major aims of food web studies is to produce food webs of complexity comparable to that of real ones, using relatively simple models that account for dynamics of the involved populations. Here results of the research conducted on community modules enter the food web stage: mechanisms expected to hold complex food webs together recruit from those that tend to stabilize dynamics of module-forming populations. Mechanisms shown to stabilize population dynamics on simple food webs vary and include adaptivity of consumers with respect to their diet composition[40,41], type III functional responses[51,60], degree of immigration[4], and intraspecific interference in consumers[54,32]. Other mechanisms enhancing food web stability may manifest themselves only at the level of food webs as a whole and include proportion and distribution of weak links[50,3] or allometric body size relationships[9]. One may also be interested in the impacts on food web stability of mechanisms that destabilize population dynamics in simple community modules, such as Allee effects[7,18], including Allee effects due to enhanced exploitation of rare species[17]. Surprisingly, no study conducted so far has explored effects on food web stability of two or more of these mechanisms operating simultaneously. Do they act in synergy or do they suppress one another? Although Ref. [47] studied the effects of type III functional responses and predator interference, they did not consider the effects of their interaction. A brief study of this kind is carried out in Sec. 4.1 of this paper, showing that although any of the examined mechanisms has a positive effect on food web stability, their interaction is far from clear-cut. To explore population dynamics on complex food webs, theoretical ecologists either combine structural food web models with models of population dynamics (the "top-down" approach) or let complex food webs evolve from an initially small number of species by combining population dynamic models with dynamic processes of speciation and alien species invasion (the "bottom-up" approach). In what follows, we discuss each of these approaches in more detail.

3.1. *Top-down approach*

The top-down approach is apparently a legacy of a considerable amount of research on structural properties of real food webs, summarized in a brilliant and comprehensive review by Ref. [21]. The approach consists of generating an "initial" food web structure via a structural food web model, on which a model of population dynamics is superimposed and run. After the dynamics attain a steady state, the "eventual" food web structure is recorded. Interestingly, virtually all studies of this kind conducted so far reported that initial food webs collapsed to eventually result in impoverished, persistent food webs[4,37,8] , with structural properties different from those of the initial food webs[47]. The top-down approach thus represents a very useful tool to explore constraints set up by food web structure on community dynamics, and vice versa. More specifically, it allows us to seek for properties of species (functional response, degree of migration, carrying capacity in the case of basal species) or food webs (connectance, network type, proportion of basal, intermediate and top species) that make food webs more resistant to species extinctions and thus more robust to biodiversity loss[47]. It is just the top-down approach that has been used to study the impacts of potentially stabilizing mechanisms on food web persistence, in terms of the number or proportion of species persisting in the eventual food web. A variety of mechanisms have indeed been shown to enhance food web persistence, including type III functional responses and functional responses with consumer interference[60,47], adaptivity of consumers with respect to their diet composition[8,37,39], preference of omnivores for lower trophic-level prey[60], or increased immigration rates of the involved species from nearby source populations[4]; see also Sec. 4.1. The top-down approach has also been repeatedly used to address the still controversial complexity-stability issue: are more complex food webs more stable or less stable than less complex ones? This issue is obviously a legacy of Robert May's work[49], and consists in showing whether the number or proportion of species persisting in the eventual food web increases (enhanced stability) or decreases (reduced stability) with increasing initial food web connectance C. Most of the studies addressing this issue considered adaptive foraging as a key mechanism, showing that whether we observe positive or negative complexity-stability relationship depends on the choice of structural food web model, population-dynamic model, and the way basal species are treated[8,37,38,39,57]. Positive relationships between the number or proportion of species persisting in the eventual food web and initial food web connectance has recently been observed also in models without adaptive foraging[35] and in models of competitive communities[26]. Published studies using the top-down approach differ in a number of methodological aspects. These include the structural food web model (random, cascade, niche and/or other), model used to describe consumer-resource interactions (Lotka-Volterra model, models involving type II, type III, or consumer-density-dependent functional responses, bioenergetic model sensu Ref. [62], an individual-based model), consideration or not of adaptive foraging, the way parameter values are chosen (fixed and species-independent, randomly generated from some pre-specified ranges, related to body size through allometric relationships), threshold density below which populations are considered extinct, or the moment(s) at which this threshold is applied (continuously or only at the end of simulation runs). Such a variety of modelling alternatives prevents us from any judicious synthesis and rather calls for additional studies that would "fill the gaps" (see Sec.5). An interesting question is, for example, how would the May's relationship, and its equivalent based on an individual-based model[36], would look like if random interaction matrices are replaced by those generated by some structural food web models. The observation that eventual, persistent food webs have structural properties different from those of food webs from which they depart[47] becomes an issue, since the initial food web structures are generated through one or more of the structural food web models which claim to satisfactorily capture complexity of many real food webs[21]. If this is true, any collapsed, eventual food web no more seems to be a good representation of reality. Should not we rather solve an inverse problem then: what food web structure to start with to end up with a food web structure comparable to those generated by the structural food web models? Or, alternatively, what (stabilizing) mechanisms our population-dynamic model should consider and combine to prevent any extinction in the initial food web? On the other hand, given that many structural food web models have been based on data coming from cumulative food webs, or snapshot food webs with limited resolution, we cannot perhaps expect dynamic persistence of all species present in the corresponding initial web. What if the eventual food webs are closer to reality than the initial ones? Obviously, many problems thus remain open and in need to be addressed.

3.2. *Bottom-up approach*

Using bottom-up approach, one builds food webs "from below". Starting with one or two species (a consumer-resource pair), new species are added one by one at a rate and at a small abundance. Their dynamics relative to those of the established food web then decide on whether or not these new species succeed in augmenting the food web and whether or not they subsequently cause extinctions of some species established earlier. One may distinguish two augmentation frameworks. In the "speciation" framework, new species appear as mutants of the species already in the web, where mutation events are usually separated by long enough time for the perturbed dynamics to settle in a new ecological attractor (equilibrium point, limit cycle or chaotic attractor)[20]. In the "invasion" framework, new species need not have anything in common with the species already in the web, and usually arrive much more frequently, at times comparable in scale to population dynamics[58]. Models defined within the invasion framework are sometimes referred to as assembly models with "free" structure. This is to distinguish them from assembly models with a pre-specified trophic structure[48,43]. As with the top-down approach, published studies using the bottom-up approach differ in such a variety of aspects that prevents us from any sound synthesis at the moment. Apart from the diverse methodological aspects listed above for the top-down approach (excluding the choice of a structural food web model, obviously), one of the most important steps in developing food webs from below is to define how new species that appear through speciation or invasion will become linked to those already in the web. A number of ingenious techniques have already been used: (i) linking based on species body sizes [58] , (ii) linking based on a "matrix of scores" that allows for features possessed by individual species[11,20,19,53], (iii) linking based on the technique underlying the structural niche model[29], or (iv) assuming that the new species always becomes a consumer with a fixed number of already present species taken randomly as resources[4]. Bottom-up models also exist that work with individuals rather than populations and where individuals are represented by a sort of genome allowed to mutate upon reproduction[31,1,6]. Despite that variety, a common theme of all these studies is how complex can be the generated food webs and how their complexity varies with specific model elements. Thus, for example, communities assembled through invasions have many properties different from those of stable communities of equal size taken at random[58], evolving food webs allowing consumers to change their diet adaptively are much more complex than those in which consumers have fixed preferences for resources[29], or complexity of evolving food webs increases with increasing (fixed) amount of external resources and decreasing intensity of interspecific interference[20]. However, as no pair of studies is directly comparable, one has to be cautious about generality of the predictions that regard food web dynamics. On the other hand, wide heterogeneity in the currently available techniques of modelling food web dynamics using the bottom-up approach clearly demonstrates what is possible. The question of to what extent food webs generated via the bottom-up approach are comparable, in terms of structural properties, to real food webs, food webs generated by structural food web models, or food webs generated through the top-down approach, has not been addressed properly yet. It would also be interesting to examine whether or not different model assumptions lead to comparable food webs, and why. Therefore, as with the top-down approach, much remains to be done before any sound synthesis can be made.

4. Some specific models and some novel results

In this section, we illustrate the top-down and bottom-up approaches with some specific models of population dynamics on complex food webs, and use them to present some novel results.

4.1. *Simultaneous effect of stabilizing mechanisms*

4.1.1. *Model*

Here we explore, using the top-down approach, whether and how several ecological mechanisms known to stabilize consumer-resource dynamics in small community modules – disproportionately reduced feeding on rare resources (type III functional response), continuous immigration of consumers from nearby source populations and a sort of intraspecific interference in consumers – contribute to food web persistence, i.e., to the proportion of surviving species from the initial food web. Previous studies considered diverse mechanisms only individually[4,37,47]. Here we explore how they act both in isolation and simultaneously, i.e., how they interact. An initial food web structure is built first, using the niche model with $S = 30$ species and connectance $C = 0.15$; see also Fig. 2. The following population model is then superimposed and run on it. All basal species are assumed to independently obey logistic dynamics in the absence of consumers, and all non-basal (i.e., intermediate and top) species die out exponentially in the absence of resources and no immigration. Consumption is driven by either a type II (saturating) or type III (sigmoid) functional response. Also, we allow for a proportion p_M of non-basal species to immigrate from unspecified nearby source populations, and a proportion p_I of non-basal species to face intraspecific interference. The rate of change of density x_i of population i ($i = 1, \ldots, S$) is therefore

$$\frac{dx_i}{dt} = (r_i - s_i x_i)x_i - (d_i + f_i x_i)x_i + m_i$$
$$+ \frac{\sum_{j=1}^{S} a_{ji} e_{ji} \lambda_{ji} x_j^n}{1 + \sum_{j=1}^{S} a_{ji} h_{ji} \lambda_{ji} x_j^n} x_i - \sum_{j=1}^{S} \frac{a_{ij} \lambda_{ij} x_i^n}{1 + \sum_{k=1}^{S} a_{kj} h_{kj} \lambda_{kj} x_k^n} x_j \tag{1}$$

The matrix $A = (a_{ij})$ is the interaction matrix generated by the niche model ($a_{ij} = 1$ if species j consumes species i and $a_{ij} = 0$ otherwise), and n defines the type of functional response (type II for $n \in (0, 1)$ and type III for $n > 1$). By (our) definition, basal species have $d_i = f_i = m_i = 0$ and non-basal species have $r_i = s_i = 0$. Finally, population model (1) is parameterized as follows. Parameter values corresponding to individual species are generated as uniform random deviates from some pre-specified parameter ranges, a procedure common to the top-down approach[49,37,33]: we choose the range [0.5, 1.5] for the basal species growth rate r, [0.1, 0.5] for the non-basal species death rate d, [0.1, 1] for a measure of negative density dependence in basal species s, [0, 1] for immigration rate m, encounter rate λ and conversion efficiency e, [0, 20] for intensity of intraspecific interference f, and [0, 0.2] for consumer handling time h. Initial densities of the basal species are set to their respective carrying capacities r/s, while those of the non-basal species are generated uniformly randomly from the range [0.001, 0.1]. We use three different values for each of the mechanisms examined for its contribution to food web persistence, that is, for the proportion of non-basal species able of immigration ($p_M = 0$, 0.2 and 0.5), for the proportion of non-basal species facing intraspecific interference ($p_I = 0$, 0.5 and 1), and for the type of functional response ($n = 1$ (type II), 2 and 5 (both type III)). We use the full factorial design for further processing, so that we have 27 combinations in total of which the one with type II functional response, no interference and no immigration is referred to as the baseline scenario further on. Altogether, 50 simulation replicates were conducted for each of the 27 combinations, each run for 5000 time units. Only species that reached density $> 10^{-6}$ at the end of simulations were considered surviving.

4.1.2. *Results*

We start by statistically assessing an impact of each of the stabilizing mechanisms and their interactions on the proportion of surviving species from the initial food web (Table 1). We observe that any of the stabilizing mechanisms has a positive significant effect on food web persistence. Quite surprisingly, all the second-order interactions (of which only the interaction between type of functional response and proportion of consumers with interference is statistically significant) are negative, suggesting that rather than acting in synergy, the examined mechanisms suppress one another.

Table 1. Statistical analysis of the importance of different stabilizing mechanisms on the proportion of species surviving from the initial food web. The "General Regression Models" routine of the Statistica package (StatSoft, Inc., ver. 8) was used; no transformation was applied to the dependent variable as it complied with normal distribution.

Effect	Estimate	SE	t-value	P
Intercept	9.934	0.458	21.683	0.000
Functional response n [A]	0.982	0.145	6.775	0.000
Prop. migration p_M [B]	16.828	1.474	11.420	0.000
Prop. interference p_I [C]	4.319	0.710	6.085	0.000
A x B	-0.0034	0.466	-0.0072	0.994
A x C	-0.845	0.224	-3.764	0.00017
B x C	-1.482	2.283	-0.649	0.516
A x B x C	-0.304	0.722	-0.421	0.674

Figure 3A then exemplifies how the initial food web due to the niche model collapses if all consumers follow the baseline scenario (type II functional response, no interference, no immigration). The resulting food web is of the pyramid type. On the other hand, the best combination out of those examined (in terms of the maximum proportion of surviving species; maximum scenario) resembles the initial food web structure, as measured by B, I and T, with the number of intermediate species slightly reduced. While the best scenario corresponds to $n = 2$, $p_M = 0.5$ and $p_I = 0$, the scenario in which all the mechanisms are set at their maximum (maximum scenario) is $n = 5$, $p_M = 0.5$ and $p_I = 1$. This again suggests that some mechanisms indeed suppress some other. To really substantiate an antagonistic operation of some stabilizing mechanisms, we explore how food web persistence changes with the relative order in which individual mechanisms are incorporated in the population-dynamic model, for the maximum scenario with $n = 5$, $p_M = 0.5$ and $p_I = 1$ (Fig. 3B). We can see, for example, that while the effect of interference is positive for three right-most situations in Fig. 3B, it is negative for the first and third ones. Also, if the effect of a mechanisms is always positive, such as in the case of immigration, its contribution to food web persistence may differ depending on when it is added relative to the other mechanisms (Fig. 3B).

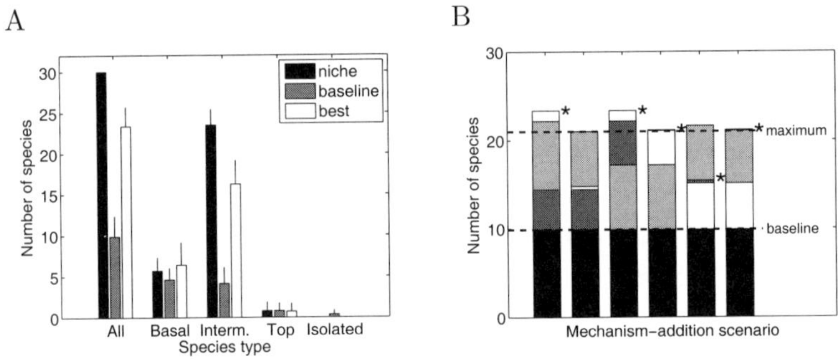

Figure 3. Effects of immigration, intraspecific interference and functional response on food web persistence. A. Comparison of the initial food web structure due to the niche model (niche), the food web with type II functional response, no immigration and no interference (baseline), and the scenario with $n = 2$, $p_M = 0.5$ and $p_I = 0$ for which the maximum proportion of surviving species is attained among all those examined (best). Isolated species are those that have neither resources nor consumers. B. Effects of adding the three examined mechanisms one by one, for the maximum scenario with $n = 5$, $p_M = 0.5$ and $p_I = 1$: black – baseline, dark grey – type III functional response with $n = 5$, light grey – immigration with $p_M = 0.5$, white – interference with $p_I = 1$. Note that some mechanisms can both increase and decrease species richness depending on when they are incorporated relative to the others. Asterisks denote negative increment. Vertical strokes represent respective values of standard deviation.

4.2. *Impacts of speciation vs. invasion on food web structure*

4.2.1. *Model*

The second model we develop in this paper builds food webs from below (bottom-up approach), and we are interested in the extent to which food webs generated via speciation (evolutionary food webs) and invasion (assembled food webs) might deviate. While the former process is typical of mainlands and spans very long timescales (typically millions to hundreds of millions of years), the latter is more adequate to islands and runs on much shorter timescales (typically tens to thousands of years). The developing food web starts with an external resource x_1, growing logistically with an intrinsic growth rate r and a carrying capacity K in the absence of any other species, and an autotroph x_2 consuming this resource. This two-species system is described by the following equations typical of many predator-prey models:

$$\frac{dx_1}{dt} = rx_1\left(1 - \frac{x_1}{K}\right) - \frac{\lambda x_1^n}{1 + h\lambda x_1^n}x_2$$
$$\frac{dx_2}{dt} = -dx_2 - fx_2^2 + \frac{e\lambda x_1^n}{1 + h\lambda x_1^n}x_2$$

All model parameters are explained in Table 2. The resource and consumer densities are initiated at the carrying capacity and a small value x^0, respectively. The body size b_1 of the resource is set to 1 and that of the consumer, b_2, is generated from a pre-specified range, spanning six orders of magnitude (Table 2). We distinguish two processes through which the food web can grow in size: speciation and invasion. Events corresponding to these processes occur at rates r_s and r_i, respectively. Times of their actual occurrence are thus generated via exponentially distributed random deviates with these rates as their respective means. Any new species x_i begins with a small density x^0, and is assumed to follow dynamics specified by model (1), with $r_i = K_i = m_i = 0$ (i.e., we assume no immigration). The involved processes operate as follows:

(1) **Speciation.** Upon a speciation event occurrence, a species is randomly chosen from the set of currently present species (excluding the external resource), with probability proportional to its actual density. Both its body size and other parameters are slightly changed relative to its parent species' values by adding to each a value generated from a normal distribution with the zero mean and a standard deviation σ (common to all parameters). Provided that any of the parameters would fall below zero, it is set back to a (minimal) value 0.01. The mutated body size may cause a change in linking of the new species to the others, relative to its parent species. We use here the algorithm proposed by Virgo et al. (2006): species i with body size b_i is assigned to the diet of species j with body size b_j with probability

$$P_L = \alpha \exp\left[-\left(\log_{10}\left(\beta\frac{b_i}{b_j}\right)/\gamma\right)^2\right] \tag{2}$$

This formula says that there is a most preferred ratio of consumer size to resource size, that consumers substantially larger or smaller than a resource do not practically consume that resource, and that consumers can also feed on a slightly larger resource; see Table 2 for an explanation of the involved parameters and Fig. 4 for a graphical representation of formula (2). The new species becomes cannibalistic with probability P_C if its parent species is not, and ceases to be cannibalistic with the same probability in the opposite case.

(2) **Invasion.** Upon an invasion event occurrence, a new species invades from an outer environment. Its body size and other parameters are uniformly randomly generated from some pre-specified ranges (Table 2). Linking of the new species to the others follows the procedure described above for speciation. The invading species is cannibalistic with probability P_C.

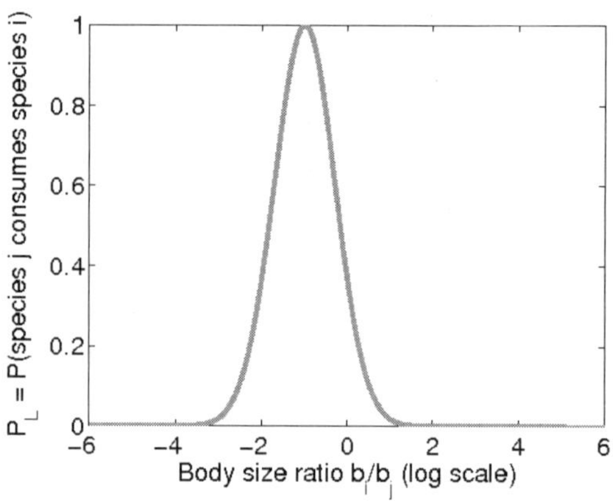

Figure 4. Graphical representation of formula (2) that defines diet of a consumer species; see the main text for more details.

Simulations are run until a maximum time $t_{\max}$ is reached, after which the model can be run for another time interval t_{final} in which none of these processes acts. This can serve as a test of whether the resulting community is stable or is only maintained by the extinction-addition balance. Once density of a species falls below an extinction threshold ϵ (here 10^{-10}), that species is instantly removed from the food web.

4.2.2. *Results*

Speciation and invasion have contrasting effects on the resulting (at time $t_{\max}$) food web size and structure. Speciation causes the food web either to

Table 2. Model parameters of the bottom-up model of food web dynamics.

Par.	Meaning	Value/range
r	Resource intrinsic growth rate	1
K	Resource carrying capacity	10^6
d	Consumer per capita mortality rate	[0.1, 0.5]
e	Consumption efficiency	[0, 1]
f	Degree of intraspecific interference	[0, 10]
λ	Measure of predator-prey encounter rate	[0, 1]
h	Handling time	[0, 0.2]
n	Functional response "degree"	2
b	Body size	[1, 10^6]
x^0	Initial density of new species	0.01
r_s	Speciation rate	1/500
r_i	Invasion rate	1/5
σ	Standard deviation of mutated parameters	0.1
P_C	Probability of becoming cannibalistic (or not)	0.05
α	Probability for most preferred body size ratio	1
β	Defines most preferred body size ratio	10
γ	Range of body sizes consumed (diet breadth)	1
$t_{\max}$	Simulation time (speciation/invasion)	100000/1000
t_{final}	Extra simulation time (no speciation/no invasion)	20000/200

steadily grow to a relatively large size or does not cause any marked change relative to the initial food web (Fig. 5A), depending on the initial model parameters (the exact way how has yet to be determined). On the other hand, an island food web community formed by repeated invasions shows a fluctuating pattern in species richness in which a relatively steady growth is interspersed with dramatic collapses (Fig. 5B). Also, not considering the external resource, the food web formed through invasions consists of basal, intermediate as well as top species (here $S = 9$, $C = 0.17$, $B = 3$, $I = 3$, $T = 3$), while in that growing through speciation events all species are of the intermediate type (here $S = 38$, $C = 0.25$, $B = 0$, $I = 38$, $T = 0$). We note that the food web structure might differ at different times because of the species turnover. Once the food web augmentation procedure is stopped, the actual food web responds differently as to whether we consider speciation or invasion (Fig. 5). There is no effect of cessation of species additions in the case of speciation. Hence, the evolved community appears stable. Contrary to that, the actual food web collapses once further invasions are prevented. The assembled community is thus "only" maintained by the extinction-invasion balance; see also Ref. [4] for a similar result.

The above results provide just a glimpse at what we might expect in different situations; a more elaborated analysis is needed to demonstrate how these patterns vary with changes in other model elements, such as the extinction threshold ϵ (see also Sec. 5.3). The presented model can also be

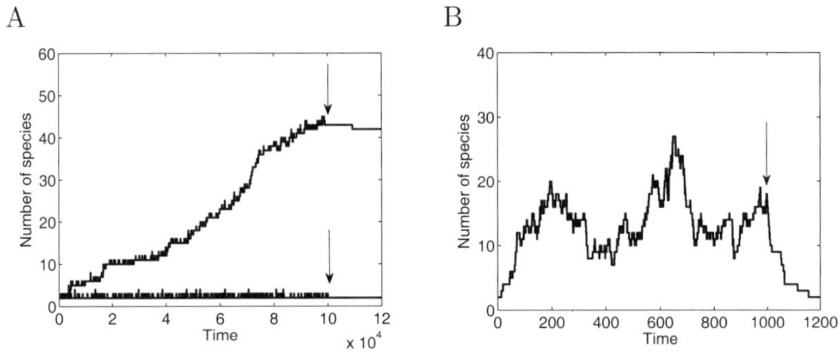

Figure 5. Effects of different food web augmentation processes on species richness. A. Speciation. B. Invasion. Two qualitatively different trajectories are observed in the speciation case, depending on the initial model parameters. Arrows indicate time instants beyond which there is no further speciation or invasion.

used to address a bunch of other questions, such as how do the resulting food webs

- vary with different rates of speciation and invasion,
- look like when food web dynamics are interspersed with external catastrophes (instant extinction of a number of randomly chosen species or of a clade in the case of evolution through speciation),
- look like when after some time of only invasions, only speciation events occur, and vice versa.

5. Discussion

In this paper, we presented two widely used, complementary ways of how to model and study population dynamics on complex food webs, the top-down approach and the bottom-up approach. The former approach starts with a complex food web structure on which a model of population dynamics is superimposed and run. On the contrary, the bottom-up approach builds complex food webs from below, starting with a simple food web and repeatedly adding new species; the success or failure of these new species is determined by a model of population dynamics. In addition, we illustrated each of these approaches by a specific model. Using the top-down approach, we studied effects of simultaneous operation of several stabilizing mechanisms, showing, somewhat surprisingly, that they might suppress each other. Through the bottom-up approach, we examined an extent to which food webs formed via speciation vs. invasion events differ, showing that they differ in size, structure as well as propensity to collapse.

In our mini-review, we also showed that a variety of models have already been developed, using each approach, which are quite heterogeneous as regards both underlying assumptions and addressed questions. Using each approach, food webs of different size and structure can be obtained, depending on the specific processes incorporated in the model. Given that both approaches have the same ultimate aim – generate food webs persistent with respect to population dynamics and of complexity comparable to food webs observed in nature – we might ask to what extent food webs resulting from both approaches are comparable. Surprisingly, this question has not been addressed yet. In what follows, we discuss three methodological issues that have grasped the attention of food web modellers: how to model (dynamics of) basal species, whether and how to account for body size of the involved species, and how to deal with simulation time and extinction threshold.

5.1. *Modelling (dynamics of) basal species*

Using the top-down approach, modellers repeatedly face the situation where many species present in the initial food web go eventually extinct[4,37,47]; see also our example in Sec. 4.1. This is a somewhat unwished result given that initial food webs are generated by structural food web models aimed at reproducing complexity of real food webs. Also, using the bottom-up approach, the evolved or assembled food webs are either much smaller[58] or much larger[20] relative to what is commonly observed in nature. In addition, the complexity-stability relationship explored using the top-down approach was shown to change from negative to positive once averaging over all generated food webs with given size and connectance was replaced by averaging only over a subset of the webs that contained a fixed number of basal species[38,39]. All these observations suggest that basal species are likely to play a major role in driving food web persistence. According to Ref. [47], "the importance of basal species for persistence emphasizes the need for high quality data resolved evenly at all trophic levels". Empirical food webs have been found to have either a "pyramid" form[10] – basal species most numerous, followed by intermediate species and then top species – or a "hump-shaped" form[59] – some basal species, many intermediate species and a few or no top species. Any structural food web model based just on species richness and connectance thus cannot cover both these food web types simultaneously. But perhaps many hump-shaped food webs could admit a pyramid form once those high quality data are collected. Different authors have approached the issue of modelling (dynamics of) basal species differently. They (i) considered for further processing just a subset of initial food webs that had the same number of basal species[38,39], (ii) fixed the number of basal species a priori[58,5,57], or (iii) added a null layer to the food web, comprising one or two species referred to as the "external resource(s)" or "environment" on which all basal species feed and whose abundance is often assumed time-invariant[4,20,19]. In the latter case, making the external resource abundance fixed and high enough, almost anyhow large food web might be generated[4,20,19]. Explicit modelling of an external resource naturally allows the basal species to compete with one another, an issue largely neglected so far [47] . But this does not happen if the external resource is indepletable, that is, fixed in density. Even when thought of as sunlight as in Ref. [20], grasses compete with bushes and trees of different height compete with one another for light due to one shading another (not to speak of space, water or nutrients). Thus, even the external resource should have its own dynamics; the only study we know of that accounted for this possibility, in the form of logistics, is that of Ref. [4]. Quantitatively, accounting for basal species competition might not only reduce the proportion of basal species in the persistent food web relative to when competition is absent, but also the total number of species, i.e., food web persistence as such. We therefore think that a fruitful study would be to explore how does food web persistence change with different ways of modelling basal species (dynamics).

5.2. *Body size and allometric relationships*

Given a food web structure and a model of population dynamics, the question is how to sensibly choose model parameters and/or initial population densities. This is a tricky issue as well, as the number, type and stability of system steady states may vary from one parameter set to another, and different states can be reached from different initial densities. Moreover, this kind of information about steady states is virtually always impossible to gain because of high dimensionality and/or strong non-linearity of many population-dynamic models; researchers typically restrict model exploration to numerical simulations. The available models corresponding to the top-down approach coped with this issue differently: they used (i) fixed, species-independent parameters sometimes motivated by empirical data[8], (ii) parameters uniformly randomly generated from some prespecified ranges[33], (iii) or a combination of these methods[37]; similar strategies were used for setting initial population densities. The situation is somewhat simpler for speciation models of the bottom-up approach (but not for the invasion models), as here new species are added at a low density and model parameters corresponding to these new species are some mutated values of their parents. Any of these ways is related to a number of questions. Namely, how does food web persistence vary if the fixed parameter values are changed or the parameter ranges shrunk or extended? What we actually gain by averaging simulation results over many replicates, each run with different model parameters (and different food web structure which gets randomized in both the top-down and bottom-up approaches)? And, above all, how to set up a representative set of model parameters for a given problem? A promising avenue to address the latter question that gets increasingly popular is to relate as much of model parameters as possible to species body size through the so-called allometric relationships[61,52]. For example, body size has been shown to affect respiration rate, maximum ingestion rate, consumer-resource interaction strength, production efficiency, and mortality rate[62,25,44,58,9]. Body-size-dependent parameters in food web models were shown to enhance food web persistence[9] and prevent any extinction due to population dynamics superimposed on a real food web[25], but also to have no significant effect[29]. As (many) model parameters might be linked to body size, an accompanying question is how to assign body sizes to individual species in a food web generated through, the niche model. Location of species on the niche axis scaled to body size can be a promising candidate[61]. Alternatively, we might first generate body sizes corresponding to individual species and only then generate a food web structure, via formula (2), in the way akin to that through which Ref. [5] generated food web structures by means of calculating optimal feeding preferences for each individual species. On the other hand, the role of body size as a unique predictor of model parameters has not to be overrated.

5.3. *Simulation time and extinction threshold*

Adopting one approach or another, there are a number of technical issues one has to deal with before running simulations. We consider two of them here: the length of simulations and the value of extinction threshold. How long should we run a simulation? Not an easy question, as we usually have no idea of how the system behaves asymptotically (i.e., how many and what type of steady states it has, and whether these are (locally) stable or not), so that we always face a risk of stopping the simulation prematurely while it is still in a transient phase. This is especially important when two or more scenarios (dynamics based on two types of functional response) or two parameter sets are to be compared, as times the system needs to "settle" may vary between them. For example, if the steady state is sort of a locally stable limit cycle in terms of species richness, as in our bottom-up model with invasions, simulations need to be run sufficiently long to identify it. Also, we used different lengths of simulations for speciation and invasion scenarios, respecting different timescales on which these processes act. Unfortunately, algorithms allowing for assessing system permanence (i.e., no extinctions in the future) are available only for the consumer-resource interactions of the most simple, Lotka-Volterra type[58]. The second issue we discuss here concerns the extinction threshold, i.e., a population density below which populations are considered effectively extinct. Whereas in individual-based models this is irrelevant as a species goes simply extinct when its last individual dies out[36], in most analytical (i.e., equation-based) models populations will never reach zero density, so that an extinction threshold needs to be set up and applied. But what is too low and what is enough? Specific values used in the literature vary greatly, over many orders of magnitude, from as low numbers as 10^{-308}, through 10^{-1547} and 0.001 [57], up to 1^{20}. The relevant question therefore is how sensitive is food web persistence to this "external" parameter? Apparently, the lower value we choose the more species will persist, but should this dependence worry us? The second aspect related to an extinction threshold is when should we apply it. Although we intuitively feel that removing extremely rare populations is more appropriate to do already in the course of simulations, as we know from the analysis of many consumer-resource models that too rare a population may still recover later on if not removed immediately, many food web studies apply it only after the simulation terminates[9,47]; some studies even do not mention the applied procedure at all. Ref. [9] claimed that their results did not change qualitatively with different choices of the extinction threshold, but how general is this observation? And what if one's focus is on quantitative rather than qualitative predictions (recall that we would like to come up with food webs of realistic complexity)? Sensitivity analysis of this kind should therefore become a part of any modelling study on food web dynamics, unless the used values are clearly substantiated.

5.4. *Concluding remarks*

Real food webs are characterized by diverse types of consumer-resource relationships (such as predator-prey or host-parasitoid), diverse types of functional forms (such as functional responses) within these types, and diverse, non-random parameters within these latter types. The question therefore is how to cope with all these intricacies, keeping a sort of generality. In other words, what the models discussed in this paper are good for and what form should they take to be successful in this endeavour? We hope we have convinced the reader that studying population dynamics on complex food webs is of distinct importance for our understanding of biodiversity, including our ability to assess threats to it due to ubiquitous negative impacts of many anthropogenic activities. We may also ask many more specific questions, some of which have been addressed and even more formulated throughout this paper. Virtually no pair of studies conducted so far on food web dynamics is entirely comparable. On one hand, such a heterogeneity in the addressed questions and modelling assumptions is welcome, as it shows possibilities and opens new research avenues (and, last but not least, facilitates publication). On the other hand, however, this heterogeneity is sort of impending, as it prevents any sound synthesis on the relationship between food web structure and dynamics. Potent food web theory can only be developed if additional and potentially less attractive studies are composed that "fill the gaps" remaining behind the "at-the-edge-of-science" works. By filling the gaps, we mean here combining components of diverse models and testing the published results for broader robustness. This diligent effort might also help us identify an ideal model of food web dynamics, if there is any, as a cocktail composed of ingredients of many models developed so far. These will certainly include a mixture of a number of stabilizing mechanisms, consideration of external resource with its own dynamics, and body-size-related model parameters, respecting at the same time the question addressed by such a model. Using the words of Ref. [4], "the strategy is to look for the simplest set of rules which appear sensible and which allow to derive the observed statistical patterns of biodiversity". Then, using either the top-down approach or the bottom-up approach, one may hope that the generated food webs will be of size and complexity comparable to those observed in real food webs.

Acknowledgments

I acknowledge funding by the Institute of Entomology (Z50070508), the Grant Agency of the Academy of Sciences of the Czech Republic (IAA100070601), and the Faculty of Science (MSM6007665801).

References

1. Anderson, P. E. and H. J. Jensen. 2005. Network properties, species abundance and evolution in a model of evolutionary ecology. Journal of Theoretical Biology 232:551–558.
2. Bascompte, J. and C. J. Melián. 2005. Simple trophic modules for complex food webs. Ecology 86:2868–2873.
3. Bascompte, J., C. J. Melián and E. Sala. 2005. Interaction strength combinations and the overfishing of a marine food web. Proceedings of the National Academy of Sciences of the USA 102:5443-5447.
4. Bastolla, U., M. Lassig, S. C. Manrubia and A. Valleriani. 2001. Diversity patterns from ecological models at dynamical equilibrium. Journal of Theoretical Biology 212:11–34.
5. Beckerman, A. P., O. L. Petchey and P. H. Warren. 2006. Foraging biology predicts food web complexity. Proceedings of the National Academy of Sciences of the USA 103:13745–13749.
6. Bell, G. 2007. The evolution of trophic structure. Heredity 99:494-505.
7. Berec, L., E. Angulo and F. Courchamp. 2007. Multiple Allee effects and population management. Trends in Ecology and Evolution 22:185–191.
8. Brose, U., R. J. Williams and N. D. Martinez. 2003. Comment on "Foraging adaptation and the relationship between food web complexity and stability". Science 301:918b.
9. Brose, U., R. J. Williams and N. D. Martinez. 2006. Allometric scaling enhances stability in complex food webs. Ecology Letters 9:1228–1236.
10. Byrnes, J. E., P. L. Reynolds and J. J. Stachowicz. 2007. Invasions and extinctions reshape coastal marine food webs. PLoS ONE 2(3):e295.
11. Caldarelli, G., P. G. Higgs and A. J. McKane. 1998. Modelling coevolution in multispecies communities. Journal of Theoretical Biology 193:345–358.
12. Camacho, J., R. Guimera and L. A. N. Amaral. 2002a. Analytical solution of a model for complex food webs. Physical Review E 65:article number 030901(R).
13. Camacho, J., R. Guimera and L. A. N. Amaral. 2002b. Robust patterns in food-web structure. Physical Review Letters 88:article number 228102.
14. Cattin, M.-F., L.-F. Bersier, C. Banašek-Richter, R. Baltensperger and J. P. Gabriel. 2004. Phylogenetic constraints and adaptation explain food-web structure. Nature 427:835–839.
15. Cohen, J. E. and C. M. Newman. 1985. When will a large complex system be stable? Journal of Theoretical Biology 113:153–156.
16. Cohen, J. E., F. Briand and C. M. Newman. 1990. Community food webs: data and theory. Springer-Verlag, Berlin, Germany.

17. Courchamp, F., E. Angulo, P. Rivalan, R. J. Hall, L. Signoret, L. Bull and Y. Meinard. 2006. Rarity value and species extinction: the anthropogenic Allee effect. PLoS Biology 4:e415.
18. Courchamp, F., L. Berec and J. Gascoigne. 2008. Allee effects in ecology and conservation. Oxford University Press, Oxford.
19. Drossel, B., A. K. McKane and C. Quince. 2004. The impact of nonlinear functional responses on the long-term evolution of food web structure. Journal of Theoretical Biology 229:539–548.
20. Drossel, B., P. G. Higgs and A. J. McKane. 2001. The influence of predator-prey population dynamics on the long-term evolution of food web structure. Journal of Theoretical Biology 208:91–107.
21. Dunne, J. A. 2006. The network structure of food webs. Pages 27–86 *in* M. Pascual and J. A. Dunne, eds. Ecological networks: linking structure to dynamics in food webs. Oxford University Press, Oxford, UK.
22. Dunne, J. A., R. J. Williams and N. D. Martinez. 2002. Food-web structure and network theory: the role of connectance and size. Proceedings of the National Academy of Sciences of the USA 99:12917-12922.
23. Dunne, J. A., R. J. Williams and N. D. Martinez. 2004. Network structure and robustness of marine food webs. Marine Ecology Progress Series 273:291-302.
24. Ebenman, B., R. Law and C. Borrvall. 2004. Community viability analysis: the response of ecological communities to species loss. Ecology 85:2591–2600.
25. Emmerson, M. C. and D. Raffaelli. 2004. Predator-prey body size, interaction strength and the stability of a real food web. Journal of Animal Ecology 73:399–409.
26. Fowler, M. S. 2009. Increasing community size and connectance can increase stability in competitive communities. Journal of Theoretical Biology 258:179–188.
27. Goldwasser, L. and J. Roughgarden. 1993. Construction and analysis of a large Carribean food web. Ecology 74:1216–1233.
28. Goodland, R. J. 1975. The tropical origin of ecology: Eugen Warming's jubilee. Oikos 26:240–245.
29. Guill, C. and B. Drossel. 2008. Emergence of complexity in evolving niche-model food webs. Journal of Theoretical Biology 251:108–120.
30. Holt, R. D. 1997. Community modules. Pages 333–350 *in* A. C. Gange and V. K. Brown, eds. Multitrophic interactions in terrestrial systems. Blackwell Science, London, UK.
31. Hraber, P. T. and B. T. Milne. 1997. Community assembly in model ecosystem. Ecological Modelling 103:267–285.
32. Huisman, G. and R. J. De Boer. 1997. A formal derivation of the "Beddington" functional response. Journal of Theoretical Biology 185:389–400.
33. Ives, A. R. and B. J. Cardinale. 2004. Food-web interactions govern the resistance of communities after non-random extinctions. Nature 429:174–177.
34. Janssen, A., A. Pallini, M. Venzon and M. W. Sabelis. 1998. Behaviour and indirect interactions in food webs of plant-inhabiting arthropods. Experimental & Applied Acarology 22:497-521.

35. Kartascheff, B., C. Guill and B. Drossel. 2009. Positive complexity-stability relations in food web models without foraging adaptation. Journal of Theoretical Biology 259:12–23.
36. Keitt, T. H. 1997. Stability and complexity on a lattice: coexistence of species in an individual-based food web model. Ecological Modelling 102:243–258.
37. Kondoh, M. 2003a. Foraging adaptation and the relationship between food web complexity and stability. Science 299:1388–1391.
38. Kondoh, M. 2003b. Response to comment on "Foraging adaptation and the relationship between food web complexity and stability". Science 301:918c.
39. Kondoh, M. 2006. Does foraging adaptation create the positive complexity-stability relationship in realistic food-web structure? Journal of Theoretical Biology 238:646–651.
40. Křivan, V. 1997. Dynamic ideal free distribution: Effects of optimal patch choice on predator-prey dynamics. American Naturalist 149:164–178.
41. Křivan, V. and A. Sikder. 1999. Optimal foraging and predator-prey dynamics II. Theoretical Population Biology 55:111–126.
42. Lawlor, L. R. 1978. A comment on randomly constructed model ecosystems. American Naturalist 112:445–447.
43. Lockwood, J. L., R. D. Powell, M. P. Nott and S. T. Pimm. 1997. Assembling ecological communities in time and space. Oikos 80:549–553.
44. Loeuille, N. and M. Loreau. 2006. Evolution of body size in food webs: does the energetic equivalence rule hold? Ecology Letters 9:171–178.
45. Martinez, N. D. 1993. Artifacts of attributes? Effects of resolution on the Little Rock Lake food web. Ecological Monographs 61:367–392.
46. Martinez, N. D. and L. J. Cushing. 2006. Additional model complexity reduces fit to complex food-web structure. Pages 87–89 *in* M. Pascual and J. A. Dunne, eds. Ecological networks: linking structure to dynamics in food webs. Oxford University Press, Oxford, UK.
47. Martinez, N. D., R. J. Williams and J. A. Dunne. 2006. Diversity, complexity, and persistence in large model ecosystems. Pages 163–185 *in* M. Pascual and J. A. Dunne, eds. Ecological networks: linking structure to dynamics in food webs. Oxford University Press, Oxford, UK.
48. May, R. M. 1971. Stability in multispecies community models. Mathematical Biosciences 12:59–79.
49. May, R. M. 1972. Will a large complex system be stable? Nature 238:413–414.
50. McCann, K., A. Hastings and G. R. Huxel. 1998. Weak trophic interactions and the balance of nature. Nature 395:794–798.
51. Murdoch, W. W. and A. Oaten. 1975. Predation and population stability. Pages 1–131 *in* A. Macfadyen, ed. Advances in Ecological Research. Academic Press.
52. Pascual, M., J. A. Dunne and S. A. Levin. 2006. Challenges for the future: integrating ecological structure and dynamics. Pages 351–371 *in* M. Pascual and J. A. Dunne, eds. Ecological networks: linking structure to dynamics in food webs. Oxford University Press, Oxford, UK.
53. Quince, C., P. G. Higgs and A. J. McKane. 2005. Deleting species from model

food webs. Oikos 110:283–296.

54. Ruxton, G. D. 1995. Short term refuge use and stability of predator-prey models. Theoretical Population Biology 47:1–17.
55. Sanders, N. J., N. J. Gotelli, N. E. Heller and D. M. Gordon. 2003. Community disassembly by an invasive species. Proceedings of the National Academy of Sciences of the USA 100:2474–2477.
56. Stouffer, D. B., J. Camacho, R. Guimera, C. A. Ng and L. A. N. Amaral. 2005. Quantitative patterns in the structure of model and empirical food webs. Ecology 86:1301–1311.
57. Uchida, S. and B. Drossel. 2007. Relation between complexity and stability in food webs with adaptive behavior. Journal of Theoretical Biology 247:713–722.
58. Virgo, N., R. Law and M. Emmerson. 2006. Sequentially assembled food webs and extremum principles in ecosystem ecology. Journal of Animal Ecology 75:377–386.
59. Williams, R. J. and N. D. Martinez. 2000. Simple rules yield complex foodwebs. Nature 404:180–183.
60. Williams, R. J. and N. D. Martinez. 2004. Stabilization of chaotic and nonpermanent food-web dynamics. European Physical Journal B 38:297–303.
61. Woodward, G., B. Ebenman, M. Emerson, J. M. Montoya, J. M. Olesen, A. Valido and P. H. Warren. 2005. Body size in ecological networks. Trends in Ecology and Evolution 20:402–409.
62. Yodzis, P. and S. Innes. 1992. Body size and consumer-resource dynamics. American Naturalist 139:1151–1175.

NEW ZEALAND PALÆODEMOGRAPHY: PITFALLS & POSSIBILITIES

C. E. M. PEARCE, S. N. COHEN, J. TUKE

School of Mathematical Sciences,
The University of Adelaide,
Adelaide, SA 5005, Australia
E-mail: charles.pearce, samuel.cohen , simon.tuke@adelaide.edu.au

Following Atholl Anderson, there has been a growing acceptance of a late (12th century) first settlement date for New Zealand. This is sometimes advanced to AD 1280. For a modest initial number of settlers, a very rapid and sustained population growth is then required to match the population extant around AD 1800. The most detailed New Zealand palæodemographic study to date, that of Brewis, Molloy and Sutton, does not support this scenario. Their data analysis hangs on several significant assumptions and approximations. We show how to proceed from weaker assumptions and also provide confidence intervals for our estimates. Our analysis does not depend on radiocarbon or other dating. We conclude by considering the implications for New Zealand prehistory.

1. Introduction & setting

Palæodemography is concerned with such issues as estimating demographic parameters from past populations (primarily from skeletons in an archæological context) and/or making deductions regarding the health of individuals in those populations. The most basic parameters are age–specific fertility in women and age–specific survival rates of both sexes, from which life tables can be constructed.

Osteologists face considerable problems with the development of reliable age indicators in skeletons to relate skeletal morphology to chronological ages. The determination of age–specific fertility rates for women depends on the fact that macro–anatomical alterations may occur at the pelvic bone, in the form of pitting, as a consequence of pregnancy and childbirth. The number of pregnancies/births may be estimated from the pitting at two sites, the posterior pubic symphysis and the preauricular groove of the ilium. Age–specific fertility rates are then derived by relating the total number of pregnancies/births over the life of a woman to her estimated age at death. This method is not without difficulties. Pitting can arise in nullipara and further its traces can gradually become less pronounced over the years following a birth. However a computer simulation study by Brewis[4] tended to confirm osteologically based estimates.

There are further difficulties in the use of statistical techniques for handling both indicators relating to survival rates and indicators relating to fertility rates. A good account of some of the ideas, techniques and problems is provided in Hoppa and Vaupel[9].

There can be subtle effects relating to the nature of the sampling involved. Suppose a population is *stable*, that is, that the distribution of ages of individuals in the population is independent of time. Denote by $S(x)$ the probability that an individual lives to age x years or more. If the population is stable and growing at a rate r *per annum* (positive, negative or zero), then at any instant of time the proportion $G(x)$ of the population that has survived to age x years or more is proportional to $S(x)\exp(-rx)$. In general $S(x) \neq G(x)$ unless $r = 0$, that is, unless the population size is static. If a survival function is determined by observing relative frequencies of different ages at death over a short time, then a function close to $G(x)$ will be obtained rather than $S(x)$. Thus Sattenspiel and Harpending[24] working with Swedish data found the mean age of death over the period 1778–1782 (a quantity here essentially governed by G) was 32.7 years while the life expectancy at birth (a quantity entirely governed by S) was 38.5 years. This situation arose in a growing population ($r = 0.006$ or 0.6% *per annum* increase).

However the mode of sampling entailed in obtaining a prehistoric skeletal population can be quite different. The skeletons may derive over a period of some centuries and the end effects arising over short intervals are of small order. The age at death for a skeleton picked at random will then be very close to S rather than to G.

For the remainder of this chapter we shall reserve the notation $S(x)$ exclusively for the survival function for females.

One difficulty in estimating $S(x)$ is that the skeletons of infants (children under one year of age) are often under–represented in real–world samples (see Buikstra and Konigsberg[6], Moore *et al.*[17], Peterson[20]) and the extent to which they are under–represented may need to be guessed. A concomitant problem is that gender may be difficult to assign accurately for children under the age of 15 years so that survival rates for males and females are not distinguished below age 15. Let $y = x - 15$ represent age beyond 15 and p the (often unknown) probability of survival to at least age 15. It is convenient to parameterize $S(x)$ for $x \geq 15$ as

$$S(x) = pS_{15}(y) \quad \text{for } y \geq 0.$$

We now introduce the age–specific fertility $f(x)$. This is the density (per woman per year) of children born to women aged x years. If $F(x)$ denotes the corresponding cumulative function, that is, the expected number of children born to a woman by the time she reaches age x years, then $f(x) = F'(x)$. Let L, U be respectively the lower and upper bounds to the ages at which a woman reproduces. The argument of Section 3 suggests that the choice $L = 15$ is realistic. This provides a convenient match with the age at which gender assignment to skeletons becomes easier. Suppose that q is the probability that a birth is female. Then a result of Lotka[15] gives for a stable population that r is determined by

$$C_p := 1/p = q \int_{15}^{U} \exp(-rx) S_{15}(x - 15) f(x)\, dx. \tag{1}$$

2. New Zealand Palæodemography

Central to New Zealand palæodemography are studies of a very limited number of Māori skeletons from disparate times and locations. Estimations of fertility in prehistoric New Zealand were made by Houghton[11], Phillipps[21], Sutton[29] and Visser[33]. Progress was been aided by studies by Houghton[10] of scoring of pelvic bones. Estimates of average age at death were given by Houghton[11] and by Simpson[25] and Sutton[28]. The most detailed New Zealand palæodemographic study published to date is by Brewis, Molloy and Sutton[5], which builds on the above. It called on an unusually large cohort of 172 skeletons.

As noted by the Brewis *et al.*, the date of first colonization was widely believed to be AD 750-950 when their study was published. This date has since moved several centuries towards the present. Following a seminal paper by Atholl Anderson[1], there has been wide acceptance of a late (12th century) first settlement date for New Zealand. This has largely arisen from a dismissal of many radiocarbon dates on the grounds of a lack of adequate chronometric hygiene. There is a perceived need for stricter controls against sample contamination and for more careful on–site stratigraphy.

By contrast with this late advent, the most credible population estimates for 1801 (the dawn of European settlement) are perhaps 155,000 and 166,000 (suggested by Rutherford) and 175,000 proposed by Urlich[32]. See Lewthwaite[14] for an overview and a discussion of other estimates. For a modest initial number of settlers (prehistorians usually posit 300 or fewer), a very rapid and sustained population growth is then required to meet the population extant around AD 1800. Assuming an initial population of 300 in 1200 and a natural balance in numbers of the sexes, the 155,000 figure for 1801 corresponds to an average rate of growth of 1.045% p.a., while the 175,000 figure gives 1.065% p.a.

Even with initial settlement dated at AD 750, such a group of 300 migrants would need the not inappreciable growth rates of 0.596% per annum and 0.608% per annum to reach the stated 1801 levels.

What light does the demographic work of Brewis *et al.* shed on this contrast of paradigms? They found a population *decline* of 0.414% p.a. in association with an improbably low infant mortality rate of 0.035. In a sample of 172 individuals, there were 6 infant deaths and 141 individuals reaching age 15. This corresponds to $C_p = 172/141$ in the notation of the previous section. If infant deaths were under–represented by 24, infant mortality would have been 15.3% and $C_p = 196/141$. Brewis *et al.* examined this scenario and obtained a corresponding population *decline* of 1.52% p.a.

The abstract to their paper begins:

"Skeletal and comparative evidence of mortality is combined with fertility estimates for the precontact Māori population of New Zealand to determine the implied rate of precontact population growth. This rate is found to be too low to populate New Zealand within the time constraints of its prehistoric sequence, the probable founding population size, and the probable population size at contact. Rates of growth necessary to populate New Zealand within the accepted time span are calculated. The differences between this minimum necessary rate and the skeletally derived rate are too large to result solely from inadequacies in the primary data."

The authors proposed four alternative explanations:

(a) the skeletal evidence of mortality is highly inaccurate;
(b) the skeletal evidence of fertility severely underestimates actual levels;
(c) there was very rapid population growth up to AD 1150 for which no skeletal evidence is currently available;
(d) the prehistoric sequence of New Zealand may have been longer than generally accepted.

After some discussion, Brewis *et al.* concluded that a combination of (c),

(d) was the most probable. In that light, they proposed, without reference to the skeletal evidence, the unrealistically high population growth rate of 0.875% p.a. sustained over the 919 years from AD 850 to AD 1769 with a small initial settlement as feasible in the context of the then current thinking about prehistoric demography in New Zealand.

In this chapter we readdress the issue of the growth rate of prehistoric New Zealand, employing modern statistical techniques to obtain refinements of the results obtained by Brewis *et al.* from skeletal analysis and in particular to derive confidence intervals for point estimates of rates of growth. In the following section we examine the assumptions underlying their analysis. In Section 4 we provide a new analysis and in Section 5 present our results. We end in Section 6 with discussion and conclusions.

3. Assumptions in Brewis *et al.*

The skeletal analysis by Brewis *et al.* employed a life table approach based on inferences from skeletal data providing information on ages at death and the total number of children produced by a woman during her life. The authors utilized the following assumptions:
(i) except for infants, the distribution of age at death as indicated by skeletal data represents age–specific mortality rates of the constituent age cohorts;
(ii) the raw proportion (3.5%) of skeletons under age one constitutes an under–represention. Two alternative adjustments were proposed: from 3.5% to 15.3% and from 3.5% to 29.7%;
(iii) a proportion $q = 0.488$ of births were female;
(iv) each child is born when the mother is 25 years of age;
(v) the level of the total fertility of a Māori woman over her reproductive life was adopted from conclusions reached earlier by Phillipps[21].

Assumption (i) was made for convenience of calculation. Citing Sattenspiel and Harpending[24] and Buikstra *et al.*[7], Brewis *et al.* argued that (i) is an approximation because "distributions of age at death generated by osteological analysis are not necessarily representative of age–specific mortality rates of the constituent age cohorts". See also Sutton and Molloy[30].

However the mode of sampling entailed in obtaining the New Zealand prehistoric skeletal population involves skeletons from a period of some centuries. In accordance with our discussion in Section 1 there should be *ceteris paribus* very little, if any, bias in the use of (i) in the present context.

Table 1. Number of births by final age of mother

age at death	no. of births (posterior pubic symphysis)	no. of births (preauricular groove of ilium)	births estimate
20	0	0	0
21	0	1	0.5
21	0–1	2	1
26	1	1	1
26	1–2	2	2
26	2	3	2.5
27	0	1	0.5
32	2	2	2
32	2–3	3–4	3
33	1	1	1
34	3	4–5	4
35	3	3–4	3
36	1	1	1
36	1	2	1.5
36	1–2	2–3	2
36	4–5	4	4
36	5	4–5	5
39	3	2–3	3
39	3	3	3
39	4	4	4
39	4–5	4–5	4.5
43	2–3	3–4	3
43	3–4	3–4	3.5
43	4	4–5	4
47*	3	0	1.5
47	3	3	3
47	4	3–4	4
47	4–5	3	4
47	4	4	4
47	3	4–5	4
47	5	4	4.5
47+	5	5	5
50	3–4	4	4

Here the caveat is for burial practices for different age groups to be equal, etc.

In respect to assumption (ii), Brewis *et al.* note that the 3.5% mortality of infants as indicated by the skeleton population appears unrealistically low. Weiss[34] states that the healthiest and most successful prehistoric and small scale populations still had an infant mortality rate of about 10%. At the other extreme Māori infant mortality following the advent of infectious diseases with European colonization has been estimated to have been 36.5–41.9% (Pool[22]), which may suggest a maximum infant mortality of about 30% in prehistoric New Zealand (see Brewis[3]).

A recent study by Navara[18] notes lattitude as a significant factor in the proportion of births that are female. This study and the references therein support the choice $q = 0.488$ in (iii) as being accurate.

Assumption (iv) is biologically unrealistic and was made for simplicity of calculation. In view of the highly nonlinear way in which r depends on F and S, we can expect this to introduce bias.

We now turn to assumption (v). Phillipps states that Māori women gave birth at a fairly uniform rate from age 20 to age 34 (and then reached menopause), with an average of 3.4 children when they lived that period out fully. See also Sutton[29] for a discussion of her results. Her conclusion derives from estimates of the number of children produced by each of a sample of 33 women during her life (estimated from two sites of pelvic pitting), coupled with corresponding estimates for age at death. The two pelvic counts were close in all but one of the skeletons, indicated by * in Table 1. This one we have set aside from our calculations. In Table 1 we list the data and propose average/concensus figures for the numbers of children produced by each of the women concerned.

In Table 2 we group the data of Table 1 into five-year intervals.

Table 2. Average number of births by final age cohort of mother

age at death	no. of women	average births
20–24	3	0.5
25–29	4	1.5
30–34	4	2.5
35–39	10	3.1
40–44	3	3.5
45–50	8	4.06

The successive age groups of women have successively higher average birth counts. This makes a *prima facie* case against the Phillipps conclusion that the childbearing years end in the mid thirties. The table further indicates age–specific fertility was highest in the late 20s and early 30s (one child per five–year interval), with reduced fertility (0.5 children per interval) reaching down to the late teens (the three women in the age interval 20–24 all died at age 20 or 21) but also extending well into the forties for women who lived that long. In particular, childbearing does not occur uniformly over the childbearing years. This suggestion appears well in keeping with human biology, in which $f(x)$ is typically a unimodal function with a pronounced peak in the late twenties/early thirties.

4. Refined estimation

Lifetables involve the determination of an appreciable number of parameters, which is an inefficient use of small data sets such as those used by Brewis *et al.* and Phillipps – our present database. This is notably the case if the object is to determine the rate of change r of a population, as this is a second–order parameter that is found indirectly. One way to combat this problem is to make use of prior information.

A considerable literature has indicated that good fits can be obtained to empirical data with the survival function $S(x)$ and the cumulative fertility function $F(x)$ to age x for women in a population by suitable canonical functions involving only two or three parameters. With $x \geq 0$ and $y \geq 0$ we use the respective canonical forms

$$F(x) = \frac{a}{1 + e^{b-cx}}, \qquad S_{15}(y) = \exp\left[\frac{\alpha}{\beta}\left(1 - e^{\beta y}\right)\right].$$

These forms also serve the useful role of smoothing data.

The data for F was entered into R and analyzed using nonlinear least squares through the nls package[23]. The results are presented in Table 3.

Table 3. Parameterization of F

	estimate	std error	t value	$Pr(>\|t\|)$	
a	4.36259	0.66952	6.516	3.91e-07	⋆⋆⋆
b	4.81899	1.64917	2.922	0.00667	⋆⋆
c	0.15344	0.05954	2.577	0.01532	⋆

Here the significance codes are 0 "$\star\star\star$"; 0.001 "$\star\star$"; 0.01 "$\star$". The correlations of the parameter estimates are $corr(a,b) = -0.70$, $corr(a,c) = -0.82$, $corr(b,c) = 0.98$.

Interestingly enough, a linear model gave a very slightly better fit to the data, which might appear to be supporting a uniform age–specific fertility function of an extended Phillips type. See Figure 1. It might also be interpreted as saying that, in the data we are observing, the women have not reached menopause and thus infertility and so we are not observing the expected asymptotic behaviour.

More importantly, to choose the straight line in place of the curve because it has better fit or is simpler would be to disregard prior information available about the shape of $f = F'$ which is less visually obvious in its integrated form F, which is smoothed. The sharper f, illustrated in Figure 2, offers all we might desire in an age–specific fertility curve, with unimodal form and pronounced peak in the late twenties and early thirties.

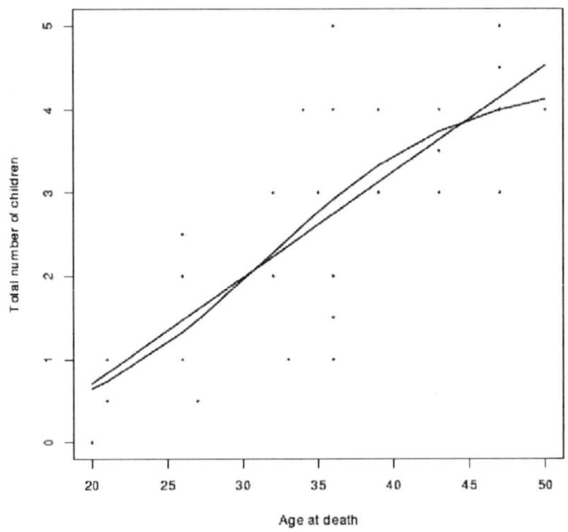

Figure 1. Observations of numbers of children vs age of death with estimated fertility curves

Recall that there is difficulty in distinguishing gender for deaths under age 15 and that Brewis *et al.* used pooled gender parameters for survival up to age 15. The form of S proposed in the Introduction is particularly

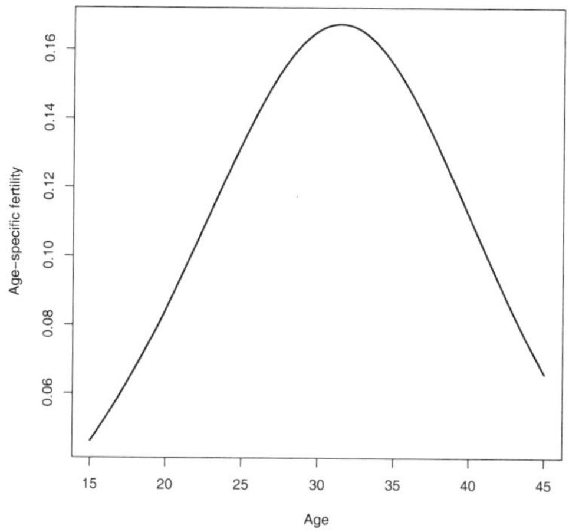

Figure 2. Estimated age–specific fertility density function

convenient for combining pooled data (up to age 15) and unpooled data (beyond age 15). The data for F was given in years and that for S only in five–year intervals. We used a maximum likelihood procedure to make best use of the latter.

The form of S_{15} may be compared with the Gompertz–Makeham function

$$\exp\left[-\gamma y + \frac{\alpha}{\beta}\left(1 - e^{\beta y}\right)\right]$$

(*cf.* Konigsberg and Herrman[12]). We experimented with the latter and found that, for our data, a maximum likelihood estimation fit gave a value of γ that was not significant. We then reverted to the basic Gompertz function with $\gamma = 0$ and made a maximum likelihood estimation of α, β using data for S relating to ages over 15.

Put $x_1 = 15$, $x_2 = 20$, etc. and suppose there were n_i deaths between x_i and x_{i+1}. In view of the independence of individuals, the probability/likelihood of observing the data we have, given the curve S, is

$$L = \prod_i \{S(x_i) - S(x_{i+1})\}^{n_i}.$$

Given the parametric form for S in terms of α and β, we can write the log–likelihood as

$$\ell = \sum_i n_i \ln[S(\alpha, \beta; x_i) - S(\alpha, \beta; x_{i+1})],$$

which can be fitted for α, β using regular methods.

By standard Fisher information theory, the variance–covariance matrix for the maximum likelihood sample estimates is estimated by the negative of the Hessian of the log–likelihood function evaluated at the maximum likelihood values and the distribution of the sample parameter estimates is asymptotically multivariate normal. This provides confidence intervals for estimates. See, for example, Louis[16].

Table 4. Parameterization of female survival curve S_{15}

	estimate	std error
α	0.01853884	0.004556504
β	0.08159386	0.012690147

Table 5. Variances & covariances for α, β

	α	β
α	2.076173e-05	-5.144155e-05
β	-5.144155e-05	1.610398e-04

We experimented with five other approaches to curve fitting related to this canonical form, including use of

$$S(x) = \exp\left[\frac{\alpha}{\beta}\left(1 - e^{\beta x}\right)\right] \qquad \text{for } x \geq 0,$$

but fitted only by data with $x \geq 15$. This approach involved modelling the implivit truncation of the data available. These seemed less satisfactory theoretically in terms of making full use of information without involving additional assumptions, but gave broadly comparable results. See Figure 3 for a maximum likelihood estimation, plotted with each sample point, age being taken as the centre of its age interval.

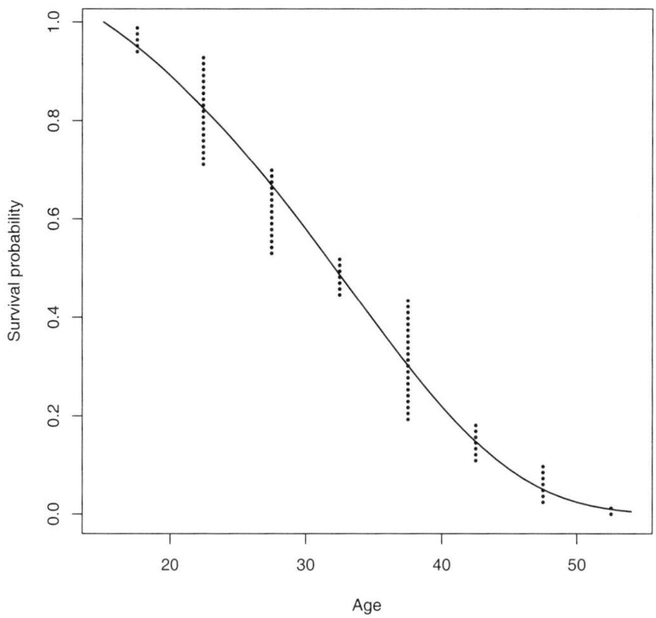

Figure 3. Observed mortality data and estimated survival curve

Suppose we choose $U = 50$, which is consonant with human biology and Table 2. By (1)

$$C_p = 0.488 \times \int_{15}^{50} \exp(-rx) S_{15}(x - 15) f(x)\, dx. \tag{2}$$

We recall that the term C_p may be interpreted as the reciprocal of the probability at birth that an individual will live to age 15 or more. The integral can be evaluated numerically to high precision, obviating the errors of numerical analysis inherent in a discrete interval approach such as is used in life tables. Equation (2) was used to obtain the population rate of change r for a variety of values C_p. These were chosen to correspond to various levels of infant mortality.

As noted, Fisher information theory applied to S_{15} supplied a joint approximately bivariate normal distribution for the parameter values α, β. Similarly we obtained a joint approximately trivariate normal distribution for the values of the parameters a, b, c involved in f. For each specific choice of p, Monte Carlo methods were used. A sequence of joint values was generated for the parameters α, β, a, b, c by sampling from the two underlying joint distributions. For each set in the sequence, (2) was solved for r. Thus a suitably large set of values of r was generated from which a point estimate and 95% confidence interval was derived. This was done in turn for each of the values p of interest. To simulate the data for the confidence intervals for r we employed the mvtnorm package in R[8].

5. Results

Table 6. Estimates for the population growth rate r

C_p	infant mortality	point estimate of r	confidence interval	estimate of Brewis *et al.*
172/141	0.035	-0.00915	(-0.0355, 0.0087)	-0.00414
189/141	0.122	-0.01251	(-0.0357, 0.0058)	
193/141	0.140	-0.01325	(-0.0360, 0.0020)	
196/141	0.153	-0.01380	(-0.0374, 0.0020)	-0.0152
201/141	0.174	-0.01469	(-0.0394, 0.0017)	
207/141	0.198	-0.01573	(-0.0408, 0.0009)	
221/141	0.249	-0.01804	(-0.0416, -0.0020)	
236/141	0.297	-0.02035	(-0.0446, -0.0041)	

The results are presented in Table 6. Values for r, with associated 95% confidence intervals, are given for several values of C_p and cover representative infant mortality rates. In common demographic practice, these figures would often be presented as percentages. Thus the leading entry 0.035 would be given as a 3.5% infant mortality and the corresponding value of r as a 0.915% p.a. population decrease. The rates obtained by Brewis *et al.* are appended in a final column for comparison. We remark that while our estimates differ somewhat from theirs, the latter lie within the relevant 95% confidence intervals for r corresponding to the relevant infant mortality level.

For an easily assimilated overview, Figure 4 displays the graph of r as a function of C_p.

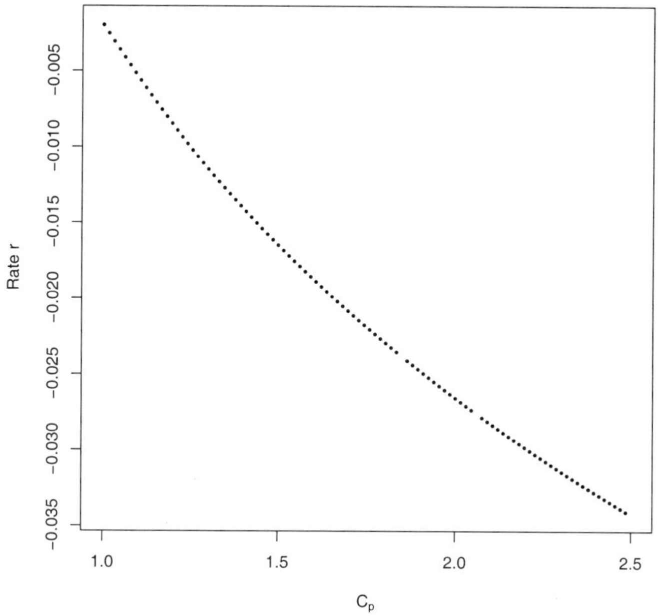

Figure 4. Estimated population growth rate as a function of C_p

At this point we have achieved the primary goal of the Chapter, to derive as sharp an estimate as possible of the prehistoric New Zealand growth rate using only primary skeletal source materials. Because of the common problem of likely infant skeleton under–representation, we have made our estimates in terms of a parameter C_p, the reciprocal of the probability of an individual living to age 15 or more. This provides a flexible range of scenarios for discussion and further analysis.

To reach this point, we have needed to characterize the survival function $S_{15}(x - 15)$, the probability that a woman will reach age $x > 15$ or more, given that she has already reached age 15. Determining this probability does not require knowledge of the level of infant mortality. That is, the answer is not scenario dependent. The probability at birth that a woman will reach age $x > 15$ or more is scenario dependent and is given by $pS_{15}(x - 15)$.

The existing data permits a parallel analysis for the derivation, for $x > 15$, of a function $T_{15}(x - 15)$ giving the probability that a man reaches age x or more, given he has reached age 15. Put $y = x - 15$. We seek a representation

$$T_{15}(y) = \exp\left[\frac{\alpha^*}{\beta^*}\left(1 - e^{\beta^* y}\right)\right] \qquad \text{for } y \geq 0.$$

Tables 7 and 8 summarize our results.

Table 7. Parameterization of male survival curve T_{15}

	estimate	std error
α^*	0.02701351	0.007267688
β^*	0.06882682	0.014693966

Table 8. Variances & covariances for α^*, β^*

	α^*	β^*
α^*	5.281929e-05	-9.054212e-05
β^*	-9.054212e-05	2.159126e-04

6. Discussion and conclusions

At the outset we note that we have deemed all pregnancies/births associated with pitting to be births (possibly stillbirths). Any resultant error will, of course, give an understatement of the extent of the decline rate of the population.

Although we have weakened the assumptions of the models of Phillipps and Brewis *et al.*, our work does, of course, still build on a model. Most notably, we assume that fertility and survival parameters were stable over a sufficient period for a stable population regime to become established "without shocks", that is, abrupt changes in demographic parameters. We assume that the survival and fertility functions are uncorrelated. Further, we assume homogeneity in the model across New Zealand. Each of these is a simplification.

The life span of a woman will be positively correlated with her health and, where lifespan was short, as in prehistoric New Zealand, both lifespan and health may be correlated with the number of children produced. Thus the survival and fertility functions may be correlated. The possible extent of such a correlation is unclear. In an overview paper, Bongaarts[2] states that while famine and starvation reduce fertility significantly, moderate chronic malnutrition results in only a very small decrease on fertility. See also the review of Spielmann[27].

On the other hand, while Phillipps states that for the skeletons available "the maximum number of pregnancies/births was only five", Māori tradition gives examples of rather larger numbers of children for some high–ranking and therefore better nourished women from pa areas in Little Ice Age New Zealand. Thus Nuakaiwhakahua of Ngāti Ruanui is recorded as having nine children (Sole[26]).

Demographic parameters can be expected to vary somewhat in both space and time. Visser[33] has plausibly argued that there is evidence of regional variations of fertility. A relevant indicator is the Māori *pa*, a type of earthwork fortification the presence of thousands of which characterized the best resourced, densely populated areas of New Zealand. Given that most of the skeletons on which this and earlier analyses are based have provenance outside the pa areas, it is possible that the force of population decline was mitigated in much of the country. These issues are taken up in a related work (Pearce and Pearce[19]).

Qualitative evidence has been adduced for variation over time, in particular in relation to the Mediæval Climatic Optimum and the Little Ice Age (see Sutton[28] and Visser[33]). And transitions between different climatic regimes can be expected to produce "shocks". Our analysis can be viewed as describing an "average" behaviour. A careful validation is beyond the scope of this chapter. A key property is that of *weak ergodicity*. In our present context, this refers to the phenomenon that (under suitable assumptions) two populations that are arbitrarily different in their age structure tend with time to adopt the same (time–dependent) age structure distribution if they have the same time–dependent regimes of mortality and fertility. See Tuljapurkar[31], Le Bras[13] and Zeifman[35]. In broad terms, we can expect to be able to weaken the assumption of a stable population in analysis provided shocks are not too large or frequent relative to the rate of convergence occurring with weak ergodicity.

The removal of assumptions and approximations in the palæodemographic analysis and the determining of standard errors puts the estimates for prehistoric growth rates on a new and sounder footing. While there may have been periods of population increase, there would seem to have been an overall pronounced decline of population in New Zealand over several centuries during prehistoric times. The consistent rapid population growth required for the Anderson late–arrival scenario now appears very unlikely. A rethinking of the possibility of a rather longer span for New Zealand prehistory would seem appropriate. This question is addressed at length from several perspectives by Pearce and Pearce[19].

References

1. A. Anderson, The chronology of colonization in New Zealand, *Antiquity* **65**, 767–791 (1991).
2. J. Bongaarts, Does malnutrition affect fecundity? A summary of evidence, *Science (New Series)* **208, no. 4444** (1980)
3. S. Brewis, Assessing infant mortality in prehistoric New Zealand: a life table approach, *New Zealand J. Archæology* **10** (1988), 73–82.
4. A. Brewis, Reconstructions of prehistoric fertility: the Maori case, *Man and Culture in Oceania* **5** (1989), 21–36.
5. A. A. Brewis, M. A. Molloy and D. G. Sutton, Modeling the prehistoric Maori population, *Amer. J. Phys. Anthrop.* **81**, 343–356 (1990).
6. J. Buikstra and L. Konigsberg, Paleodemography: critiques and controversies, *Amer. Anthrop.* **87**, 316–333 (1985).
7. J. E. Buikstra, L. W. Konigsberg and J. Bullington, Fertility and the development of agriculture in the prehistoric Midwest, *Amer. Antiquity* **51**, 528–546 (1986).
8. A. Genz, F. Bretz, T. Miwa, X. Mi, F. Leisch, F. Scheipl and T. Hothorn, *mvtnorm: Multivariate Normal and t distributions* manual Genz(2009), R-package version 0.9–5 (2009). http://CRAN.R-project.org/package=mvtnorm.
9. R. D. Hoppa and J. W. Vaupel (eds), *Paleodemography: age distributions from skeletal samples*, Cambridge University Press, Cambridge (2002).
10. P. Houghton, The bony imprint of pregnancy, *Bull. N. Y. Acad. Med.* **51**, 655–661 (1975).
11. P. Houghton, *The First New Zealanders*, Hodder and Stoughton, Auckland (1980).
12. L. W. Konigsberg and N.P Herrman, Markov chain Monte carlo estimation of hazard model parameters in paleodemography, in R. D. Hoppa and J. W. Vaupel (eds), *Paleodemography: age distributions from skeletal samples*, Cambridge University Press, Cambridge (2002), 222–242.
13. H. Le Bras, *The Nature of Demography*, Princeton University Press, Princeton N.J. (2008).
14. G. Lewthwaite, Rethinking Aotearoa's human geography, *Soc. Sci. J.* **36** (4) (1999), 641–658.

15. A. J. Lotka, Relation between birth rates and death rates (1922), in *Mathematical Demography: Selected Papers*, Eds D. Smith and N. Keyfitz, Springer–Verlag, Berlin (1977).
16. T. A. Louis, Finding the observed information matrix when using the EM algorithm, *J. Roy. Statist. Soc., Ser. B* **44(2)** 226–233 (1982).
17. J. Moore, A. Swedlund and G. Armelagos, The use of life tables in paleodemography, *in* A. Swedlund (Ed.), Population Studies in Archaeology and Biological Anthropology, a Symposium, *Amer. Antiquity* **40 (2, part 2)**, Memoir **30** (1975).
18. K. J. Navara, Humans at tropical lattitudes produce more females, *Biol. Lett.* **5** (2009), 524–527.
19. C. E. M. Pearce and F. M. Pearce, *Oceanic Migration: Oceanography and Global Climate Change as Keys to the Paths, Sequence, Timing and Range of Prehistoric Migration in the Pacific and Indian Oceans*, Springer Verlag (to appear).
20. W. Peterson, A demographer's view of prehistoric demography, *Current Anthrop.* **16**, 227–246 (1975).
21. M. A. L. Phillipps, An estimation of fertility in prehistoric New Zealanders, *N. Z. J. Archaeol.* **2**, 149–167 (1980).
22. D. I. Pool, *The Maori Population of New Zealand, 1769–1971*, Auckland University Press (1977).
23. R Development core team, *R: a Language and Environment for Statistical Computing*, *R* Foundation for Statistical Computing, Vienna (2008). http://www.R-project.org
24. L. Sattenspiel and H. Harpending, Stable populations and skeletal age, *Amer. Antiquity* **48**, 489–498 (1983).
25. A. I. F. Simpson, *An Assessment of Health in the Prehistoric Inhabitants of New Zealand and the Chatham Islands*, B. Med. Sci. Thesis, Otago Medical School (1979).
26. T. Sole, *Ngāti Ruanui: a History*, Huia, Wellington, New Zealand (2005).
27. K. L. Spielmann, A review: dietary retrictions on hunter–gatherer women and the implications for fertility and infant mortality, *Human Ecology* **17(3)** (1989), 321–345.
28. D. G. Sutton, The prehistoric People of Palliser Bay, *Bull. Nat. Mus. N. Z.* **21**, 185–203 (1979).
29. D. G. Sutton, Maori demographic change 1769–1890: The inner workings of "a picturesque but illogical simile," *J. Poly. Soc.* **95**, 291–339 (1986).
30. D. G. Sutton and M. A. Molloy, Deconstructing Pacific palaeodemography: a critique of density dependent causality, *Archaeol. Oceania* **24**, 31–36 (1989).
31. S. D. Tuljapurkar, Population dynamics in weak environments. IV. Weak ergodicity in the Lotka equation, *J. Math. Bio.* **14(2)** (1982), 221–230.
32. D. U. Urlich, *The Distribution and Migrations of the North Island Maori Population about 1800–1840*, M.A. Thesis, Auckland University, Appendix II (1969).
33. E. P. Visser, *Fertility among Prehistoric New Zealanders: An Analysis of General and Regional Patterns*, MA Thesis, University of Auckland (1986).

34. K. Weiss, Demographic models for anthropology, *Amer. Antiquity* ***38(2), Pt 2***, *Memoir* **27** (1973).
35. A. I. Zeifman, On the weak ergodicity of nonhomogeneous continuous–time Markov chains, *J. Math. Sci.* **93(4)** (1999), 612–615.

REGULATION BY OPTIMAL TAXATION OF AN OPEN ACCESS SINGLE-SPECIES FISHERY CONSIDERING ALLEE EFFECT ON RENEWABLE RESOURCE*

A. ROJAS-PALMA

Instituto de Matemáticas, Universidad Austral de Chile
E-mail: alejandro.rojas.p@mail.ucv.cl

E. GONZÁLEZ-OLIVARES

Grupo de Ecología Matemática, Instituto de Matemáticas,
Pontificia Universidad Católica de Valparaíso, Chile
E-mail: ejgonzal@ucv.cl

In this work, a bioeconomic model of an open access single-species fishery is analyzed, using a catch-rate function suggested by W. C. Clark and considering Allee effect in the exploited resource. The harvesting effort is considered to be a dynamic variable (a function of time) and also it is assumed that the exploitation of the fishery is regulated by an agency by imposing a tax per unit of landed biomass. The main objectivesare to establish the maximization of the monetary social benefit as well as to prevent the extinction of the resource. i.e., an optimal control problems is obtained, which is solved by means of the Pontryagin's Maximum Pinciple.

1. Introduction

If harvesting by individuals of a region is causing severe damage of the ecosystem of a determined region, in particular when a exploited population can become to extinction, then the governing authority of this region should plan a regulating policy which would keep the damage to the ecosystem minimal[1].

To avoid this the regulating authority levies a tax on the catch of the harvesting agency. This acts as a deterrent to the fisher and helps the renewable resource to grow, which can be an incentive to the fisher, when the tax takes the form of a subsidy[15].

*This work was partially supported by DII-PUCV project number124720/2009

In this work, a bioeconomic model of an open access single-species fishery is analyzed using a catch-rate function suggested by Clark[4]. This catch-rate function is more flexible and realistic than the usual catch-per-unit-effort hypothesis which is used in Schaefer's model[4].

The growth of the exploited population is affected by the *Allee effect* [5,18] or *depensation*[16]. The harvesting effort E is considered to be a dynamic variable (a function of time), i.e., $E = E(t)$ and also it is assumed that the exploitation of the fishery is regulated by an agency by imposing a tax τ per unit biomass of landed fish.

The net economic revenue to the fishermen (perceived rent) is the difference between of incomes and the cost[4] and the gross rate at which capital is invested at any time is assumed to be proportional to the "perceived rent" at that time[9].

Moreover, the net economic revenue to the society can be considered as the sum of the net economic revenue to the fishermen (perceived rent) plus the economic revenue to the regulatory agency.

To establish the optimal taxation, firstly the positive steady state of this system is determined and conditions for its existence and stability are obtained. Lately, we find the proper taxation policy which would give the best possible benefit through harvesting to the community while preventing the extinction of the predator, which is studied invoking the Pontryagin's Maximum Principle of Control Theory[6,8]. This form of control is different from the usual optimal harvesting policy trying the maximization of the net economic revenue to the fishermen.

Economists are particularly attracted to taxation because a competitive fisheries can be better maintained under taxation rather than other regulatory methods. However, little attention has been paid to study the dynamics of fishery resources using taxation as a control instrument[2,9,13].

2. Model construction

The populational dynamics of the fishing resource is modeled by the equation

$$\frac{dx}{dt} = F(x) - h(x, E) \tag{1}$$

where $x = x(t)$ represents the populational biomass in the time, $F(x)$ is the natural growth function and $h(x, E)$ represents the harvesting rate in the time.

In this work we consider

$$F(x) = r\left(1 - \frac{x}{K}\right)(x - m)\,x \tag{2}$$

a modification of the logistical equation with the factor $(x - m)$ representing the Allee effect. If $m > 0$, we have the *strong Allee effect*, meanwhile if $m = 0$, it has the *weak Allee effect*[20].

The harvesting function is of the sigmoid form[19] given by

$$h(x, E) = \frac{qx^2 E}{x^2 + a^2} \tag{3}$$

where $E = E(t)$ represents the fishing effort in time and q is the catchability coefficient. This function marks a difference with the function commonly used in the Schaefer hypothesis[4], expressed by

$$h(x, E) = qEx$$

since it implies that the capture is limited although the biomass is increased, whenever the effort and is limited, that is to say, $h \to qE$ when $x \to \infty$, this function expresses effects of saturation with respect to the abundance of the stock. Any function of realistic harvest must exhibit this behavior.

The harvesting agency's aim is to obtain as much revenue as possible through its activity, whereas the community needs the food through harvesting and is also keen on protecting the resource from extinction. Thus, the benefit to the community consists of the revenue through the harvest and the retained resource population.

Then, the problem of optimization of the community's benefit is a conditional optimal control problem in the sense that the revenue is to be maximized subject to the condition that the population is larger than a positive quantity as $t \to \infty$. In order to achieve this goal a regulating agency has to curb arbitrary growth of harvesting. This is done by levying a tax on the catch (which can also be a subsidy).

Tax (or subsidy) makes the harvesting effort a dynamic variable. The net benefit, from the harvesting, to the society is the revenue before the deduction of tax, obtained by the harvesting agency. The controlling agency, like the government, levies a tax τ on the harvesting agency. The purpose of the tax (which may be a subsidy) is to regulate the harvesting effort.

The fishing effort E is governed by the equations

$$E(t) = \alpha Q(t) \quad 0 < \alpha \leq 1 \tag{4}$$

$$\frac{dQ(t)}{dt} = I(t) - \gamma Q(t) \tag{5}$$

where $I(t)$ represents the rate of gross investment, $Q(t)$ represents the amount of capital inverted in the fishery in the time and γ represents the rate of depreciation of the capital, the fishing effort E at any time t is assumed like a proportion of the instantaneous amount of inverted capital, following the idea described in[9].

If $Q(t)$ is measured in standardized units of fishing $[SDF]$, for example, amount of vessels available by the fishery, then, is reasonable to consider that $E(t)$ must be a proportion of $Q(t)$ and the maximum capacity of effort must be equal to the number of vessels available ($\alpha = 1$).

The case in that $\alpha = 0$ indicates the situation in which a fishing effort is not developed, although vessels available exist. This can reflect a situation of fishing over-exploitation, the fishing fleet are due to maintain non-operative because the fishing stock has been exhausted.

The economic revenue of the fisherman (perceived rent) can be interpreted as the difference between the incomes of the harvest, and the cost of realizing the effort

$$G(x, E) = \left(\frac{(p-\tau)qx^2E}{x^2+a^2} - c\right)E$$

where p represents the price by biomass unit, which we will consider constant, c represents the cost of harvest by biomass unit and the price by unit of landed biomass is punished by the tax of the regulatory agency.

If it is assumed that the rate of gross investment is proportional to the perceived rents, is obtained

$$I = \beta G(x, E) \qquad 0 \leq \beta \leq 1 \tag{6}$$

Equation (6) says that the rate of maximum investment at any time agrees with the rents perceived for the case in that $\beta = 1$.

The case $\beta = 0$ can only happen when a situation of perceived rent exists and is negative, that is to say, when it is not feasible to stop investing in fixed actives[4]. If the fishery lost importance and the socials capitals are manageable, the sole owner of the fishery is beneficiary, allowing a lost of continuous investment of fixed actives. in this case $I < 0$ and $\beta > 0$ where the negative values of investment are possible to be interpreted like lost of investment.

Replacing the equation (6) in the equation (5) and deriving the equation (4) the differential equation obtained is

$$\frac{dE}{dt} = \alpha\left(\left(\beta\frac{(p-\tau)qx^2}{x^2+a^2} - c\right) - \gamma\right)E \tag{7}$$

The fisherman and the regulatory agency is two components different from the society. Therefore the income obtained by them are income gained from the society of the fishery. The fishery revenue, also call social revenue, is the sum of the revenue of the fisherman and the revenue obtained by the regulatory agency. Denoting by $\pi(x, E, t)$ to the function of fishery revenue, then

$$\begin{aligned}\pi(x,E,t) &= \left(\tfrac{(p-\tau)qx^2E}{x^2+a^2} - c\right)E + \tau\left(\tfrac{qx^2E}{x^2+a^2}\right) \\ &= \left(\tfrac{pqx^2}{x^2+a^2} - c\right)E\end{aligned} \tag{8}$$

Considering the growth population rate of renewable resource to be described by the more usual equation to Allee effect and variation of fishing effort in time is the difference between the gross investment rate and the amount of capital invested in the fishery at time t affected be a (constant) rate of depreciation of capital, the following model is obtained

$$X_\mu : \begin{cases} \frac{dx}{dt} = r\,(1-\frac{x}{K})(x-m)\,x - \frac{q\,x^2}{x^2\,+\,a}y \\ \frac{dE}{dt} = \left(\alpha\beta\left(\frac{(p-\tau)q\,x^2}{x^2\,+\,a^2} - c\right) - \gamma\right)E \end{cases} \tag{9}$$

where $x = x(t)$ indicates the population size of exploited resource and $E = E(t)$ represents the fishing effort as variable in time. All the parameters are positive, i. e., $\mu = (r, K, m, q, a, \alpha, \beta, p, \tau, c, \gamma) \in \mathbb{R}_+^{11}$ and have different bioeconomic meanings.

r is the intrinsic growth rate of the resource

K is the carryng capacity of the exploited population

m is the Allee threshold (minimum size of viable population)

q is the catchability coefficient

a is the biomass quantity necessary to reach half of q

α is the proportionality coefficient of the instantaneous amount of inverted capital

β is the proportionality coefficient of the rate of gross incomes

p is the price by unit of landed biomass

τ is the tax demanded by the regulatory agency

c is the harvesting cost by biomass unit

γ is the depreciation rate of the capital.

For biological reasons a, $m < K$, but m can be zero (weak Allee effect). System (9) or vector field X_μ corresponds to a Gause-type predator-prey model[7] in which the predator is represented by the action of men through the fishing effort.

The system is defined in the first quadrant, this is

$$\Omega = \{(x, E) \in \mathbb{R}^2 / \; x \geq 0, E \geq 0\} = \mathbb{R}_0^+ \times \mathbb{R}_0^+.$$

The equilibrium points of the system are $O = (0,0)$, $P_m = (m,0)$, $P_K = (K,0)$ and $P_e = (\overline{x}, \overline{E})$, where

$$\begin{aligned} \overline{x} &= \sqrt{\tfrac{a^2(\gamma+\alpha\beta c)}{\alpha\beta((p-\tau)q-c)\ -\gamma}} \\ \overline{E} &= \tfrac{r}{q\overline{x}}\,(1-\tfrac{\overline{x}}{K})(\overline{x}-m)\left(\overline{x}^2+a^2\right) \end{aligned}$$

If the model considers taxes, $\tau > 0$, is natural to assume that this tax must be smaller than the price of the landed, i. e., $0 < \tau < p$, Then, the equilibrium point P_e lies to the interior of the first quadrant whenever the following conditions are satisfied

$$\alpha\beta\left((p-\tau)\,q-c\right)\ -\gamma > 0 \tag{10}$$

$$m \leq \overline{x} \leq K \tag{11}$$

The condition (10) defines a level superior for the tax, given by

$$0 < \tau < p - \tfrac{(\gamma+\alpha\beta c)}{\alpha\beta q} \tag{12}$$

when the resource is on the edge of the extinction, must be under the protective control of a regulatory organism, in this case it is possible to be decided on a high rate of tax but that the price of landed of the resource, provided the fishermen stop realizing an excessive exploitation. This case can be considered by means of

$$\tau > p > 0$$

In order to simplify the calculus, the methodology used in[11,17] is followed, changing the variables and making a time rescaling given by the function

$$\varphi : \bar{\Omega} \times \mathbb{R} \longrightarrow \Omega \times \mathbb{R}$$

such that

$$\varphi(u,v,s) = (Ku, \tfrac{rK^2}{q}v, \tfrac{u^2\,+\,\frac{a^2}{K^2}}{rK}s) = (x,E,t)$$

and we have that

$$\det D\varphi(u,v,\tau)\ = \tfrac{(Ku)^2+a^2}{q} > 0$$

then, φ is a diffeomorphism[3], for which the vector field X_μ in the new coordinate system, is topologically equivalent to the vector field $Y_\eta = \varphi \circ$

X_μ, which takes the form $Y_\eta = P(u,v)\frac{\partial}{\partial u} + Q(u,v)\frac{\partial}{\partial v}$ and the associated polynomial differential equations system is given by

$$Y_\eta : \begin{cases} \frac{du}{ds} = \left((1-u)(u-M)\left(u^2 + A\right) - uv\right)u \\ \frac{dv}{ds} = B\left(u^2 - C\right)v \end{cases} \tag{13}$$

with $A = \left(\frac{a}{K}\right)^2 < 1$, $B = \frac{\alpha\beta((p-\tau)q-c)-\gamma}{rK}$, $C = \sqrt{A\frac{(c\alpha\beta+\gamma)}{(\alpha\beta((p-\tau)q-c)-\gamma)}}$ and $M = \frac{m}{K} < 1$, where $\eta = (A,B,C,M) \in \Delta =]0,1[\times\mathbb{R}\times\mathbb{R}_+\times[0,1[$ and the system (13) is defined in

$$\bar{\Omega} = \{(u,v) \in \mathbb{R}^2 /\ u \geq 0,\ v \geq 0\}$$

the equilibrium points of the system (13) are $O = (0,0)$, $Q_M = (M,0)$, $Q_1 = (1,0)$ and $Q_e = (C, \frac{1}{C}(1-C)(C-M)(\ A + C^2))$. Clearly, Q_e is at interior of the first quadrant if and only if, $M < C < 1$. The jacobian matrix of the system is

$$D(u,v) = \begin{pmatrix} a_{11} & -u^2 \\ 2Buv & B\left(u^2 - C^2\right) \end{pmatrix}$$

with

$$a_{11} = -5u^4 + (4+4M)u^3 + (-3M-3A)u^2 + (2A+2MA-2v)u - MA$$

3. Main Results

System (13) was already studied in Ref. [17], the main results on local stability and qualitative behavior of the system are given next, omitting their proofs. In order to see details with respect to the proofs we can make reference to the mentioned work.

3.1. *Strong Allee effect*

For system (13), with $m > 0$ we have the following results:

Lemma 3.1. *a) The set* $\tilde{\Gamma} = \left\{(u,v) \in \mathbb{R}^2/0 \leq u \leq 1, v \geq 0\right\}$ *is an invariant region.*

b) The solutions are bounded.

c) The equilibrium point $Q_1 = (1,0)$ *is*

c.1) a hyperbolic attractor, if and only if, $C > 1$ *and* (C, L) *lies at the four quadrant*

c.2) a hyperbolic saddle point, if and only if, $C < 1$ *and* (C, L) *lies at the first quadrant c.3) a non-hyperbolic attractor, if and only if,* $C = 1$.

d) The point $Q_M = (M, 0)$ *is*

d.1) a hyperbolic saddle point, if and only if, $C > M$

d.2) a hyperbolic repellor, if and only if, $C < M$

d.3) a non-hyperbolic repellor, if and only if, $C = M$.

e) The equilibrium point $O = (0,0)$ *is an hyperbolic attractor for any set of parameter values.*

Lemma 3.2. *Assuming* $0 < M \leq C \leq 1$, *let* $W^s(M,0)$ *and* $W^u(1,0)$ *the stable and unstable manifold of the equilibrium point* $(M,0)$ *and* $(1,0)$ *respectively, then there exists a heteroclinic curve joining those points, that is, there exists a subset of parameter values for which* $W^s(M,0) = W^u(1,0)$.

Theorem 3.1. *Let* $(u^*, v^s) \in W^s(M,0)$ *and* $(u^*, v^u) \in W^u(1,0)$ *and assuming that* $M < C < 1$, *the equilibrium point* (C, L) *is at the interior of the first quadrant*

1) Supposing that $v^s > v^u$ *the singularity* (C, L) *is*

a) a hyperbolic local attractor, if and only if,

$$-3C^4 + 2(1+M)C^3 - (M+A)C^2 + AM < 0,$$

and

a1) If

$$B < \frac{\left(-3C^4 + 2(1+M)C^3 - (M+A)C^2 + AM\right)^2}{8AC^2(1-C)(C-M)},$$

is an attractor node.

a2) If

$$B > \frac{\left(-3C^4 + 2(1+M)C^3 - (M+A)C^2 + AM\right)^2}{8AC^2(1-C)(C-M)},$$

is an attractor focus.

b) a hyperbolic repellor, if and only if,

$$-3C^4 + 2(1+M)C^3 - (M+A)C^2 + AM > 0$$

and

b1) If

$$B > \frac{\left(-3C^4 + 2(1+M)C^3 - (M+A)C^2 + AM\right)^2}{8AC^2(1-C)(C-M)},$$

is a repellor focus surrounded by a limit cycle.

b2) If

$$B < \frac{\left(-3C^4 + 2(1+M)C^3 - (M+A)C^2 + AM\right)^2}{8AC^2(1-C)(C-M)},$$

is a repellor node.

2) Supposing that $v^s < v^u$ *then* (C, L) *is repellor and the equilibrium point* $(0,0)$ *is the* $\omega -$ *limit of all trajectories of system.*

Theorem 3.2. *Considering that* $v^s > v^u$. *The positive equilibrium point* (C, L) *is a two order weak focus, if and only if,*

$$-3C^4 + 2(1+M)C^3 - (M+A)C^2 + AM = 0.$$

In Figure 1 it is shown the two limit cycles.

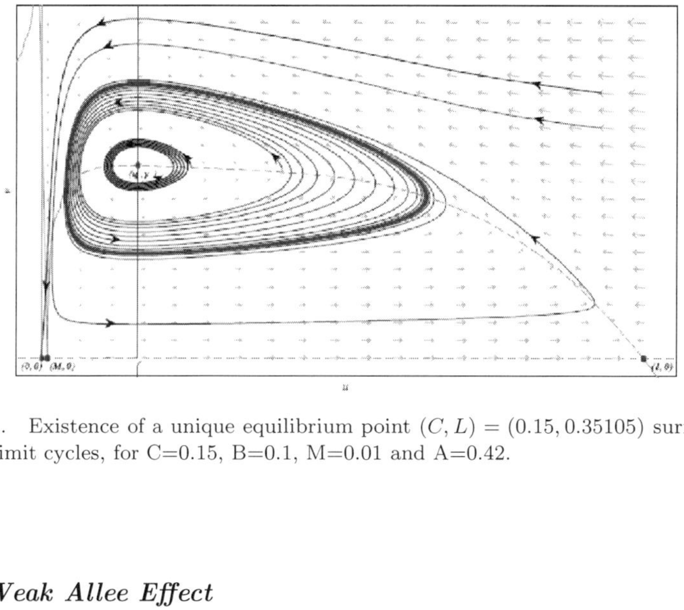

Figure 1. Existence of a unique equilibrium point $(C, L) = (0.15, 0.35105)$ surrounded by two limit cycles, for C=0.15, B=0.1, M=0.01 and A=0.42.

3.2. *Weak Allee Effect*

For the case when $M = 0$ we obtain similar results for the following system

$$Y_\lambda : \begin{cases} \frac{du}{d\tau} = \left((1-u)(\ A + u^2) - v\right)\ u^2 \\ \frac{dv}{d\tau} = B(u^2 - C^2)v \end{cases} \tag{14}$$

where $\lambda = (A, B, C) \in \Delta_1 = \,]0,1[\, \times \mathbb{R}_+ \times \mathbb{R}_+$. System (14) or vector field Y_λ has three equilibrium points $(0,0)$, $(1,0)$ and (C, L') with $L' = (1-C)(\ A + C^2)$.

Theorem 3.3. *a) The origin is non-hyperbolic attractor point and there exist a separatrix curve* Σ *that divides the behavior of trajectories. For any trajectory passing by a point upper the curve* Σ, *the equilibrium* $(0,0)$ *it is their* $\omega - limit$.

b) The equilibrium point $(1\,,0)$ *is*

b1) a hyperbolic attractor, if and only if, $C > 1$ *and* (C, L') *lies at the four quadrant*

b2) a hyperbolic saddle point, if and only if, $C < 1$ *and* (C, L') *lies at the first quadrant*

b3) a non-hyperbolic attractor, if and only if, $C = 1$.

c) Let $W^u\,(1,0)$ *the unstable manifold of the equilibrium point* $(1,0)$, *then there exists a heteroclinic curve joining the points* $(0,0)$ *and* $(1,0)$, *that is, there exists a subset of parameter values for which*

$$\Sigma = W^u\,(1,0)\,.$$

Theorem 3.4. *Let* $(u^*, v^s) \in \Sigma$ *and* $(u^*, v^s) \in W^u\,(1,0)$

1) Supposing that $v^s > v^u$, *then the equilibrium point* (C, L') *is*

1.1) a hyperbolic local attractor, if and only if,

$$A > 2C - 3C^2.$$

Moreover,

a) If

$$B < \frac{C\left(A - 2C + 3C^2\right)^2}{8A(1-C)},$$

is an attractor node.

b) If

$$B > \frac{C\left(A - 2C + 3C^2\right)^2}{8A(1-C)},$$

is an attractor focus.

1.2) a hyperbolic repellor if and only if,

$$A < 2C - 3C^2.$$

Moreover

c) If

$$B < \frac{C\left(A - 2C + 3C^2\right)^2}{8A(1-C)},$$

is a repellor node.

d) If

$$B > \frac{C\left(A - 2C + 3C^2\right)^2}{8A(1-C)},$$

is a repellor focus surrounded by a limit cycle.

2) Supposing $v^s < v^u$, then the equilibrium point (C, L') is repellor and the equilibrium $(0,0)$ is globally asymptotically stable.

Theorem 3.5. *The singularity (C, L') is a two order weak focus, if and only if,*

$$A = (2 - 3C)\,C.$$

The diagram of bifurcations for Weak Allee effect is shown in Figure 2.

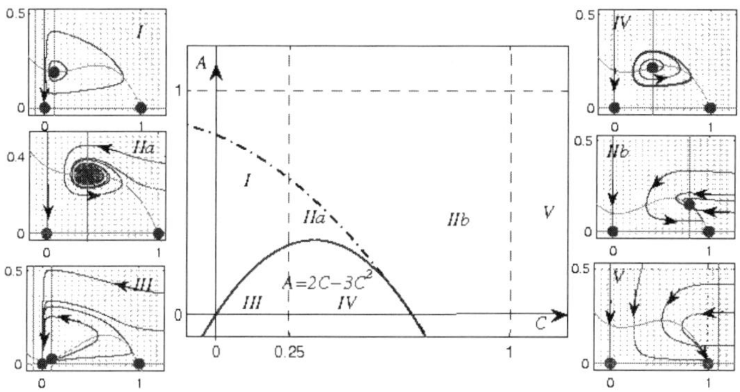

Figure 2. Bifurcation diagram for the *weak Allee effect.* For the vector field Y_λ we have five subsets in the AC – *plane* of the parameter space, in which the systems are not topologically equivalent. The parameter B determines mainly whether the positive equilibrium point is node or focus.

4. Optimal taxation policy

The optimal taxation policy introduces the maximization of a fishery in a time interval, so stable equilibrium conditions are avoided[9]. However, profit is not the concept to maximize but the *net present value* (NPV) of the fishery[4]. The formulation proposed by Clark[4] includes the discount rate δ into the economic balance through time. This parameter is used to discount net benefits that will accrue in the future compared with net benefits that can be achieved today[4]. The first approach is to maximize the net present value into a non limited time interval which is described by the equation:

$$J = \int_0^\infty e^{-\delta t} \left(\frac{pqx^2}{x^2+a^2} - c \right) E dt$$

Then, the objective of the regulatory agency is to find the optimal taxation policy which gives maximum benefit to the society[2,10,13], i.e., a taxation policy which gives the maximum benefit the tax $\tau = \tau(t)$ as the control variable to maximize J subject to the state equations (1) and the control constraints

$$\tau_{\min} < \tau < \tau_{\max}$$

As the social benefit includes preservation of population of the renewable resource we should maximize J such that $x(t)$ approaches a nonzero value as $t \to \infty$. Then, the problem of optimal control is:

Maximize:

$$J(x, E, t) = \int_0^\infty e^{-\delta t} \left(\frac{pq\, x^2}{x^2+a^2} - c \right) E dt$$

subject to:

$$X_\mu : \begin{cases} \frac{dx}{dt} = r\,(1 - \frac{x}{K})(x - m)\, x - \frac{qx^2 E}{x^2+a^2} \\ \frac{dE}{dt} = \left(\alpha\beta \left(\frac{(p-\tau)qx^2}{x^2+a^2} - c \right) - \gamma \right) E \end{cases}$$

$$\tau_{\min} < \tau(t) < \tau_{\max}$$

which can be solved applying the Pontryagin's Maximum principle.

Clearly, this is a linear control problem on infinite horizon. Hence, the solution will be a combination of bang–bang and singular controls. First we study the singular solution for the optimization problem. The Hamiltonian for the above control problem is given by

$$\begin{aligned} H(\lambda_1, \lambda_2, x, E, \tau, t) = {} & \left(\frac{pq\, x^2}{x^2+a^2} - c \right) E e^{-\delta t} \\ & + \lambda_1 \left(r\,(1 - \frac{x}{K})(x - m)\, x - \frac{q\, x^2 E}{x^2+a^2} \right) \\ & + \lambda_2 \left(\alpha\beta \left(\frac{(p-\tau)q\, x^2}{x^2+a^2} - c \right) - \gamma \right) E \end{aligned} \quad (15)$$

where $\lambda_1 = \lambda_1(t)$ and $\lambda_2 = \lambda_2(t)$ are costate functions or adjoint variables[10].

The Hamiltonian (15) must be maximized for $\tau \in [\tau_{\min}, \tau_{\max}]$, from the first order condition for maximum it is obtained

$$\frac{\partial H}{\partial \tau} = 0$$
$$-\frac{\lambda_2 \alpha\beta q\ x^2 E}{x^2+a^2} = 0$$

where $E(t) > 0$, it follows that $\lambda_2(t) = 0$. This gives the condition for the singular control.

The adjoint equation $\lambda_2(t)$ is

$$\frac{d\lambda_2(t)}{dt} = -\frac{\partial H}{\partial E}$$
$$-\left(\frac{pq\ x^2}{x^2+a^2} - c\right) e^{-\delta t} - \lambda_1 \frac{q\ x^2}{x^2\ +\ a^2} = 0$$

the equation is reduced to

$$\lambda_1(t) = \left(p - \frac{c\left(x^2+a^2\right)}{q\ x^2}\right) e^{-\delta t} \tag{16}$$

and this indicates that the shadow price throughout the singular path is given by

$$\lambda_1(t)\, e^{\delta t} = p - \frac{c\left(x^2+a^2\right)}{q\ x^2} \tag{17}$$

The adjoint equation for $\lambda_1(t)$ is

$$\frac{d\lambda_1(t)}{dt} = -\frac{\partial H}{\partial x}$$

where

$$\begin{aligned}\frac{\partial H}{\partial x} &= \frac{2a^2pqx}{(a^2+x^2)^2} E e^{-\delta t} \\ &+\lambda_1\left(-\frac{r\left(3x^2-2(K+m)x+Km\right)}{K} - \frac{2a^2qxE}{(a^2+x^2)^2}\right) \\ &+\lambda_2 \frac{\alpha\beta(p-\tau)2a^2qEx}{(a^2+x^2)^2}\end{aligned}$$

with the help of the system (9), the equation (15) and some algebraic calculations, the last equation can be written in the following form

$$\begin{aligned}&\left(p - \frac{c}{q}\right) + \frac{2ca^2}{qx^2} r\ (1 - \frac{x}{K})(x-m)\ - \delta\left(p - \frac{c\left(x^2+a^2\right)}{q\ x^2}\right) \\ = &\left(p - \frac{c\left(x^2+a^2\right)}{q\ x^2}\right) \frac{r\left(3x^2-2(K+m)x+Km\right)}{K}\end{aligned}$$

this equation that also can be written as

$$\frac{(pq-c)\ x^2+2ca^2r\ (1-\frac{x}{K})(x-m)}{(pq-c)x^2\ -ca^2}-\frac{r\left(3x^2-2(K+m)x+Km\right)}{K}=\delta \tag{18}$$

eq. (18) is called the *fundamental equation*[4] and it can be described by

$$\frac{3r(c-pq)x^4-2r(c-pq)(K+m)x^3+\left(cra^2+K(\delta+mr)(c-pq)\right)x^2+Ka^2c(\delta-mr)}{Kqx^2}=0 \tag{19}$$

then, to assure, at least, the existence a positive root of the equation (18) the following conditions must satisfy simultaneously

$$\frac{c}{p}\left(1+\frac{ra^2}{K(\delta+mr)}\right)<q \tag{20}$$

$$m<\frac{\delta}{r} \tag{21}$$

If we call $x=x^*$ to the positive solution of the equation (18), then, the optimal equilibrium values will be

$$x=x^*$$

$$E=E^*=\frac{r}{qx^*}\ (1-\frac{x^*}{K})(x^*-m)\left((x^*)^2+a^2\right)$$

$$\tau=\tau^*=p-\frac{(\gamma+\alpha\beta c)\left((x^*)^2+a^2\right)}{\alpha\beta q(x^*)^2}$$

Obtained the singular solution, it is important to define a tax policy that allows the trajectories of the system to come near at the level of optimal equilibrium population as fast as it is possible, maintaining as well the levels of optimal effort and tax. We remark that for such an optimal policy to be practicable, it is necessary that the optimal equilibrium point be at least locally asymptotically stable[15].

We consider the optimal equilibrium point (x^*, E^*) be such that there exists an admissible tax policy $\tau(t)$ that it fulfills (9) and such that the equation $(x(t_1), E(t_1))=(x^*, E^*)$ has a strictly positive solution under the policy $\tau(t)$ with $t>t_1$; let T be the set of all such policies and $\tau^*\in T$ the optimal tax policy obtained previously. We now consider the problem of approaching (x^*, E^*) most rapidly, i.e., the most rapid approach[4,12]. Let $\overline{\tau}(t)$ be such a policy and t_f be the shortest time in which the optimal tax is obtained. The optimal tax policy will be

$$\tau(t)=\begin{cases}\overline{\tau}(t)\ \text{if}\ 0<t<t_f\\ \tau^*\ \ \text{if}\quad t\geq t_f\end{cases} \tag{22}$$

We can consider the period $[0, t_f]$ like a period of planning, in this case we will consider it as the most short period to approach to the optimal population level.

Let (x_0, E_0) a initial condition contained in the region of attraction of the interior equilibrium point (x^*, E^*) obtained previously. We can propose the following optimal time control problem:

Minimize:

$$\int_0^{t_f} dt = t_f(\tau)$$

$$\tau \in [\tau_{\min}, \tau_{\max}]$$

subject to :

$$X_\mu : \begin{cases} \frac{dx}{dt} = r\,(1 - \frac{x}{K})(x - m)\,x - \frac{q\,x^2 E}{x^2\,+\,a^2} \\ \frac{dE}{dt} = \left(\alpha\beta\left(\frac{(p-\tau)qx^2}{x^2+a^2} - c\right) - \gamma\right) E \end{cases}$$

$$(x(0), E(0)) = (x_0, E_0)$$
$$(x(t_f), E(t_f)) = (x^*, E^*)$$

$$0 \leq t \leq t_f$$

This time we would be looking for non-singular control of bang–bang type. Using Pontryagin's Maximum principle[6,12], the Hamiltonian for this problem is

$$H(\mu_1, \mu_2, x, E, \tau, t) = 1 + \mu_1\left(r\,(1 - \tfrac{x}{K})(x - m)\,x - \tfrac{q\,x^2 E}{x^2\,+\,a^2}\right) + \mu_2\left(\alpha\beta\left(\tfrac{(p-\tau)qx^2}{x^2+a^2} - c\right) - \gamma\right) E$$

The adjoint equations are

$$\frac{d\mu_1}{dt} = -\frac{\partial H}{\partial x}$$
$$\frac{d\mu_2}{dt} = -\frac{\partial H}{\partial E}$$

they allow to define the system of equations

$$\begin{cases} \frac{d\mu_1}{dt} = \mu_1\left(\frac{r\left(3x^2-2(K+m)x+Km\right)}{K} + \frac{2a^2qxE}{(a^2+x^2)^2}\right) - \mu_2\left(\frac{2\alpha\beta a^2(p-\tau)qxE}{(a^2+x^2)^2}\right) \\ \\ \frac{d\mu_2}{dt} = \mu_1\left(\frac{qx^2}{x^2+a^2}\right) - \mu_2\left(\alpha\beta\left(\frac{(p-\tau)qx^2}{x^2+a^2} - c\right) - \gamma\right) \end{cases} \tag{23}$$

Now, by following the methodology of Ref. [14], we derive the conditions on $\mu_1 = \mu_1(t)$ and $\mu_2 = \mu_2(t)$; they are given by the equations defined in the Pontryagin's minimum principle[6,12]

$$\max_{\tau\in[\tau_{\min},\tau_{\max}]} H(\mu_1, \mu_2, x^*, E^*, \tau, t) = H(\mu_1, \mu_2, x^*, E^*, \tau^*, t) \tag{24}$$

and

$$H(\mu_1, \mu_2, x^*, E^*, \tau^*, t) = 0 \tag{25}$$

The equation (23) implies that

$$H(\mu_1, \mu_2, x^*, E^*, \tau^*, t) \leq H(\mu_1, \mu_2, x^*, E^*, \tau, t)$$

therefore

$$\mu_2\left(\frac{-q\alpha\beta(x^*)^2E^*}{a^2+(x^*)^2}\right)\tau^* \leq \mu_2\left(\frac{-q\alpha\beta(x^*)^2E^*}{a^2+(x^*)^2}\right)\tau \tag{26}$$

now, the equation (25) indicates that our problem is

$$\min_{\tau\in[\tau_{\min},\tau_{\max}]} \mu_2\left(-\frac{q\alpha\beta(x^*)^2E^*}{a^2+(x^*)^2}\right)\tau \tag{27}$$

and it is obtained

$$-\frac{q\alpha\beta(x^*)^2E^*}{a^2+(x^*)^2} < 0$$

thus, the minimum values will be reached in the extremes of the interval.

Then, the control policy $\overline{\tau}$ is determined by

$$\overline{\tau}(t) = \begin{cases} \tau_{\max} \text{ if } \mu_2(t) > 0, & 0 \leq t \leq t_f \\ \tau_{\min} \text{ if } \mu_2(t) < 0, & 0 \leq t \leq t_f \end{cases}$$

Finally, the optimal taxation policy, as combination of the singular and non-singular controls, determined by

$$\tau(t) = \begin{cases} \tau_{\max} \text{ if } \mu_2(t) > 0, & 0 \leq t \leq t_f \\ \tau_{\min} \text{ if } \mu_2(t) < 0, & 0 \leq t \leq t_f \\ \tau^* \text{ if } & t > t_f \end{cases}$$

5. Discussion

The economic goal of this work is to maximize the monetary benefit to the community and ecologically we wish to keep the population resource from extinction. The regulation instrument used to drive the system towards such a state is tax. The existence of Allee effect or depensation mathematically causes a region in which all initial condition will tend to the origin of the system, this implies that as much the fishing resource in the long term as the fishing effort will tend to fall. In order to avoid this, one proposes a condition for the existence of a point of stable equilibrium, which will be the objective of the optimization problem, and an optimal regulation policy.

Bio-economically, we have looked for an optimal tax policy and an interior equilibrium corresponding to this tax policy. Next, we drove this system to this interior equilibrium in the shortest possible time and most beneficially. The optimal tax policy is a combination of bang–bang and singular control policies. Because we reach the interior equilibrium in the shortest time possible, the number of switches in the bang–bang control are decreased and thus we have a smooth tax policy. This and the optimality combine to make the policy obtained, the most efficient tax policy. Mathematically, a multigoal optimization problem is reduced to a sequence of optimization problems. One on the infinite horizon and one on a finite horizon which is a time optimal control problem.

The optimal policy are obtainable only if an interior equilibrium exists and is sufficiently rough. Constraints on tax policies have to be imposed to ensure the existence and stability of the interior equilibrium. These mathematical constraints allow sensible bio-economic conclusions and thus are validated. One interesting conclusion is that the existence of interior equilibrium strongly depends on the tax level, but its stability may not depend on tax level if the population of fish resource has sufficiently large recouping quotient. Use of a nonconventional harvest function helps in obtaining such conclusions.

References

1. L. G. Anderson and D. R. Lee. Optimal governing instrument, operation level and enforcement in natural resource regulation: the case of fishery, American Journal of Agricultural Economics, 68, 678-690 (1986).
2. K. S. Chaudhuri, A bioeconomic model of harvesting a multispecies fishery, Ecological Modelling 32, 12 (1986).
3. C. Chicone , *Ordinary differential equations with applications*, (2nd edition), Texts in Applied Mathematics 34, Springer (2006).
4. C. W. Clark, *Mathematical Bioeconomic: The optimal management of renewable resources*, (2nd edition). John Wiley and Sons, (1990).
5. F. Courchamp, L. Berec and J. Gascoigne, *Allee effects in Ecology and Conservation*, Oxford University Press (2008).
6. A. K. Dixit, *Optimization in Economic Theory*, Oxford University Press, (1990).
7. H. I. Freedman. *Deterministic Mathematical Model in Population Ecology.* Marcel Dekker, (1980).

8. R. V. Gamkrelidze, *Principles of Optimal Control Theory*, Mathematical concepts and methods in science and engineering, Plenum Press, New York, (1978).
9. S. Ganguly and K.S. Chaudhuri, Regulation of a single-species fishery by taxation, Ecological Modelling 82, 51-60 (1995).
10. B-S Goh. *Management and Analysis of Biological Populations.* Elsevier Scientific Publishing Company, (1980).
11. E. González-Olivares, J. D. Flores and J. Mena-Lorca, Metastability in an open access fisheries model with multiple Allee effects on the exploited population, submitted to Mathematical Biosciences (2009).
12. D. Grass, J. P. Caulkins, G. Feichtinger, G. Tragler, D. A. Behrens, *Optimal Control of Nonlinear Processes.* Springer-Verlag, Berlin Heidelberg, (2008)
13. T. K. Kar, Conservation of a fishery through optimal taxation: a dynamic reaction model, Communications in Nonlinear Science and Numerical Simulation 10, 121–131 (2005).
14. D. E. Kirk, *Optimal Control Theory. An Introduction.* Dover Publications, Inc. 1970.
15. S. V. Krishna, P. D. N. Srinivasu and B. Kaymacalan, Conservation of an Ecosystem through Optimal Taxation, Bulletin of Mathematical Biology 60, 569–584 (1998).
16. M. Liermann and R. Hilborn, Depensation: evidence, models and implications, Fish and Fisheries 2, 33-58 (2001).
17. A. Rojas-Palma, E. González-Olivares and B. González-Yañez, Metastability in a Gause type predator-prey models with sigmoid functional response and multiplicative Allee effect on prey, In R. Mondaini (Ed.) Proceedings of International Symposium on Mathematical and Computational Biology, E-papers Serviços Editoriais Ltda., 295-321 (2007).
18. P. A. Stephens, and W. J. Sutherland. Consequences of the Allee effect for behaviour, ecology and conservation. Trends in Ecol. Evo., Vol. 14 N° 10 401-405 (1999).
19. J. Sugie, K. Miyamoto and K. Morino. Absence of limits cycle of a Predator-prey system with a sigmoid functional response, Applied Mathematical Letters 9, 85-90 (1996).
20. G. A. K. van Voorn, L. Hemerik, M. P. Boer and B. W. Kooi, Heteroclinic orbits indicate overexploitation in predator–prey systems with a strong Allee effect, Mathematical Biosciences 209, 451–469 (2007).

EFFECT OF MASS MEDIA ON THE CULTURAL DIVERSITY OF AXELROD MODEL OF SOCIAL INFLUENCE

J. F. FONTANARI

Instituto de Física de São Carlos
Universidade de São Paulo
Caixa Postal 369, 13560-970 São Carlos SP, Brazil
E-mail: fontanari@ifsc.usp.br

A surprising feature of Axelrod's model for culture dissemination or social influence is the existence of many multicultural absorbing states, despite the fact that the local rules that specify the agents interactions are explicitly designed to decrease the cultural differences between agents. In particular, Axelrod's model has two control parameters, namely, the number F of culture features that characterize the culture of an agent, and the number q of values that each feature can take on. The agents are placed on the sites of a 2-dimensional lattice of linear size L. For $F > 2$ the model exhibits a discontinuous non-equilibrium phase transition between a homogeneous regime for which all agents share the same culture and a completely disordered regime for which the number of cultures is maximum, q^F. Here we re-examine the problem of introducing a global interaction - the mass media effect - in the interaction rules of Axelrod's model: in addition to their nearest-neighbors, each agent has a certain probability p to interact with a virtual neighbor whose cultural features are the average cultural features of the entire population. Most surprisingly, this apparently homogenizing effect actually increases the cultural diversity of the population. We show that, contrary to previous claims in the literature, even a vanishingly small value of p is sufficient to destabilize the homogeneous regime, so there is no phase transition when a global mass media effect is taken into account in Axelrod's model.

1. Introduction

Why do people have different opinions given that the natural tendency would be the emergence of some common sense after repeated interactions and conversations? Why are there different cultures given that the media has transformed the planet in a global village[a]? These are the issues

[a]The expression 'global village' was coined by Marshall McLuhan[1], who is also responsible for the celebrated phrase 'the medium is the message'.

addressed by Axelrod's model for the dissemination of culture or social influence[2], which can be viewed as the paradigm for idealized models of collective behavior which seek to boil down a collective phenomenon to its functional essence[3]. In fact, building on just a few simple principles, Axelrod's model provides a highly nontrivial answer to those questions. In Axelrod's model, an agent[b] is represented by a string of F cultural features, where each feature can adopt a certain number q of distinct traits. The interaction between any two agents takes place with probability proportional to their cultural similarity, i.e., proportional to the number of traits they have in common. According to our intuitive expectation, the result of such interaction is the increase of the similarity between the two agents, as one of them modifies a previously distinct trait to match that of its partner.

Notwithstanding the built-in assumption that social actors have a tendency to become more similar to each other through local interactions[4,5], Axelrod's model does exhibit global polarization, i.e., a stable multicultural regime[2]. More importantly, however, at least from the statistical physics perspective, is the fact that the competition between the disorder of the initial configuration and the ordering bias of the local interactions produces a nontrivial threshold phenomenon (more precisely, a nonequilibrium phase transition) which separates in the space of parameters of the model the globally homogeneous from the globally polarized regimes[6,7].

A feature that sets Axelrod's model apart from other lattice models which exhibit nonequilibrium phase transitions[10,11], is the fact that all stationary states of the dynamics are absorbing states, i.e., the dynamics freezes in the long time regime[6]. This is so because, according to the rules of Axelrod's model, two neighboring agents who do not have any cultural trait in common cannot interact and agents who share all cultural traits interact but the interaction does not change their cultural features. Hence at equilibrium we can safely predict that, regarding their cultural features, any neighbor of a given agent is either identical to or completely different from it. This is a double-edged sword: on the one hand, we can easily identify the stationary regime, which is a major problem in the characterization of nonequilibrium phase transitions; on the other hand, the dynamics can take an arbitrarily large time to freeze for some parameter settings and initial conditions[6,7,8,9].

It is indeed the rule that prohibits the interaction between completely different agents (i.e., agents which do not have a single cultural trait in

---

[b]The social agents can be thought of as individuals or as culturally homogeneous villages.

common) which is the key ingredient for the existence of a stable globally polarized state. This was first pointed out by Kennedy[12] who relaxed this rule and permitted interactions regardless of the similarity between agents (see Section 2). As a result, the system evolved until all agents became identical, i.e., the only absorbing states were the homogenous ones[c]. In addition, Klemm and colleagues[13] have shown that the introduction of external noise to the dynamics so that a single trait of an arbitrarily chosen agent was changed at random ends up destabilizing the polarized state. Moreover, expansion of communication modeled by increasing the connectivity of the lattice[14,15] or by placing the agents in more complex networks[16] (e.g., small-world and scale-free networks) also results in cultural homogenization.

It should be mentioned, however, that other models of social influence seem to yield a more robust polarized state. For instance, the frequency bias mechanism[17,18] for cultural or opinion change assumes that the number of people holding an opinion is the key factor for an agent to adopt that opinion, i.e., people have a tendency to espouse cultural traits that are more common in its social environment[d]. Parisi and colleagues[20] have replaced the rules of Axelrod's model by the frequency bias mechanism (essentially, a majority rule) and found a stable polarized state for small lattices. Since similarity plays no role in the agents' interactions, the frequency bias mechanism is naturally robust to noise. As we stress in this contribution, however, the difficulty is to find a polarized state for lattices of infinite size. To the best of our knowledge, there is no study of the finite-size effects of the frequency bias mechanism which would validate such a claim for arbitrarily large lattices.

The impression is then that the globally polarized (multicultural) state is very frail, being disrupted by any (realistic or not) extension of the original model. In view of this feeling, it came as a big surprise the finding by Shibanai and colleagues[21] that the introduction of a homogeneous media effect (i.e., it is the same for all agents) aiming at influencing the agents' opinions actually favors polarization. This finding is at odds with the common sense view that mass media, such as newspapers and television, are devices that can be effectively used to control people's opinions and so homogenize society[e].

---

[c]There are q^F distinct absorbing homogenous configurations.
[d]This is then the standard voter model of Statistical Physics[19].
[e]The effect of media in real personal networks is much more complicated and seem to

Although this counterintuitive effect of the mass media has been reproduced (and investigated) by several independent groups[23,24,25,26] there is still no first-principles explanation to it. Those works have focused on the search for a threshold of the intensity of the media effect such that above the threshold the population becomes polarized and below it, the population becomes culturally homogeneous. The main goal of this contribution is to argue that such threshold is in fact an artifact of finite lattices: when a careful finite-size effects analysis is carried out we find that even a vanishingly small media effect is sufficient to destabilize the culturally homogeneous regime.

The rest of this paper is organized as follows. In Section 2 we describe the original Axelrod's model, discuss at some length the basic assumptions of the model and introduce the effect of media according to the original suggestion by Shibanai and colleagues[21]. In Section 3 we present an efficient algorithm to simulate Axelrod's model. The simulation results as well as a discussion of our main results are also in that section. Finally, in Section 4 we present our concluding remarks.

2. Model

In Axelrod's model each agent is characterized by a set of F cultural features which can take on q distinct values. Hence an agent is represented by a string of symbols, e.g. 13211 in the case of $F = 5$ and $q = 3$. Clearly, for this parameter setting there are only $q^F = 243$ different agents. The agents are fixed in the sites of a square lattice with open boundary conditions (i.e., the lattice is surrounded by walls) and can interact only with their nearest neighbors: agents in the corners interact with two neighbors, agents in the sides with three, and agents in the bulk with four nearest neighbors. The initial configuration is completely random with the features of each agent given by random integers drawn uniformly between 1 and q. At each time we pick an agent at random (this is the target agent) as well as one of its neighbors. These two agents interact with probability equal to their cultural similarity, defined as the fraction of common cultural features. For instance, assuming that the target agent is described by the string 13211 and its neighbor by 13331, the interaction occurs with probability 3/5. Explicitly, we first generate a uniformly distributed random number r in

follow the so-called 'two-step flow of communication' in which the media affect opinion leaders first, who then influence the rest of the population[22]. In fact, personal networks seem to serve as a buffer for the media effect.

$[0, 1]$ and then compare it with the similarity between the two agents. In our example, if $r < 3/5$ the agents are allowed to interact, otherwise we pick another target agent at random and repeat the procedure. An interaction consists of selecting at random one of the distinct features, and changing the target agent's trait on this feature to the neighbor's corresponding trait. Returning to our example, if the third feature is chosen the target agent becomes 13311 and its neighbor remains unchanged. This procedure is repeated until the system is frozen in an absorbing configuration.

The basic assumption of Axelrod's model is that similarity is a main requisite for social interaction and, as a result, exchange of opinions. This is the 'birds of a feather flock together' hypothesis which states that individuals who are similar to each other are more likely to interact and then become even more similar[5]. Recent empirical evidence in favor of this assumption comes from the analysis of Web 2.0 social networks[27]. Study of a population of over 10^7 people indicates that people who chat with each other using instant messaging are more likely to have common interests, as measured by the similarity of their Web searches, and the more time they spend talking, the stronger this relationship is. We note, however, that this assumption is disputed by other researchers[28] who have shown that people are attracted to others who resemble their ideal, rather than their actual selves.

To introduce the effect of a global media following the seminal paper by Shibanai and colleagues[21], we need first to define a virtual, consensus agent whose cultural traits reflect the opinion of the majority of the population. Hence each cultural feature of this virtual agent, which plays the media role, has the trait which is the most numerous in the population (in case of degeneracy, we choose the consensus trait randomly among the degenerate options). For example, consider a population comprised of the agents 13211, 13331, 22121 and 31222 which inhabit the sites of a square lattice of linear size $L = 2$. The consensus agent is then 13221, which happens to be different from all four real agents. We note that this virtual agent is a global (non-local) property of the population, which can change whenever a single real agent changes its state.

Next, we need to specify how the media interact with the real agents. To do that we introduce a new control parameter $p \in [0, 1]$, which measures the strength of the media influence. As in the original Axelrod's model, we begin by choosing a target agent at random, but now it can interact with the media with probability p or with its neighbors with probability $1 - p$. Since we have defined the media as a virtual agent, the interaction follows exactly the same rules as before. The original model is recovered for $p = 0$, provided we properly define the halting criterion of the dynamics, as discussed in the next section.

3. Results

To simulate efficiently Axelrod's model we make a list of the active agents. An active agent is an agent that has at least one feature in common and at least one feature distinct with at least one of its nearest neighbors. Clearly, since only active agents can change their cultural features, it is more efficient to select the target agent randomly from the list of active agents rather than from the entire lattice. Note that the randomly selected neighbor of the target agent may not necessarily be an active agent itself. In the case that the cultural features of the target agent are modified by the interaction with its neighbor, we need to re-examine the active/inactive status of the target agent as well as of all its neighbors so as to update the list of active agents. The dynamics is frozen when the list of active agents is empty. This is the halting criterion we mentioned in the last section. The important point here is that the media do not enter the procedure to determine whether an agent is active or not; otherwise the dynamics would never freeze (except in the uniform regime).

A feature that sets our results apart from those reported previously in the literature is that our data points represent averages over at least 10^3 independent runs. This requires a substantial computational effort, especially in the regime where the number of cultures decreases with the lattice size since then the time for absorption can be as large as $10^6 \times A$ where $A = L^2$ is the lattice area or, equivalently, the number of agents. In the figures presented in the following, the error bars are smaller or at most equal to the symbol sizes.

Before considering the effect of media, let us review briefly the main findings regarding the nonequilibrium phase transition in the original Axelrod's model (i.e., for $p = 0$). According to Castellano and colleagues[6] this threshold phenomenon results from the competition between the disorder of the initial configuration and the ordering bias of the local interactions. Hence, it is instructive to calculate the number of cultures in the totally disordered initial configuration, in which the $A = L^2$ agents are assigned one of the any q^F cultures. This is a classical occupancy problem discussed at length in Feller's book[29]. In this occupancy problem, the probability that exactly m cultures are not used in the assignment of the A agents to the q^F cultures is

$$P_m\left(A,q^F\right)=\binom{q^F}{m}\sum_{\nu=0}^{q^F-m}\binom{q^F-m}{\nu}(-1)^\nu\left(1-\frac{m+\nu}{q^F}\right)^A, \quad (1)$$

which in the limit where A and q^F are large reduces to the Poisson distribution

$$p\left(m;\lambda\right)=\mathrm{e}^{-\lambda}\frac{\lambda^m}{m!} \quad (2)$$

where $\lambda=q^F\exp\left(-A/q^F\right)$ remains bounded[29]. Hence the average cultural diversity C_r resulting from the random assignment of agents to cultures is simply $q^F-\langle m\rangle$, which yields

$$C_r=q^F\left[1-\exp\left(-A/q^F\right)\right]. \quad (3)$$

This quantity is always a monotonically increasing function of A which grows linearly in the regime $A\ll q^F$ and tends to the maximum diversity value q^F when $A\gg q^F$.

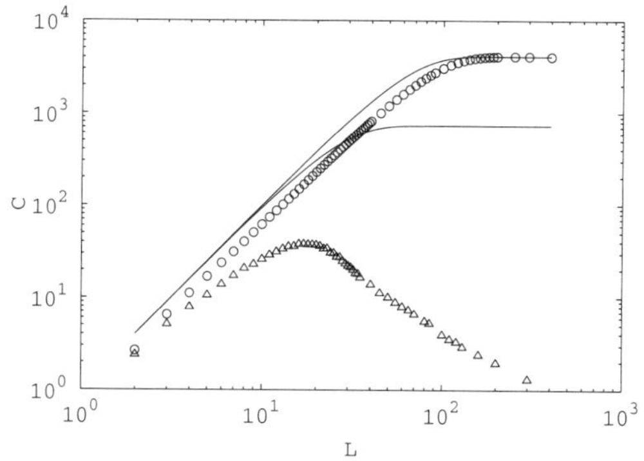

Figure 1. Logarithmic plot of the number of distinct cultures C as function of the linear size L of the lattice for $p=0$, $F=3$ and $q=9\,(\triangle)$ and $q=16\,(\circ)$. The solid lines are the results of Eq. (3) which yield the average number of cultures in the initial random configuration. In the limit $L\to\infty$ there are two distinct regimes: either $C\to 1$ $(q<16)$ or $C\to q^F$ $(q\geq 16)$.

In Fig. 1 we show the number of distinct cultures C at equilibrium as function of the linear size L of the lattice for $F=3$ and two values of the number of traits q, which correspond to different asymptotic regimes. We note that in the calculation of C the cultural diasporas, which occur when regions with specific cultural features are disconnected from other regions

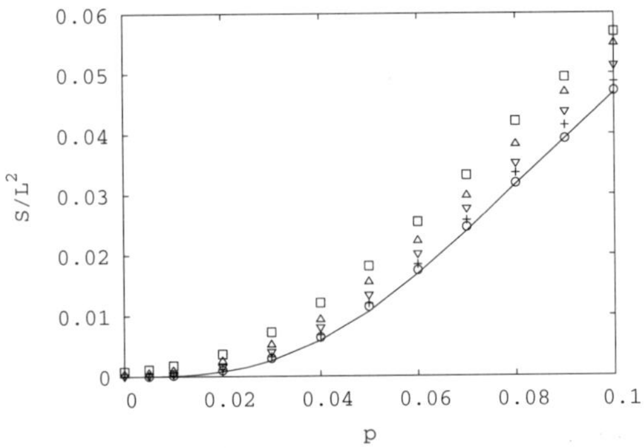

Figure 2. Ratio between the number of cultural domains and the lattice are as function of the strength of the media influence for $F = 5$, $q = 10$ and $L = 40\,(\square)\,, 60\,(\triangle)\,, 100\,(\nabla)\,, 200\,(+)\,200, 300\,(\circ)$. The solid line is the result of the extrapolation of the data to the limit $L \to \infty$.

with the same cultural features[14], are counted as a single culture. This figure illustrates an important point already mentioned: it is easy to find models that yield polarized states for finite lattices, the difficulty is to maintain the multicultural configurations when the lattice size goes to infinity. In fact, only in that limit we can observe a genuine threshold phenomenon as illustrated in Fig. 1: for $q < q_c$ we find $C \to 1$ (homogeneous regime) whereas for $q \geq q_c$ we have $C \to q^F$ (polarized regime where all possible cultures are present). The threshold value $q_c = q_c\,(F)$ is an increasing function of F and for $F = 3$ we find $q_c = 16$. We note that the transition from the uniform to the polarized regime is discontinuous (C jumps from 1 to q^F at $q = q_c$). In addition, the solid curves in Fig. 1 show that the random occupancy hypothesis, Eq. (3), yields a good qualitative description of the data in the polarized regime ($q > q_c$), although it consistently overestimates the values for the cultural diversity. This is expected as the effect of the local interactions in Axelrod's model is to decrease the cultural differences between neighboring agents. A counterintuitive result exhibited in Fig. 1 is the non-monotonic dependence of the number of cultures on the size of the lattice for $q < q_c$. The expectancy here would be a monotonic increase of C with the area of the territory, as in the case of the species-area relation[7].

We turn now to the analysis of the effect of media. To do that we need to define an 'order parameter' which remains finite (and nonzero) when the lattice size goes to infinity. A convenient choice is the ratio between the number of clusters (or cultural domains) S and the lattice area L^2.

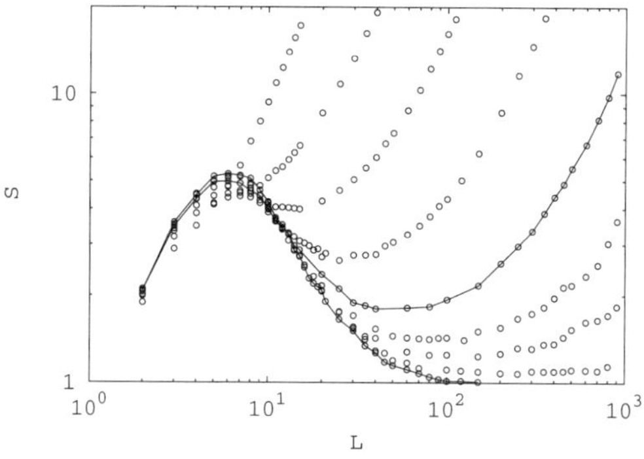

Figure 3. Number of cultural domains as function of the linear size of the lattice for $F = 5$, $q = 10$ and (bottom to top at $L = 10^2$) $p = 0, 0.001, 0.002, 0.003, 0.005, 0.01, 0.02, 0.04$ and 0.1. The solid lines connecting the symbols for $p = 0$ and $p = 0.005$ are guides to the eye.

A cluster is simply a bounded region of uniform culture. In the case of diasporas, the two or more cultural domains (which are characterized by the same culture) are counted separately. We note that whereas the number of cultures C has the upper bound q^F, S is bounded by L^2, so that $S/L^2 < 1$. Figure 2 exhibits this order parameter as function of the strength of the media influence p for different lattice sizes and fixed values of q and F. The suitability of the order parameter is demonstrated by the fact that the data converge to well-defined values (solid line in Fig. 2) as the lattice size is increased. In other words, S scales as L^2 for p not too small. Indeed, from that figure it seems that the ratio S/L^2 vanishes for small p which indicates the existence of a minimum strength value p_c, above which the uniform regime is destabilized[23,24,25,26]. For the data shown in Fig. 2 we find $p_c \approx 0.01$ by visual inspection. (The determination of p_c in those papers[23,24,25,26] was based on the analysis of lattices of linear size up to $L = 60$.)

However, a more careful analysis reveals a different story, as shown in Fig. 3. In fact, consider the data for $p = 0.005$ (the symbols are connected to single out the curve for this parameter in Fig. 3). An analysis of lattices of sizes up to $L = 60$ indicates a clear tendency of convergence towards the uniform regime (i.e., S decreases with increasing L), but this trend changes completely when lattices of sizes greater than $L = 100$ are considered: S increases with L and our analysis indicate that S/L^2 tends to a nonzero value when $L \to \infty$ for any $p > 0$. More pointedly, for the data shown

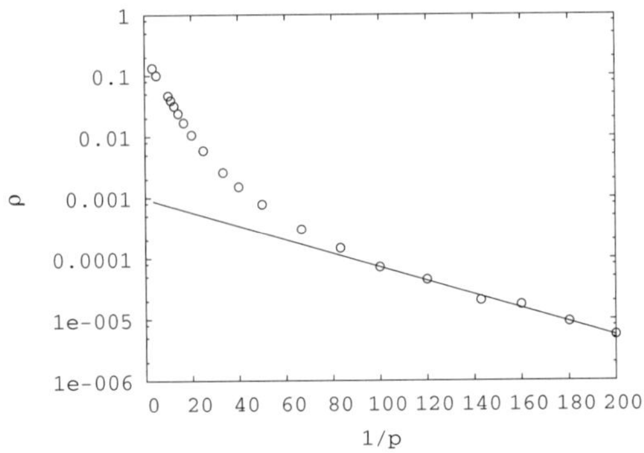

Figure 4. The ratio between the number of cultural domains and the lattice area for $L \to \infty$ as function of the reciprocal of the strength of the media influence for $F = 5$ and $q = 10$. The straight line is the fitting given by Eq. (4).

in Fig. 3 we calculate the ratio S/L^2 for a fixed p and different values of L and then extrapolate the results to $L \to \infty$. The resulting ratio, $\rho \equiv \lim_{L\to\infty} S/L^2$, is then plotted against $1/p$ as shown in Fig. 4. For small p (i.e., large $1/p$) the data is fitted very well by the equation

$$\rho = 9.5 \times 10^{-4} \exp\left(-0.026/p\right) \tag{4}$$

as indicated in the figure. This finding explains why the numerical simulations yielded a nonzero value for p_c: for small p it is virtually impossible to distinguish the result of Eq. (4) from zero. As a result, Axelrod's model does not exhibit a phase transition for $p > 0$: the only stable regime for infinite lattice sizes is the polarized one, regardless of the values of the control parameters q and F.

4. Conclusion

In addition to describing the dynamics of social influence, minor variants of Axelrod's model can be used to study language competition[30] and sympatric speciation[31]. The key ingredient in those applications is the requisite of similarity between individuals for the occurrence of interactions. In fact, sympatric speciation is based on a mate choice mechanism that depends on the similarity (genetic distance) between mates whereas communication can take place provided the individuals' languages are sufficiently similar. Kennedy[12] proposed a surprising interpretation of Axelrod's model as an algorithm for maximizing similarity. More generally, he demonstrated that social interaction can function as a general purpose optimization algorithm.

The idea is exceedingly simple: the strings that characterize the agents are interpreted as solutions of a given optimization problem, and interactions are allowed only with neighbors who are better solutions than the target agent[12]. In this way the best solutions are more likely to be copied and eventually they can take over the entire lattice. It is this vast range of applications that has given to Axelrod's model the status of paradigm for idealized models of collective behavior[3].

In this contribution we have revisited an important extension of Axelrod's model in which, in addition to the local interactions between agents, there is a global element – the media – that influences the agents' opinions or cultural traits. In stark contrast to the common sense opinion that the media effect is to homogenize the society, we find that the media actually promotes polarization or the diversity of opinions. In fact, this effect is so powerful that a vanishingly small influence strength is sufficient to destabilize the cultural homogenous state for very large lattices. At present we have no idea of why this is so. An analysis of the relaxation times as well as of the distribution of sizes of cultural domains may provide some clue on this counterintuitive effect. Work in this line is under way.

Acknowledgments

This work was supported in part by CNPq and FAPESP, Project No. 04/06156-3.

References

1. M. McLuhan, *Understanding Media: The Extensions of Man* (Signet Books, New York, 1966).
2. R. Axelrod, *J. Conflict Res.* **41**, 203 (1997).
3. R. L. Goldstone and M. A. Janssen, *Trends Cog. Sci.* **9**, 424 (2005).
4. B. Latané, *American Psychologist* **36**, 343 (1981).
5. S. Moscovici, *Handbook of Social Psychology* **2**, 347 (1985).
6. C. Castellano, M. Marsili and A. Vespignani, *Phys. Rev. Lett.* **85**, 3536 (2000).
7. L. A. Barbosa and J. F. Fontanari, *Theor. Biosc.* DOI:10.1007/s12064-009-0066-z (2009)
8. D. Vilone, A. Vespignani and C. Castellano, *Europ. Phys. J. B* **30**, 399 (2002).
9. F. Vazquez and S. Redner, *EPL* **78**, 18002 (2007).
10. J. Marro and R. Dickman, *Nonequilibrium Phase Transitions in Lattice Models* (Cambridge University Press, Cambridge, UK, 1999).
11. H. Hinrichsen, *Adv. Phys.* **49**, 815 (2000).
12. J. Kennedy, *J. Conflict Res.* **42**, 56 (1998).

13. K. Klemm, V. M. Eguíluz, R. Toral, M. San Miguel, *Phys. Rev. E* **67**, 045101R (2003).
14. J. M. Greig, *Conflict Res.* **46**, 225 (2002).
15. K. Klemm, V. M. Eguíluz, R. Toral, M. San Miguel, *Physica A* **327**, 1 (2003).
16. K. Klemm, V. M. Eguíluz, R. Toral, M. San Miguel, *Phys. Rev. E* **67**, 026120 (2003).
17. R. Boyd and P. J. Richerson, *Culture and the evolutionary process* (University of Chicago Press, Chicago, 1985).
18. A. Nowak, J. Szamrej and B. Latané, *Psychological Review* **97**, 362 (1990).
19. T. M. Ligget, *Interacting Particle Systems* (Springer, New York, 1985).
20. D. Parisi, F. Cecconi, F. Natale, *J. Conflict Res.* **47**, 163 (2003).
21. Y. Shibanai, S. Yasuno, I. Ishiguro, *J. Conflict Res.* **45**, 80 (2001).
22. P. Lazarsfeld, B. Berelson and H. Gaudet, *The people's choice* (Columbia University Press, New York, 1948).
23. J. C. González-Avella, M. G. Cosenza and K. Tucci, *Phys. Rev. E* **72**, 065102R (2005)
24. J. C. González-Avella, M. Eguíluz, M. G. Cosenza, K. Klemm, J. L. Herrera and M. San Miguel, *Phys. Rev. E* **73**, 046119 (2006).
25. K. I. Mazzitello, J. Candia and V. Dossetti, *Intern. J. Mod. Phys. C* **18**, 1475 (2007).
26. J. Candia and K. I. Mazzitello, *J. Stat. Mech.* P07007 (2008).
27. P. Singla P and M. Richardson, *Proceedings of the 17th International World Wide Web Conference* (ACM Press, Toronto, 2008) pp. 655–664.
28. C. G. Wetzel and C. A. Insko, *J. Exp. Soc. Psych.* **18**, 253 (1985).
29. W. Feller, *An Introduction to Probability Theory and Its Applications* (vol I, 3rd Edition, Wiley, New York, 1968).
30. C. Schulze, D. Stauffer and S. Wichmann, *Commun. Comput. Phys.* **3**, 271 (2008).
31. P. G. Higgs and B. Derrida, *J. Mol. Evol.* **35**, 465 (1992).

LESLIE MATRICES AND SURVIVAL CURVES CONTAIN THERMODYNAMICAL INFORMATION

F. R. MOMO
Instituto de Ciencias, Universidad Nacional de General Sarmiento, J.M. Gutiérrez 1150
1613 Los Polvorines, Argentina

S. DOYLE
CONICET - INEDES, Universidad Nacional de Luján, Rutas 5 y 7
6700 Luján, Argentina

J. E. URE
Instituto de Ciencias, Universidad Nacional de General Sarmiento, J.M. Gutiérrez 1150
1613 Los Polvorines, Argentina

By mean of numerical simulations we demonstrated here that Leslie matrices contain information about the way as population dissipate energy and pump entropy outside the system. We established two trade-off axes between fertility and survival and writing the corresponding matrices for each case. We found that r populations have higher entropy costs that K populations, whereas populations having Type I survival curves (high survival of larvae and low survival of adults) shows lower entropy costs than populations with Type IV survival curves. This study demonstrates that both, survival curve and projection matrix, have information about thermodynamic characteristics of populations.

1. Introduction

1.1. *Leslie matrices and survival curves*

Leslie matrices are one of the most useful tools in population ecology and are probably the best-known age-structured models. They represent populations divided into age classes of the same length as the time step[1]. A proportion s_i of individuals in age class i are assumed to survive into the next age class. The age-specific fertility is called f_i (number of female offspring in the first age class per female in age class i). The matrix build in that way allows us to calculate the abundance and age structure of one population across time. Abundances of each age-class at each time form a vector that represents the structured population. The population vector at time $t+1$ can be derived from the vector at time t from the following matrix equation:

$$\begin{pmatrix} n_1 \\ n_2 \\ \dots \\ n_k \end{pmatrix}_{t+1} = \begin{pmatrix} f_1 & f_2 & \dots & f_k \\ s_1 & 0 & \dots & 0 \\ 0 & s_2 & 0 & 0 \\ \dots & \dots & s_{k-1} & 0 \end{pmatrix} \begin{pmatrix} n_1 \\ n_2 \\ \dots \\ n_k \end{pmatrix}_t \quad (1)$$

Values of parameters in the Leslie matrix give information about the life history traits of the population and the kind of selection (i.e. r or K) under which this population evolved[2]. Similarly, survival curves, that show the percent of a cohort that survives until a given age, can be associated to different demographic traits and suggest how the population allocates the available energy. The shape of curve will describe the distribution of mortality with age. Slobodkin[3] recognized four basic curves called type I, II, III and IV; in type I, mortality acts most heavily on the old individuals; in type II, a constant number die per unit of time; in type III, the mortality rate is constant; and, in type IV, mortality acts most heavily on the young ages. Textbooks of ecology[4,5,6] coincide in relate type IV curves with populations exposed to an *r-selection* (in these populations the main life history traits are high fecundities, small body size and opportunistic strategies); in the same way, type I curves have been related with populations under *K-selection* (that is, populations with low fecundities, big body size, parental care of juveniles, and specialist strategies). Clearly, the type of survivorship curve can be represented by the values of parameters in the Leslie matrix. Type I curves correspond to matrices with low values of f_i, high values of s_i of young ages and low values of s_i of old ages; conversely, type IV curves are represented by Leslie matrices with high values of f_i, and low values s_i of young ages, and high values of s_i of older ages. However, we will demonstrate in this work that it is necessary to consider other physiological trade-offs in order to predict population dynamics using matrices and survival curves for r and K species.

1.2. *Thermodynamic functions in populations*

Recently, several ecological works[7,8,9] have established different thermodynamic approaches to populations and community ecology. Populations are dissipative systems *sensu* Prigogine[10] because off they maintain their age structure and biomass "pumping" entropy to their environment. By analogy with other thermodynamic systems we can establish the following relationship:

$$\frac{dS}{dU} = \frac{1}{T} \tag{2}$$

where S is the entropy, U is the internal energy, and T is temperature. However, this relationship is incomplete for ecological purposes because dissipative systems maintain their organization by mean of fluxes of entropy and energy; for this reason, it is more accurate to write Eq. (2) as

$$\frac{dS/dt}{dU/dt} = \frac{1}{T} = \tau \tag{3}$$

Eq (3) represents the relationship between entropy expulsion and energy exchange between the system and its environment. Both fluxes can be calculated and the magnitude $\tau = 1/T$ may be interpreted as a demographic sensibility as we will explain later. In order to calculate the structural entropy of a population we use here the following formula: $S = \sum p_i \ln p_i$, where p_i is the percent abundance of individuals of age i. However, the flux of entropy is not only the rate of variation of this magnitude because mortality of individuals implies expulsion of entropy outside the system; having the mean individual biomass of each stage or age and their survival rates, we can calculate the rate of expelled entropy by death of individuals. Moreover, using the know relationship between biomass and metabolic rate, we calculated the rate of energy dissipation and obtain values of Eq. (3). The slope of expelled entropy rate versus energy dissipation provides the new parameter τ (analog to $1/T$ in classic thermodynamic studies) that represents the entropy cost of maintenance per unit of energy flux.

2. Methods

We started establishing some trade-off between fertility and survival and writing the corresponding matrices for each case. For simplicity, we considered a three rows Leslie matrix with fecundity only in the last age. We assumed that $f = 1/(s_1 s_2)$ in order to have a $\lambda_1 > 1$ (growing populations); then, we performed different numerical simulations considering cases having more or less fertility combined with more or less survivals (that is the r-K trade-off). In addition, we changed the relationship between s_1 and s_2

in order to simulate the shift between different survival curves (from type I to IV). The used matrix were the following:

$$\begin{pmatrix} 0 & 0 & f \\ s_1 & 0 & 0 \\ 0 & s_2 & s_3 \end{pmatrix} \tag{4}$$

To simulate only the r-K axis we simulated a trade off between fertility and survival without differences between ages in survival; in consequence, a first set of simulations were performed considering $s_1 = s_2 = s$ and $f = 1/s^2$. Values of s were varied from 0.1 to 0.9; s_3 remained constant and equal to 0.5. The second set of simulations were performed taking into account the shift between type I and type IV survival curves; so, s_1 varied from 0.1 to 0.9 and, simultaneously, s_2 varied from 0.9 to 0.1 (from type IV to type I survivorship curves); the parameter s_3 remained again constant and equal to 0.5, and I were computed as $1/(s_1 s_2)$. Despite the fact of body size is also a trade-off magnitude, in order to avoid excessive complications that can shade the main effects, individual biomass were considered constant in all essays. Values of biomass for each age used were (in arbitrary units) $b_1 = 0.1$, $b_2 = 1$, and $b_3 = 5$. Having the results of simulations we plot the rate of dissipated energy versus the pumped entropy and obtained values of τ for different hypothetical populations. We examined the relationships between the population life history traits and their dynamics and thermodynamics (changes in entropy fluxes throughout time). In entropy calculations we used percent of abundance of each age, in consequence, it is clear that $\sum p_i = 100$; according this constraint we can examine the structural entropy trajectories of different populations in a phase-space of p_2 versus p_1 in order to establish if some kind of maximization exists and, if so, what kind.

3. Results and discussion

The first set of simulations shows that r populations have higher energetic costs than K populations; that is: for similar energy dissipation, K selected species dissipate more entropy per unit of time and, consequently, maintain a higher organization. This fact is evident if we compute the structural entropy as $S = \sum p_i \ln p_i$; the values of S are lower for K populations (Table I). Moreover, relaxation times of S (the time needed to rise $dS/dt = 0$) are lower in K populations than in r ones (Table I); however, the demographic sensibility (τ) of K populations is higher than the demographic sensibility of r populations; that is, for changes in energy availability, K species are more affected in their age structure than r ones.

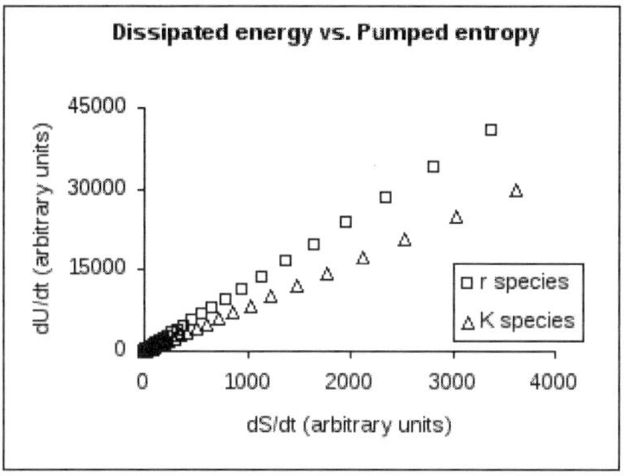

Figure 1. Rate of dissipated energy versus rate of dissipated (pumped) entropy for r and K populations.

The second set of simulations shows that populations with type IV survival curves (mortality concentrated in younger ages) dissipate more energy per unit of time respect to the entropy dissipation, but, it is notorious that populations having Type I survival curves (high survival of larvae and low survival of adults) rise higher energy and entropy fluxes (more biomass and more demographic complexity). These facts can be appreciated in Figure 2. Moreover, Type I populations have higher demographic sensibility than Type IV ones, that is higher values of τ (Table 1). Relaxation times are lower in Type I populations too. However, structural entropy has a minimum in populations having intermediate survival curves with survival rates roughly constant across ages (Table 1). This study demonstrates that both, survival curve and projection matrix, have information about thermodynamic characteristics of populations. However the most important thing is that r-K trade-off is not equivalent to Type IV Type I trade-off, despite that many ecological literature consider equivalent both life history traits axes.

Moreover, entropy dissipation is maximized by Type I populations but age structure entropy are maximized in intermediate survivorship curves. On the contrary, K selected populations maximize both, rate of pumped entropy and structural age entropy.

Demographic sensitivity τ can be a useful parameter in order to de-

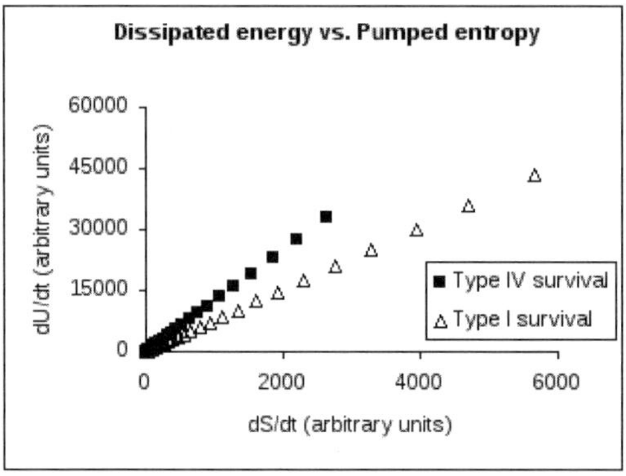

Figure 2. Rate of dissipated energy versus rate of dissipated (pumped) entropy for populations having Type IV to Type I survival curves.

scribe the response of populations to environmental changes, especially in reference to energy availability or productivity changes. It is accepted that K species tend to have big body sizes than r species. This trade-off was not examined in this work and remains as an interesting topic for future studies. Demographic entropy used here is not an entropy strictly speaking in thermodynamic sense; however it is similar to other called "entropies" used in ecological literature[9,11] and is an useful function in order to characterize the population complexity. On the other hand, entropy and energy fluxes (dS/dt and dU/dt) are the right magnitudes needed to study dissipative systems like populations. Despite our approximation is linear and dissipative structures are non linear[10] , populations are frequently in quasi-stationary regimes and our approximation may be useful to study them. Moreover, from a conceptual point of view, this paper is a forward step in the understanding of thermodynamic applications in ecology.

Table 1. Thermodynamic and dynamic life history traits for populations across r-K and IV-I trade-off axes. Entropy units have arbitrary units (see text).

Life history characterization	Demographic sensititvity (τ)	Relaxation time until $dS/dt = 0$ (generations)	Stationary demographic entropy
R	0.083	20	404
Intermediate	0.115	15	364
K	0.121	8	351
Type IV	0.081	20	405
Type III	0.112	17	368
Type I	0.130	12	374

References

1. P. Leslie, *Biometrika* **33**, 183 (1945).
2. R. MacArthur and E. Wilson, *The Theory of Island Biogeography.* Princeton University Press (1967).
3. L. Slobodkin, *Growth and regulation of Animal Populations.* Hol, Rinehart & Winston (1961).
4. S. Dodson, T. Allen, S. Carpenter, A. Ives, R. Jeanne, J. Kitchell, N. Langston and M. Turner, *Ecology.* Oxford University Press (1998).
5. E. P. Odum, *Fundamentals of Ecology.* Saunders (1971).
6. M. Begon, J. Harper and C. Townsend, *Ecology: Individuals, Populations and Communities.* Blackwell Science (1990).
7. S. Jrgensen and Y. Svirezhev, *Towards Thermodynamic Theory for Ecological Systems.* Elsevier (2004).
8. L. Demetrius, *Demography* **16**, 329. (1979).
9. H. T. Odum, *Ecol. Model.* **158**, 201 (2002).
10. I. Prigogine, *Introduction to Thermodynamics of Irreversible Processes,* Thournes. (1955).
11. L. Demetrius, V. Gundlach and G. Ochs, *Theor. Pop. Biol.* **65**, 211. (2004).

GRAPH PARTITIONING APPROACHES FOR ANALYZING BIOLOGICAL NETWORKS

NENG FAN AND P. M. PARDALOS

Department of Industrial and Systems Engineering
University of Florida
Gainesville, FL 32611, USA
E-mail: andynfan@ufl.edu, pardalos@ufl.edu

A. CHINCHULUUN

Centre for Process Systems Engineering,
Imperial College London,
London SW7 2AZ, UK
E-mail: altannar@imperial.ac.uk

This paper presents graph partitioning methods for identifying clusters in biological networks. Integer programming formulations are proposed for graph partitioning and several relaxations of these formulations are introduced for solving the problems. These relaxations include spectral methods, quadratic programming and semidefinite programming. In addition, the results of some numerical examples on biological networks are reported.

1. Introduction

1.1. *Biological Networks*

The mathematical formulations of networks were developed to represent the complex networks from biological systems based on classical graph theory. These networks include genetic, proteomic and metabolic networks, and among them, the "protein-protein interaction networks"[5] and "gene co-expression networks"[12] are most widely studied.

Grigorov[11] discussed the biological implications of the small-world[20] and scale-free[3] global topological properties of these biological networks. For example, in a kind of gene co-expression network defined as that two genes are connected when the pair of genes is involved in the same biological process[19], the small-world means that one gene can be reached from another one by a small number of genes which are connected to each other. The scale-free property of a network is that its degree distribution of nodes follows a power law[3], and this proposition is noteworthy in protein networks.

A network comprises of nodes and links connecting the nodes. Generally, in protein-protein interaction networks, proteins are nodes and a pair is connected by a link if they are known to interact with each other; and in gene co-expression networks, genes are nodes and a link indicates that the pair of genes are co-expressed over some threshold value, based on microarray experiments. We describe the general concept of clustering before introducing the objectives of analysis of biological networks. Clustering is a technique used in data mining to group large set of objects into groups (clusters) such that objects within each group behaviors similarly in some way. Thus, in a network with a set of nodes, clustering can find clusters (usually named modules in biology) and describe their properties in the process of grouping. Clustering can also be used to classify large data sets and eliminate outliers. The process of grouping data can be considered as data reduction since the nodes with similar behavior are grouped.

Identifying the modules from these biological networks has been a focus of many researchers in recent years and many methods have been proposed in the literature. Cliques in a graph are kinds of tight clusters in a network clustering. Thus, the algorithms in graph theory for cliques are used to find modules in biological networks[16], and recently by clique relaxations[2] or high density subgraphs[18,10]. Other popular methods used recently include Markov clustering[7], restricted search clustering[15], super paramagnetic clustering[4], molecular complex detection[1], and etc. An evaluation of these clustering algorithms for protein-protein interaction networks is presented by Brohee and Helden[5], and a websites for comparison results can be found on the internet[17].

In this paper, the graph partitioning model is used to identify modules in biological networks. This model is different from those based on cliques, and similar to find subgraphs. However, this method will globally classified all nodes in the meantime. Before presenting this model, the notations in graph theory are introduced.

1.2. *Graph Partitioning and Clustering*

Consider a graph $G = (V, E)$ with the set $V = \{v_1, \cdots, v_n\}$ of n vertices and the set $E = \{(i, j) : \text{edge between } v_i \text{ and } v_j, i, j \leq n\}$ of edges. There is usually a weighted matrix $W = (w_{ij})_{n\times n}$, where w_{ij} is the weight of the edge (i,j), associated with the graph such that $w_{ij} \geq 0, w_{ij} = w_{ji}, w_{ii} = 0$.

A *clique* C is a subset of vertices V such that an edge exists between every pair of vertices in C, and the induced subgraph by C is a complete graph. A clique is maximal if it is not a subset of any larger clique, while a clique is maximum if there are no larger cliques in the graph. To find cliques in a graph, we can use the related greedy maximal independent set algorithm or integer programming models[2].

A *bipartition* for $G = (V, E)$ is defined as two subsets V_1, V_2 of V such that $V_1 \cup V_2 = V$ and $V_1 \cap V_2 = \emptyset$. More generally, a k-*partition* is the collection of k subsets $V_1, V_2, \cdots, V_k$ such that $V_1 \cup \cdots \cup V_k = V$ and $V_i \cap V_j = \emptyset$ for $i, j \in \{1, 2, \cdots, k\}$ and $i \neq j$. Suppose that the vertex set V of a graph is partitioned into two disjoint subsets V_1, V_2, then the corresponding graph *cut* is defined as

$$cut(V_1, V_2) = \sum_{(i,j)\in E, i\in V_1, j\in V_2} w_{ij}.$$

For the case of k-partition, the k-cut is

$$cut(V_1, V_2, \cdots, V_k) = \sum_{1\leq i<j\leq k} cut(V_i, V_j).$$

The "nodes" and "links" are usually used in networks while correspondingly "vertices" and "edges" used in graphs.

The clique C is required to be complete while the partitioned parts V_i are just subgraphs of G and not necessary to be complete. Therefore, using graph partitioning is more general than just finding the cliques. The process of finding subgraphs by this approach can help us to understand the structure and function of proteins based on protein interaction maps of organisms, decomposes the whole protein-protein interaction network into functional modules and protein complexes and identifies known or unknown modules in a complex network.

This paper is organized as follows. Section 2 introduces graph bipartition formulations and relaxations while Section 3 presents multi-partition formulations and relaxations including spectral methods, quadratic programming and semidefinite programming. In Section 4, we present some numerical results and Section 5 concludes the paper.

2. Graph Bipartition Formulations and Relaxations

For a graph $G = (V, E)$ with weighted matrix W, the Laplacian matrix $L = (l_{ij})_{n\times n}$ is a symmetric matrix such that

$$l_{ij} = \begin{cases} d_i, & \text{if } i = j, \\ -w_{ij}, & \text{if the edge } (i,j) \text{ exists}, \\ 0, & \text{otherwise}, \end{cases}$$

where d_i is the weighted degree of vertex v_i that is $d_i = \sum_j w_{ij}$. The propositions of this matrix have been reviewed in the paper[8]. The matrix D is defined as a diagonal matrix with $d_{ii} = d_i$ and $d_{ij} = 0$ for $i \neq j$. In the following, we describe the approaches of bipartition based on the second smallest eigenvalue and corresponding eigenvector of L.

By the proposition about L[8], the cut $cut(V_1, V_2)$ of bipartition $\{V_1, V_2\}$ for $G = (V, E)$ can be expressed as $\frac{1}{4}p^T Lp$, where $p = (p_1, \cdots, p_n)^T$ is the indicator vector such that $V_1 = \{v_i : p_i = 1\}$ and $V_2 = \{v_i : p_i = -1\}$. The expression $\frac{1}{4}p^T Lp$ is called the inter similarity and $\frac{1}{4}p^T(D + W)p$ is called the intra similarity. In the graph partitioning, the vertices are highly close to each other should be decomposed into a same subgraph. Thus, the cut or inter similarity should be minimized. Therefore, the integer programming formulation for bipartition of a graph is

$$\begin{aligned} \min \quad & \frac{1}{4}p^T Lp \\ s.t. \quad & p = (p_1, \cdots, p_n)^T, p_i \in \{-1, 1\}. \end{aligned}$$

However, this objective may produce quite unbalanced parts and isolated vertices. In order to exclude such cases, two widely used cuts are defined: ratio cut and normalized cut. The ratio cut is defined with respect to the total vertices in each part

$$\frac{cut(V_1, V_2)}{|V_1|} + \frac{cut(V_1, V_2)}{|V_2|},$$

where $|V_i|$ denotes the number of vertices in V_i. The normalized cut is related to total weighted degree in each part

$$\frac{cut(V_1, V_2)}{d_{V_1}} + \frac{cut(V_1, V_2)}{d_{V_2}},$$

where $d_{V_i} = \sum_{j:v_j \in V_i} d_j$.

From theorems in the paper[8], the relaxation of integer program for bipartition based on ratio cut is

$$\begin{aligned} \min \quad & \frac{y^T L y}{y^T y} \\ s.t. \quad & y^T e = 0, y \neq 0, \end{aligned} \tag{1}$$

where y, either positive or negative, is used to classify the two groups as $p_i = 1$ and $p_i = -1$. By spectral graph theory[6], this program has the objective value as the second smallest eigenvalue and the solution as the corresponding eigenvector, say y.

Similarly, for normalized cut, the relaxation is

$$\begin{aligned} \min \quad & \frac{y^T L y}{y^T D y} \\ s.t. \quad & y^T D e = 0, y \neq 0, \end{aligned} \tag{2}$$

and, by Proposition of L, the solution of the program can be found by finding the second smallest eigenvalue and the corresponding eigenvector of the generalized eigenvalue problem $Ly = \lambda D y$.

The original binary integer programming problem for (1) and (2) is nonlinear and quite difficult to solve, but the eigenvalue problem is quite direct and can be effective for not so large problems.

Another approach for bipartition is based on quadratic programming[13]. Let $x = (x_1, \cdots, x_n)^T$ be the indicator vector and $x_i \in \{0, 1\}$. The integer programming in the form of x is

$$\begin{aligned} \min \quad & (e - x)^T W x \\ s.t. \quad & x = (x_1, \cdots, x_n)^T, x_i \in \{0, 1\}, \end{aligned}$$

or adding another part without changing the objective as

$$\min \quad (e - x)^T (W + D) x,$$

where D should be an appropriate choice. By theorem 2.1 from the paper[13], it is based on the program

$$\begin{aligned} \min \quad & (e - x)^T (W + D) x \\ s.t. \quad & 0 \leq x_i \leq 1, e^T x = m, \end{aligned} \tag{3}$$

where m is a given integer.

Theorem 2.1. *(Theorem 2.1[13]) If D in (3) is chosen such that*

$$d_{ii} + d_{jj} \geq 2 w_{ij}$$

for each i and j, then (3) *has a 0/1 solution and the partition* $\{V_1 = \{v_i : x_i = 1\}, V_2 = \{v_i : x_i = 0\}\}$ *is minimal cut. Moreover, if for each i and j,*

$$d_{ii} + d_{jj} > 2w_{ij},$$

then every local minimizer of (3) *is a 0/1 vector.*

By this theorem, the bipartition can be relaxed into a continuous quadratic programming problem.

3. Multi-partition Formulations and Relaxations

Let k be the number of parts, where $k \in \{2, 3, \cdots, n-1\}$. The *Partition Matrix* is defined as a $\{0,1\}$ rectangular $n \times k$ matrix $X = (x_{ij})_{n\times k}$, where $x_{ij} \in \{0,1\}$ and $\sum_{j=1}^{k} x_{ij} = 1$ for all $i = 1, \cdots, n$.

From the definition of partition matrix, each row corresponds to a vertex while each column to a part of the vertex set. The element x_{ij} defines that whether a vertex i is in part j (when $x_{ij} = 1$) or not (when $x_{ij} = 0$). The constraint $\sum_{j=1}^{k} x_{ij} = 1$ of X ensures that vertex i belongs to exactly one part. The column sum $\sum_{i=1}^{n} x_{ij}$ is corresponding to the number of vertices in part $j(j = 1, \cdots, k)$. Let $m_j = \sum_{i=1}^{n} x_{ij}$ and thus $\sum_{j=1}^{k} m_j = n$.

The feasible region of a partition matrix can be written as

$$\{X = (x_{ij})_{n\times k} : x_{ij} \in \{0,1\}, \sum_{j=1}^{k} x_{ij} = 1, m_j = \sum_{i=1}^{n} x_{ij}, \sum_{j=1}^{k} m_j = n\},$$

and the inter similarity can be expressed[9] in the matrix form $\frac{1}{2}tr(X^T LX)$. Thus the integer programming problem in the matrix form can be written as

$$\begin{aligned} \min \quad & \frac{1}{2}tr(X^T LX) \\ s.t. \quad & Xe_k = e_n, \\ & X^T e_n = m, \\ & m^T e_k = n, \\ & x_{ij} \in \{0,1\}, \end{aligned} \tag{4}$$

where e_k and e_n denote k and n vectors whose elements are all ones, respectively, and $m = (m_1, \cdots, m_k)^T$. For the program (4), there are many approaches for the equal partition forms where $m_1 = \cdots = m_k$, and the approaches for general form (4) may include spectral methods, quadratic programming and semidefinite programming.

Similarly as bipartition, the ratio cut and normalized cut for the multi-partition $\{V_1, \cdots, V_k\}$ are

$$R_k = \sum_{i,j:i\neq j} \left(\frac{cut(V_i, V_j)}{|V_i|} + \frac{cut(V_i, V_j)}{|V_j|} \right),$$

$$N_k = \sum_{i,j:i\neq j} \left(\frac{cut(V_i, V_j)}{d_{V_i}} + \frac{cut(V_i, V_j)}{d_{V_j}} \right),$$

respectively.

Let $X = (x_1, \cdots, x_k)$, where x_i are row vectors of matrix X, and $Y = (y_1, \cdots, y_k)$, where row vectors $y_i = x_i/m_i^{1/2}$. The ratio cut can be expressed as

$$R_k = y_1^T L y_1 + \cdots + y_k^T L y_k = tr(Y^T L Y).$$

Thus the relaxation of the program for ratio cut is

$$\begin{aligned} \min \quad & tr(Y^T L Y) \\ s.t. \quad & Y^T Y = I, \end{aligned}$$

where I is the identity matrix. By K. Fan's Theorem[6], the optimal solution is the eigenvectors $(v_1, \cdots, v_k)$ such that $Lv_i = \lambda_i v_i$ and the lower bound for objective value is $\min tr(Y^T L Y) \geq \lambda_1 + \cdots + \lambda_k$.

For the normalized cut, the indicators y_i are defined as

$$y_i = D^{1/2} x_i / \|D^{1/2} x_i\|$$

and the normalized cut can be expressed as

$$N_k = y_1^T (I - \tilde{W}) y_1 + \cdots + y_k (I - \tilde{W}) y_k = tr(Y^T (I - \tilde{W}) Y),$$

where $\tilde{W} = D^{-1/2} W D^{-1/2}$. The relaxation form of the mathematical program for the normalized cut is

$$\begin{aligned} \min \quad & tr(Y^T (I - \tilde{W}) Y) \\ s.t. \quad & Y^T Y = I, \end{aligned}$$

and similarly, by K. Fan's theorem[6], the optimal solution is the eigenvectors $(v_1, \cdots, v_k)$ such that $(I - \tilde{W}) v_i = \lambda_i v_i, (D - W) u_i = \lambda_i D u_i, u_i = D^{-1/2} v_i$ and the lower bound for objective value is $\min tr(Y^T L Y) \geq \lambda_1 + \cdots + \lambda_k$.

In the following, the quadratic programming approaches and algorithms for (4) are presented[13,14]. Since $tr(X^T D X) = \sum_{j=1}^k x_j^T D x_j = \sum_{i=1}^n d_i = \sum_i \sum_j a_{ij}$ is a fixed number related to data matrix W, $\min tr(X^T L X) =$

$\min tr(X^T(D-W)X)$ is equivalent to $\max tr(X^TWX)$ or $\max tr(X^T(W+D)X)$. The equivalent form for (4) is

$$\begin{aligned} \max \quad & tr(X^T(W+D)X) \\ s.t. \quad & Xe = e, X^Te = m \\ & X \in \Lambda = \{\in R^{nk}, x_{ij} \in \{0,1\}, 1 \leq i \leq n, j \leq j \leq k\}. \end{aligned} \tag{5}$$

In the papers[13,14], this program can be solved based on the following theorem.

Theorem 3.1. *(Theorem 6.1[13]) If D is chosen to satisfy $d_{ii} + d_{jj} \geq 2a_{ij}$ for each i and j, then the continuous problem*

$$\begin{aligned} \max \quad & tr(X^T(W+D)X) \\ s.t. \quad & Xe = e, X^Te = m \\ & X \geq 0 \ \textit{(every element of X is nonnegative)} \end{aligned} \tag{6}$$

has a maximizer contained in Λ, and hence, this maximizer is a solution of the discrete problem (5). Conversely, every solution to (5) is also a solution to (6). Moreover, if $d_{ii} + d_{jj} > 2w_{ij}$ for each i and j, then every local maximizer for (6) lies in Λ.

Another approach for solving the problem (4) is based on semidefinite programming relaxations[21] as shown in the following by a series of transformations of (4):

$$\begin{aligned} \min \quad & tr(L_AY) \\ s.t. \quad & arrow(Y) = 0, \\ & tr(D_1Y) = 0, \\ & tr(D_2Y) = 0, \\ & \mathcal{G}_J(Y) = 0, \\ & Y_{00} = 1, \\ & Y \succeq 0, \end{aligned} \tag{7}$$

where $L_A = \begin{pmatrix} 0 & 0 \\ 0 & \frac{1}{2}I \bigotimes L \end{pmatrix}$, $arrow(Y) = diag(Y) - (0, Y_{0,1:n^2})^T$, where $Y_{0,1:n^2}$ is the vector formed from the last n^2 components of the first, or 0, row of Y, $(\mathcal{G}_J(Y))_{ij} = \begin{cases} Y_{ij}, & \text{if } (i,j) \text{ or } (j,i) \in J; \\ 0, & \text{otherwise,} \end{cases}$ where $J = \{(i,j) : i =$

$(p-1)n+q, j=(r-1)n+q,$ for $p<r, p,r \in \{1,\cdots,k\}, q \in \{1,\cdots,n\}\}$, and

$$D_1 = \begin{pmatrix} n & -e_k^T \bigotimes e_n^T \\ -e_k \bigotimes e_n & (e_k e_k^T) \bigotimes I_n \end{pmatrix}, \quad D_2 = \begin{pmatrix} m^T m & -m^T \bigotimes e_n^T \\ -m \bigotimes e_n & I_k \bigotimes (e_n e_n^T) \end{pmatrix}.$$

The notation $Y \succeq 0$ means Y is positive semidefinite. We refer to paper[21] for the details of the transformation. This relaxation has no strictly feasible points[21]. Thus, in order to use semidefinite programming solving methods, a reduction of this relaxation

$$\begin{aligned} \min \quad & tr(\hat{V}^T L_A \hat{V})Z \\ s.t. \quad & \mathcal{G}_{\bar{J}}(\hat{V} Z \hat{V}^T) = \mathcal{G}_{\bar{J}}(E_{00}), \\ & Z \succeq 0, \end{aligned} \tag{8}$$

and its dual problem

$$\begin{aligned} \max \quad & W_{00} \\ s.t. \quad & \hat{V}^T \mathcal{G}_{\bar{J}}(W) \hat{V} \preceq \hat{V}^T L_A \hat{V} \end{aligned} \tag{9}$$

are considered.

We left the details[21] for (8) and (9). For primal-dual interior-point methods, the positive definite feasible points for both primal and dual feasible sets are given in Theorem 4.1 and Theorem 4.2[21].

An algorithm by a primal-dual interior-point approach is used to solve the relaxation (8). The both a lower bound for the graph partitioning problem and an appropriate solution $\bar{Y}$ for the relaxation (7) can be obtained. $\bar{Y}$ can be re-shaped to find an $n \times k$ matrix $\bar{Z}$ which satisfies the constraints of (4) except $\{0,1\}$ constraint, and an upper bound can be obtained by solving a network subproblem with $\bar{Z}$ as its adjacency matrix[21]. In this paper[21], the gap of upper and lower bounds with respect to lower bound is used to measure how close the upper bound to the optimal solution.

4. Preliminary Numerical Results

For numerical experiments, we use the spectral methods introduced in the previous section, and compare them with the existing methods from the site[a].

For the dataset "Cattle PPI (IntAct)"[b], it is IntAct protein-protein interaction network of cattle, including more than 180 interactions of more

[a] http://biit.cs.ut.ee/graphweb/index.cgi
[b] http://www.ebi.ac.uk/intact/, Pubmed Id: 10831611

than 130 proteins. From the software[17] providing many clustering methods, modules can be obtained from this network. One of the largest modules of the network obtained by connected component method is shown in Fig. 1 and another one obtained by betweenness centrality clustering method is shown in Fig. 2. From these two modules, we know that the network should have "hub" for modules.

Using spectral methods based on adjacency matrix of this network (shown in the left-hand side of Fig. 3), the corresponding eigenvector for the second smallest eigenvalue of this matrix is shown in the left-hand side of Fig. 4. Nothing can be obtained from these two figures. However, after reordering the eigenvector shown in the right-hand side of Fig. 4, three parts should be in this network, one of which is the middle stable part without big changes. As we can see, three parts of this data matrix is shown in the right-hand side of Fig. 3. By the relationships between adjacency matrix and network, the middle part shows part of itself has hubs as we have already learned from Fig. 1 and Fig. 2.

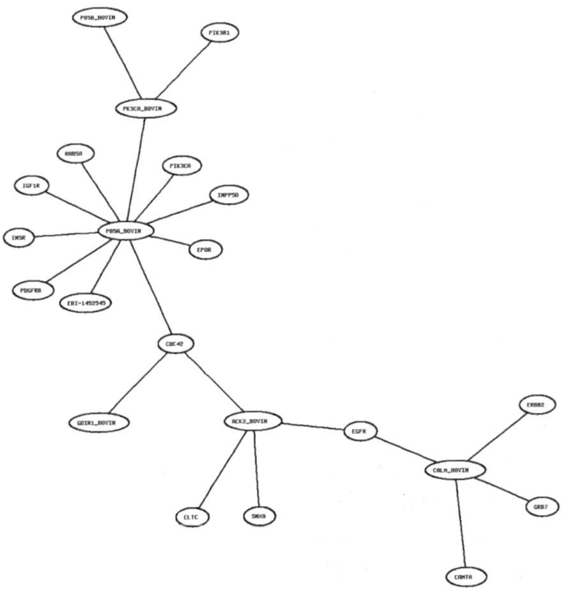

Figure 1. One of modules from Cattle PPI network by connected component method

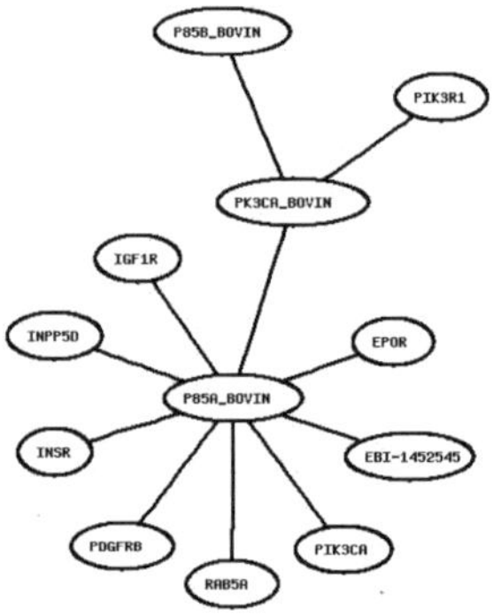

Figure 2. One of modules from Cattle PPI network by Betweenness centrality clustering method

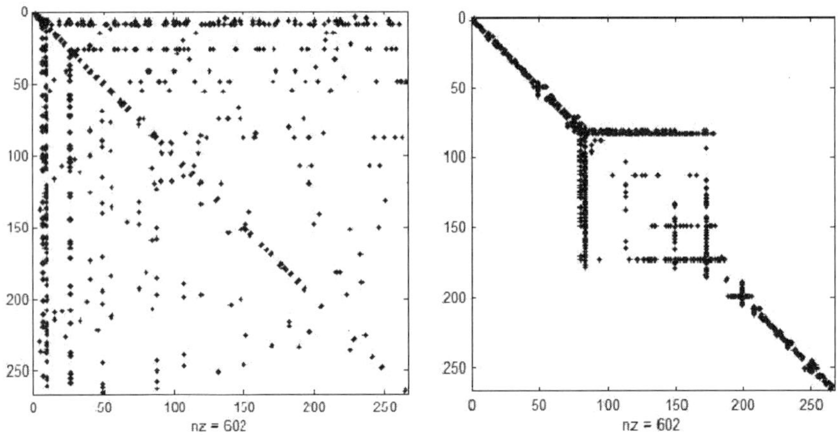

Figure 3. Data matrix and data after reordering

5. Conclusion

The use of graph partitioning approach in analysis of biological networks has not been studied well so far. For social networks, some research has began[22]. In the paper, we studied graph partitioning approaches for biological networks. The original problem (4) for graph partitioning is NP-hard, and many approaches are based on relaxations. Among these relaxations, quadratic programming and semidefinite programming have been studied for many years. The spectral methods have better applications even for more complicated cuts such as ratio cut and normalized cut. The future

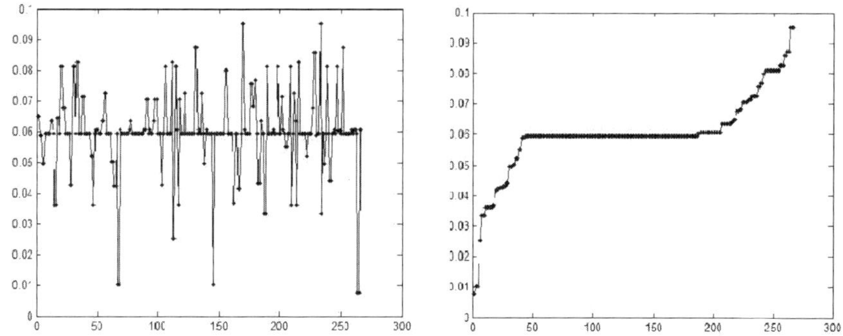

Figure 4. Eigenvectors

research in the direction of graph partitioning is to design efficient algorithms and to apply them in proper networks.

Acknowledgement

The research of the third author is partially supported by MOBILE, ERC Advanced Grant No 226462.

References

1. G.D. Bader and C.W.V. Hogue, An automated method for finding molecular complexes in large protein interaction networks, *BMC Bioinformatics* **4**, 2 (2003)
2. B. Balasundaram, S. Butenko and S. Trukhanov, Novel approaches for analyzing biological networks, *Journal of Combinatorial Optimization* **10(1)**, 23–39 (2005)
3. A. Barabasi, R. Albert, Emergence of scaling in random networks, *Science* **286**, 509–512 (1999)
4. M. Blatt, S. Wiseman and E. Domany, Superparamagnetic clustering of data, *Phys. Rev. Lett.* **76(18)**, 3251–3254 (1996)
5. S. Brohee and J. van Helden, Evaluation of clustering algorithms for protein-protein interaction networks, *BMC Bioinformatics* **7**, 488 (2006)
6. F.R.K. Chung, Spectral graph theory, *Conference Board of the Mathematical Sciences* **92**, American Mathematical Society (1997)
7. A.J. Enright, S. V. Dongen and C. A. Ouzounis, An efficient algorithm for large-scale detection of protein families, *Nucleic Acids Res.* **30(7)**, 1575–1584 (2002)
8. N. Fan, A. Chinchuluun and P. M. Pardalos, Integer programming of biclustering based on graph models, In: *Optimization and Optimal Control: Theory*

and Applications, edited by A. Chinchuluun, P.M. Pardalos, R. Enkhbat and I. Tseveendorj, Springer to appear (2009)

9. N. Fan and P. M. Pardalos, Direct approach of multi-group biclustering, in preparation, (2009)
10. J. Gagneur, R. Krause, T. Bouwmeester and G. Casari, Modular decomposition of protein-protein interaction networks, *Genome Biology* **5(8)**, R57.1 (2004)
11. M.G. Grigorov, Global properties of biological networks, *Drug Discov Today* **10(5)**, 365–372 (2005)
12. L.L. Elo, H. Jarvenpaa, M. Orei, R. Lahesmaa and T. Aittokallio, Systematic construction of gene coexpression networks with applications to human T helper cell differentiation process, *Bioinformatics* **23(16)**, 2096–2103 (2007)
13. W. Hager and Y. Krylyuk, Graph partitioning and continuous quadratic programming, *SIAM J. Discrete Math.* **12(4)**, 500–523 (1999)
14. W. Hager and Y. Krylyuk, Multiset graph partitioning, *Math. Meth. Oper. Res.* **55**, 1–10 (2002)
15. A. D. King, N. Przulj and I. Jurisica, Protein complex prediction via cost-based clustering, *Bioinformatics* **20(17)**, 3013–3020 (2004)
16. X. Peng, M.A. Langston, A.M. Saxton, N.E. Baldwin and J.R. Snoddy, Detecting Network Motifs in Gene Co-expression Networks Through Integration of Protein Domain Information, In: *Methods of Microarray Data Analysis V*, edited by P. McConnell, S. M. Lin, P. Hurban, 89–102, Springer (2007)
17. J. Reimand, L. Tooming, H. Peterson, P. Adler and J. Vilo, GraphWeb: mining heterogeneous biological networks for gene modules with functional significance, *Nucl. Acids Res.* **36**, W452–W459 (2008)
18. V. Spirin and L.A. Mirny, Protein complexes and functional modules in molecular networks, *Proceedings of the National Academy of Sciences* **100(21)**, 12123–12128 (2003)
19. L. Tari, C. Baral, and P. Dasgupta, Understanding the global properties of functionally-related gene networks using the gene ontology, *Pacific Symposium on Biocomputing* **10**, 209–220 (2005)
20. D. Watts, Networks, dynamics and the small world phenomenon, *American Journal of Sociology* **105(2)**, 493–527 (1999)
21. H. Wolkowicz and Q. Zhao, Semidefinite programming relaxations for the graph partitioning problem, *Discrete Applied Mathematics* **96-97**, 461–479 (1996)
22. Graph Partitioning at Yahoo! Research, http://research.yahoo.com/node/2368

PROTEIN-PROTEIN INTERACTIONS PREDICTION USING 1-NEAREST NEIGHBORS CLASSIFICATION ALGORITHM AND FEATURE SELECTION

M. R. GUARRACINO AND A. NEBBIA
High Performance Computing and Networking Institute,
National Research Council,
Naples, Italy
E-mail: mario.guarracino@cnr.it

A. CHINCHULUUN
Centre for Process Systems Engineering,
Imperial College London,
London SW7 2AZ, UK
E-mail: altannar@imperial.ac.uk

P. M. PARDALOS
Industrial and Systems Engineering Department,
University of Florida,
Gainesville, FL, USA
E-mail: pardalos@ufl.edu

In this paper, we consider the problem of predicting protein-protein interactions. The motivation is the prediction of interacting proteins can give greater insight in the study of many diseases like cancer, and it provides valuable information in the study of active small molecules. Here we formulate the problem as a binary classification problem and apply k-Nearest Neighbors classification technique to the classes of interacting and noninteracting proteins. A case study is analyzed to show it is possible to reconstruct a real network of thousands interacting proteins with high prediction accuracy in cross validation.

1. Introduction

Proteins are composed of amino acids arranged in a linear sequence of variable length. Protein sequences are composed of amino acids, The sequence of amino acids is defined by the nucleotide sequence of a gene and it is determined by its genetic code. The length of protein can vary from few hundreds amino acids to some thousands. The longest known human proteins are the titins, whose length is around 27,000 amino acids. Every protein in a living cell has its own function, and most proteins in a cell accomplish their task interacting with other proteins. The interactions involve some form of binding among proteins. Most proteins function cooperatively and are involved in large complex networks. Proteins can interact with the nucleic acids of a cell to regulate gene transcription, or they can bind to proteins of small nonprotein ligands, as in the case of enzyme binding to prevent catalytic activity. The interactions occurring between proteins (*Protein-Protein Interaction – PPI*) provide different functions, such as catalytic, structural, localization, cleavage, transferral, or inhibitory functions. Identifying PPIs is important to systematically understand the biological role of proteins. This is particularly important in deciphering biological mechanisms and in the identification of new molecules active against diseases.

The most used wet lab methods to detect PPIs are Yeast to Hybrid (Y2H) and tandem affinity purification-mass spectrometry (TAP-MS). Those methods have some drawbacks, mainly due to to the fact they are not capable to detect some protein complexes. More recently, other methods have been introduced. Among those, a method called protein-fragment complementation assay (PCA) seems to provide more accurate results. For a discussion about these methods, the interested reader may refer to the book edited by Li and Ng[11].

For these reasons, over the years many efforts have been devoted to devise methods to predict PPIs[20,18,17]. In a recent paper, Bock et al.[5] predict a complex network for aging. An elaboration of the network, detailed in the present paper, is shown in Figure 1.

Many databases contain information that can be used to predict PPIs. The oldest and major source of data for proteins is the Protein Data Bank (PDB). In July 2009, the number of protein structures available in PDB are over 46,700. These have been obtained with various techniques, such as X-rays, NMR and electron microscopy. More recently, the European Bioinformatics Institute (EBI) has set up the Protein Quaternary Structure database, which complements the PDB with information about the the quaternary structure of proteins, defined on their website as *the level of form in which units of tertiary structure aggregate to form homo- or hetero-multimers*.

Many databases contain manually curated, known PPIs. Among those we cite Biomolecular Interaction Network Database[1] (BIND), Database

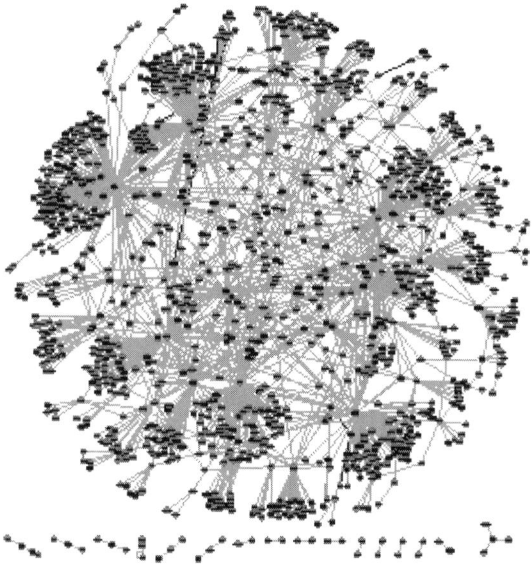

Figure 1. The aging network reconstructed from Bock et al. data

of Interacting Proteins[21] (DIP) and a Molecular Interactions Database[22] (MINT). Among those, a notable example is the Human Protein Reference Database (HPRD - www.hprd.org), which contains more that 37,000 taking place among nearly 25,000 proteins and small molecules. For a survey and a comparison on existing PPIs databases, the readers are referred to see Mathivanan's article[12].

In this work we address the problem of predicting PPIs applying machine learning methods only to sequence information. This problem has been addressed by Shen et al.[16]. Others have also tackled the same problem, obtaining less accurate results [4], [14]. All these works only use sequence information on the assumption that the sequence specifies the structure. Shen et al. approach is based on a coding of the proteins based on the frequency of the amino acids in the protein sequence. In this way, proteins, composed of a different number of amino acids, are represented by vectors of the same size. As it will be detailed in the next sections, the 20 amino acids are clustered in 7 classes, and the frequency of all possible class triples is recorded. The number of possible triples of elements drawn from the seven classes is equal to $7 \times 7 \times 7 = 343$. The problem is modeled as a binary classification task, in which the positive class (golden positive set - GPS) is composed of couples of proteins from HPRD that are known to interact in vitro or in vivo, and a negative one (GNS) that is composed of random couples that are not known to interact. Each class element is obtained by pairing the two protein vectors. In this setting, the classification model is obtained in a space with 686 dimensions.

Shen et al. report a classification accuracy, obtained on a dataset composed of 32,886 pairs, never above 86% using Support Vector Machines (SVMs), with various kernels, on a training composed of 32,486 proteins and a test set constructed of 400 protein pairs. In the same paper, authors try to reconstruct three different networks. It has to be noted that, although all interacting protein pairs of the three networks were contained in HPRD, and therefore in the training set, the use of kernel SVMs could predict a large part of the interactions, but not all of them.

In our work we could not reproduce the same experiments of Shen et al. [16], due to the computational complexity and memory footprint of the classification problem. For this reason, we directed our attention to alternative classification methods. Furthermore, we decided to use an instance based classification method, which would be exact for all pairs in the training set, without the side effect of kernel methods of overfitting the problem.

The contribution of this work consists in showing: *(1)* it is possible to gather a GNS dataset to increase prediction accuracy, and *(2)* it is possible to obtain very accurate results using a low computational complexity classification method, namely k-Nearest Neighbors (k-NN)[13], with an appropriate choice of the metric. In this new setting, accuracy on datasets of over 32,000 protein pairs can be over 90%, with an execution time in the order of tens of minutes on standard hardware.

The rest of the paper is organized as follows. In the next section, classification methods are described and compared. Then, in Section 3, the data preparation is described in details. In Section 4, classification results are discussed and, in Section 5, a case study is analyzed. Finally, in Section 6, conclusions are drawn and future work is proposed.

2. Methods

SVM[19] is state of the art technique in supervised learning and it has been successful applied for solving many scientific and technological problems[9]. In case of two linearly separable classes, it finds two parallel hyperplanes that separates the elements of the training set belonging to the two different classes. The separating hyperplanes are usually chosen to maximize the margin between the two classes. In other words, the optimization problem maximizes the distance between two parallel hyperplanes $w'x - b = \pm 1$ that leave all cases of the two classes on different sides. The classification hyperplane is $w'x - b = 0$, which is in the middle of, and parallel to, those maximizing the margin.

Let two matrices $A \in R^{n \times m}$ and $B \in R^{p \times m}$ represent the two classes in which each row is a point in the feature space. Then the following quadratic programming problem finds the optimal hyperplane (w, b):

$$\begin{aligned} \min f(w) &= \frac{w'w}{2} \\ s.t.\ (Aw + b) &\geq e \\ (Bw + b) &\leq -e, \end{aligned} \tag{1}$$

where e is the unit vector of appropriate dimension.

A point $\hat{x}$ for which the class label is unknown, is classified in class A if $(\hat{x}'w + b) \geq 0$, in class B otherwise.

For the classes which are not linearly separable, a non linear transformation method called *kernel function* can be used to project the points in the classes in a higher dimensional space, in which the two classes are linearly separable. One of the widely used kernel functions is the Radial Basis Function (RBF)

$$k(x) = exp(-\gamma \|x\|^2).$$

The class label of a new point is defined according to the image point and the classification plane in the new space.

Computational complexity of the quadratic programming problem is $O((n+p)^3)$, and $O(m^{2.1})$ when Sequential Minimal Optimization[15] is used. The number of proteins we are dealing with is in the order of 30,000 with 686 features, which cannot be handled by standard hardware with acceptable execution times. Another shortcoming of the algorithm is that it does not usually produce original class labels for the points in the training set. That is usually acceptable in standard machine learning problems since data are affected by errors and a classification model exactly reproducing all training set class labels would overfit the noisy data. However, in our case, data represent frequencies of amino acids, which are exactly known. To avoid these problems, we turned our attention to the k-Nearest Neighbors algorithm. The objective of the method is to discover k nearest neighbors for the given instance according to the majority class of k nearest neighbors. This instance is defined as the training dataset, and it is used to classify each member of the target dataset. When $k = 1$, the point is simply assigned to the same class of its closest neighbor. Different distance functions can be used to calculate how close each member of the training set is to the target data that is being examined. Here we use the *cityblock* distance, that is induced by the Minkowski metric with $l = 1$:

$$d(a,b) = \sum_{i=1}^{m} (|a_i - b_i|^l)^{1/l}, \tag{2}$$

which has has a lower computational complexity with respect to Euclidian metric.

3. Dataset preparation

The analysis of protein sequences becomes one of the most successful areas in bioinformatics, and researchers are increasingly relying on computational techniques to classify these sequences into functional and structural families based on sequence homology. One of the major challenges to predict PPIs using only sequence information is an efficient representation of the protein amino acids sequence. In protein structural class prediction tasks, Costantini and Facchiano[6] have considered the frequency of groups of t adjacent amino acids in order to preserve sequence information. When $t = 1$, the representation of a protein is in a space of dimension 20, and each component represents the relative frequency of one amino acid. When $t = 2$, the protein representation vector x has the of size $20 \times 20 = 400$, which is the total number of pairs that can be obtained with 20 amino acids. Larger tuples of amino acids produce higher dimensional vector spaces, without increasing the descriptive capabilities of the vector. Costantini and Facchiano have reported that a reasonable structural classification of proteins can be obtained with $t = 3$. In their work, the vector space has dimension of 8,000, while, Shen et al.[16] use *cojoint triads* of amino acids. They grouped amino acids into seven groups, {'A' 'G' 'V'}, { 'I' 'L' 'F' 'P'}, {'Y' 'M' 'T' 'S'}, {'H' 'N' 'Q' 'W'}, {'R' 'K'}, {'D' 'E'},{'C'}, based on biochemical considerations. With this grouping, the dimension of the vector space is reduced from 20^3 to 7^3, which makes the problem computationally tractable. For each protein i, the vector x_i with the frequencies of all 343 possible triplets is computed. Then, we can normalize the data using the formula:

$$v_i = \frac{x_i - min(x_i)}{max(x_i)}. \tag{3}$$

The normalization is used to avoid longer proteins sequences to have larger frequency values.

To prepare the positive class of the training set we have used HPRD, release 090107. This database contains 38,167 protein interactions between 25,661 proteins from the human proteome. Each interaction has been manually extracted by experts from the literature. We took a set of 16,650 pairs at random and, for each protein sequence in the pair (i, j), we computed the vectors v_i and v_j, and their concatenation composed of 686 components. For the negative GNS, we extracted at random the pair (l, m) and (k, s) from the HPRD database, and we checked that pair (l, s) was not present in HPRD. This does not mean the two proteins do not really interact, but the choice is motivated by the fact that if the two proteins are both listed in the database, and there is no evidence of their interaction, they can be supposed to be a non interacting pair.

In a way similar to Hur and Noble[10], we computed the Pearson coefficient of the corresponding frequency vectors:

$$\rho_{l,s} = \frac{\sigma_{l,s}}{\sigma_l, \sigma_s},$$

where $\sigma_{l,s}$ is the correlation, σ_l and σ_s the variance of the vectors v_l and v_s. We accept the pair as a non interacting one, if the absolute value of the Perason coefficient is less then 0.3.

The process has ended when 16,650 non interacting pairs have been computed. The total number of unique proteins is 7,652 in the negative class and 10,780 in the positive one. The number of unique proteins present in both classes is 155.

Protein-protein interactions can be represented as a network where proteins are vertices and certain types of interactions between proteins are edges. These structures are called protein-protein interaction networks. These networks play an important role in computational biology. They can be easily visualized and are are convenient for understanding the complex nature of different types of interactions between proteins by applying graph theory methods.

First attempts to study of complex networks were based on the theory of classical random networks. Erdos and Renyi[7] introduced the idea that in a random network, any two given nodes are connected with probability p, and the degrees of the nodes follow a Poisson distribution. In such a network many nodes have a number of connections closed to the average. Furthermore, in such settings, the probability of nodes with t links decreases exponentially for large values of t, since $P(t) = e^{-t}$. The latter shows that it becomes more and more unlikely to encounter nodes of with a degree that is significantly higher than the average. More recently a different model has been proposed by Barabasi and Albert[2]. In their *scale-free* model, they propose to model a biological network with a power-law relationship ($P(k) = e^{-r}$), rather than the Poisson distribution. In such a network, the most of the nodes are highly connected via hubs, with very few links between them. In Fig. 2 the sample distribution of the interactions in the HPRD (letf) and positive interaction in the training set (right) is reported. In both cases, there is a large number of proteins with few interactions and a few proteins with a large number of links. As said, this is usually accepted as representative of the behavior of proteins in cells, in which only few proteins take part to many different processes.

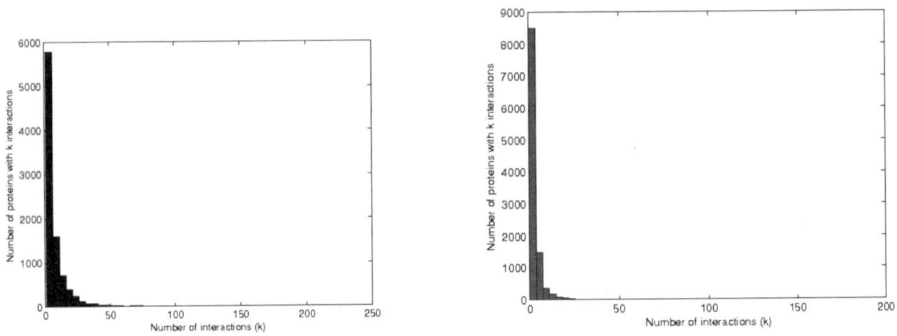

(a) HPRD (b) Interacting proteins in training set

Figure 2. Histograms of protein interactions

4. Computational experiments

All macros have been implemented with Matlab 7.3.0. Results are calculated using an Intel Xeon CPU 3.20GHz, 6GB RAM running Red Hat Enterprise Linux WS release 3.

Table 1 reports results of a 3-fold cross validation for the dataset in the previous section. Accuracy refers to the mean classification accuracy on the three folds. Sensitivity and specificity are defined as:

$$Sensitivity = \frac{TP}{TP + FN},$$

$$Specificity = \frac{TN}{TN + FP}$$

where TP is the number of true positives, FN the number of false negatives, and FP the number of false positives. Standard deviation of the accuracy on the three fold is also given.

4.1. *Comparison among metrics*

The distances used are cityblock, as defined in equation (2), euclidian and correlation, defined as one minus the correlation of the two vectors.

Table 1. Results for dataset with small overlap.

Distance	Accuracy	Sensitivity	Specificity	Std. dev.
Cityblock	94.29%	96.22 %	93.03 %	$\pm$ 0.24 e-2
Euclidean	84.64%	95.19%	79.56%	$\pm$ 0.79 e-3
Correlation	76.89%	82.21%	80.27%	$\pm$ 0.51 e-2

4.2. *Dataset consistency*

In order to understand if the accuracy results would depend on the number of proteins present in both the datesets, we repeated the experiments with a dataset with more proteins in common. We selected 16,200 pairs from HPRD with 1,957 unique proteins. We have generated 65,000 couples from those unique proteins, from which we filtered out 16,200 pairs with an absolute Pearson coefficient less than 0.3. The unique proteins in the negative class is 1,855. The results obtained with this dataset are reported in Table 2.

Table 2. Results for dataset with large overlap.

Distance	Accuracy	Sensitivity	Specificity	Std. dev.
Cityblock	88.33%	91.05%	88.09%	$\pm$ 0.43e-2
Euclidean	86.69%	92.33%	84.53%	$\pm$ 0.59e-2
Correlation	87.65%	92.26%	86.00%	$\pm$ 0.58e-3

We note that accuracy depends on the number of overlapping proteins in the two classes. Nevertheless, cityblock distance always achieves highest results in terms of accuracy.

As expected, we obtained lower accuracy results when the overlap is larger. This is due to the fact that the distance among points in the vector space is smaller.

We believe that in a real setting, the training set has to contain all information about interactions available from experiments. For the GPS, all known interacting proteins should be taken in consideration. For GNS, pairs should be selected with respect to the analysis task at hand, which means that processes, localization and functions of the proteins to be analyzed should be taken into account in the design. On the other hand, the accuracy values obtained by a random choice of the training set can still be satisfactory, as it will be shown in the following experiment, where the GPS and GNS were the ones described in section 3.

4.3. *Feature selection*

Feature selection addresses the problem of searching a minimal set of features that maximizes the discrimination among classes. If there are n features, there are 2^n possible features subsets. When n is very large, it is impossible to find the optimal features subset, therefore a suboptimal solution needs to be found. Feature transformation methods obtain a new set of features as linear combination of the original ones, that is, they linearly project points in a space of lower dimension. Feature selection methods have usually lower computational complexity with respect to feature transformation ones. Since there is no standard choice for feature selection method, and there is no clear biological understanding why certain triplet frequencies have greater discriminating power, we decided to use the one proposed by Golub et al.[8], which is reminiscent of Fisher discriminant criterion This method provides a scoring procedure with a computational complexity linear in the number of features. For each triplet frequency j, the means μ_j^+ and μ_j^- are calculated considering the classes of interacting proteins and non-interacting ones, separately. In the same way, the standard deviations σ_j^+ and σ_j^- are calculated. These values are used to evaluate the discriminant between the two classes:

$$F(j) = |\frac{\mu_j^+ - \mu_j^-}{\sigma_j^+ + \sigma_j^-}| \tag{4}$$

The best features are those with greater value of $F(j)$.
We calculate the $F(j)$ value for each gene and it sorts the features for decreasing the values of $F(j)$. Then, we choose those frequencies for which the sum of $F(j)$ is equal to a fixed percentage α of the total sum. In other words, the algorithm returns the frequencies j_k such that:

$$\sum_{k=1}^{m_\alpha} F(j_k) = \alpha \sum_{j=1}^{m} F(j), \tag{5}$$

with $0 \leq \alpha \leq 1$ and $m_\alpha \leq m$. We have decided to apply the feature selection procedure within the 3-fold cross validation process, that has been detailed earlier, on the dataset with lower overlap. This means, for each fold the most discriminating features are computed on the training set and accuracy is computed on the test set. Results are shown in Table 3.

Table 3. Classification results with feature selection for cityblock metric

α	Accuracy	Sensitivity	Specificity	Std. dev.
0.1	93.80 %	95.49 %	92.87 %	± .29e-2
0.2	94.77 %	96.96 %	93.19 %	± .59e-3
0.3	94.88 %	97.30 %	93.06 %	± .46e-2
0.4	94.98 %	97.20 %	93.33 %	± .61e-2
0.5	94.97 %	97.12 %	93.37 %	± .27e-2
0.6	94.80 %	96.93 %	93.26 %	± .40e-2
0.7	94.72 %	96.71 %	93.33 %	± .36e-2
0.8	94.70 %	96.61 %	93.38 %	± .22e-2
0.9	94.64 %	96.51 %	93.37 %	± .13e-2
1.0	94.29 %	96.22 %	93.03 %	± .24e-2

We note that the accuracy results are comparable for all choices of α, but the best results are obtained for $\alpha = 0.4$.

5. A case study

We tested the method on the protein-protein interaction network found in Bell et al.[3]. This network is composed of human homologs of proteins that have an impact on the longevity of invertebrates species. The network is composed of 175 human homologs of proteins that have been experimentally found to increase longevity in yeast, nematode, or fly, and 2,163 additional human proteins that interact with these homologs. Overall, the network consists of 3,271 binary interactions among 2,338 unique proteins. The article provides the names of interacting proteins pairs, but no accession number. From these 3,271 pairs we took the 1,740 protein names that were also present in the HPRD database. Then we found 2,062 pairs with both proteins in HPRD.

We then proceed to investigate which is the effect of the feature selection process on the accuracy of prediction. We used k-NN trained on the 33,300 protein pairs we obtained as described in section 3. In Figure 3 the bars represent the percentage of features selected with $\alpha \in [0.1, 0.2, \ldots, 1]$, while the dots represent the accuracy in the prediction for the aging network obtained in correspondence with that choice of α. We again find that

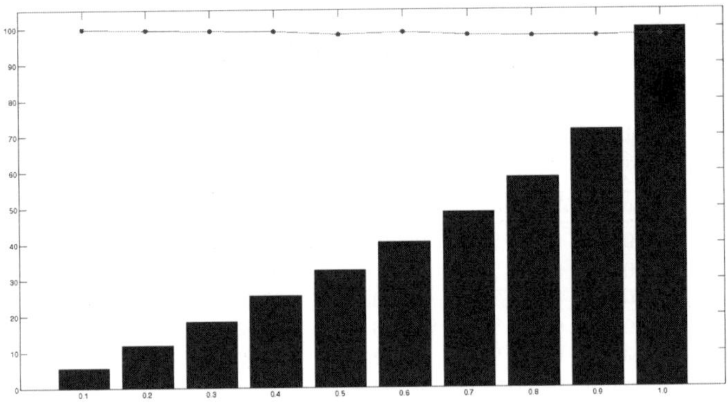

Figure 3. Effect of feature selection on accuracy prediction for the aging network

the best result is achieved with a value of $\alpha = 0.1$, with no significative difference in accuracy with respect to $alpha = 0.4$ ($p < 0.05$). Therefore, we decided to use the latter α to filter the features used in the prediction of the aging network.

The classifier was able to correctly predict the interaction of 2,023 pairs out of 2,062, with a classification accuracy of 98.11%. It is worth noting the training set contains only 32 interacting pairs present in the network. The resulting network has been processed with Cytoscape (www.cytoscape.org) and the result is depicted in Figure 1.

6. Conclusion

In this work, we propose a novel approach to predict protein protein interactions. Results on a network of 3,271 binary interactions among 2,338 unique proteins gives a prediction accuracy of 98,11%. Validation results are higher then those available in literature using different classification methods and methodologies to build the training set. Future work will be devoted to the implementation of different filtering techniques for GNS, to fully understand the capability of the prediction algorithm in connection with the training sets. Finally, it will be interesting to investigate the correctness of the classifier on the remaining 1,209 pairs of the aging network composed by the 598 proteins not found in HPRD. For all proteins contained in the network, amino acid sequences are available, but for those not in HPRD, obtaining the sequence from the name, rather than the accession number, which is not provided in the original paper, becomes a cumbersome and prone to error task.

Acknowledgement

Adriano Nebbia spent a period at ICAR-CNR as an undergraduate student. He has contributed to the present work implementing all software needed to codify the proteins frequencies for the data sets used for experiments in sections 4 and 5. This work has been partially funded by MIUR project PRIN 2007, and MOBILE, ERC Advanced Grant No 226462.

References

1. G. D. Bader, I. Donaldson, C. Wolting, B. F. Ouellette, T. Pawson, and C. W. Hogue. Bind–the biomolecular interaction network database. *Nucleic Acids Res*, 29(1):242–245, January 2001.
2. A. L. Barabasi and R. Albert. Emergence of scaling in random networks. *Science*, (286):509–512, 1999.
3. R. Bell, A. Hubbard, R. Chettier, D. Chen, J. P. Miller, P. Kapahi, M. Tarnopolsky, S. Sahasrabuhde, S. Melov, and R. E. Hughes. A human protein interaction network shows conservation of aging processes between human and invertebrate species. *Plos Genetics*, 5(3), 2009.
4. J. R. Bock and D. A. Gough. Predicting protein–protein interactions from primary structure. *Bioinformatics*, 17(5):455–460, May 2001.
5. J. R. Bock and D. A. Gough. Predicting proteinprotein interactions from primary structure. *Bioinformatics*, 17(5):455–460, 2001.
6. S. Costantini and A. M. Facchiano. Prediction of the protein structural class by specific peptide frequencies. *Biochimie*, 1-4, 2008.
7. P. Erdös and A. Rényi. On random graphs, i. *Publicationes Mathematicae (Debrecen)*, 6:290–297, 1959.
8. T. R. Golub, D. K. Slonim, P. Tamayo, C. Huard, M. Gaasenbeek, J. P. Mesirov, H. Coller, M. Loh, J. R. Downing, M. A. Caligiuri, C. D. Bloomfield, and E. S. Lander. Molecular classification of cancer: Class discovery and class prediction by gene expression monitoring. *Science*, 286:531–537, 1999.
9. M.R. Guarracino, S. Cuciniello, D. Feminiano, G. Toraldo, and P.M. Pardalos. Current classification algorithms for biomedical applications. *Centre de Recherches Mathmatiques CRM Proceedings & Lecture Notes of the American Mathematical Society*, 45(2):109–126, 2008.

10. Asa B. Hur and William Noble. Choosing negative examples for the prediction of protein-protein interactions. *BMC Bioinformatics*, 7(Suppl 1), 2006.
11. X. Li and S. Ng. *Biological Data Mining in Protein Interaction Networks.* Igi Global, 2009.
12. Suresh Mathivanan, Balamurugan Periaswamy, T. K. B. Gandhi, Kumaran Kandasamy, Shubha Suresh, Riaz Mohmood, Y. L. Ramachandra, and Akhilesh Pandey. An evaluation of human protein-protein interaction data in the public domain. *BMC Bioinformatics*, 7(Suppl 5), 2006.
13. Tom M. Mitchell. *Machine Learning.* McGraw-Hill, New York, 1997.
14. Loris Nanni. Hyperplanes for predicting protein-protein interactions. *Neurocomputing*, 69(1-3):257–263, 2005.
15. J. Platt. *Advances in Kernel Methods: Support Vector Learning*, chapter Fast training of SVMs using sequential minimal optimization, pages 185–208. MIT press, Cambridge, MA, 1999.
16. Juwen Shen, Jian Zhang, Xiaomin Luo, Weiliang Zhu, Kunqian Yu, Kaixian Chen, Yixue Li, and Hualiang Jiang. Predicting protein-protein interactions based only on sequences information. *PNAS*, 104(11):4337–4341, March 2007.
17. T. L. Shi, Y. X. Li, Y. D. Cai, and K. C. Chou. Computational methods for protein-protein interaction and their application. *Curr Protein Pept Sci*, 6(5):443–449, October 2005.
18. B. Shoemaker and A. Panchenko. Deciphering proteinprotein interactions - part ii. computational methods to predict protein and domain interaction partners. *PLoS Computational Biology*, 3(4):595–601, 2007.
19. V. Vapnik. *The Nature of Statistical Learning Theory.* Springer-Verlag, 1995.
20. A. Walker-Taylor and D. Jones. *Proteomics and protein-protein interactions : biology, chemistry, bioinformatics, and drug design*, chapter Computational methods for predicting protein protein interactions, pages 89–114. Springer, 2005.
21. I. Xenarios, D. W. Rice, L. Salwinski, M. K. Baron, E. M. Marcotte, and D. Eisenberg. Dip: the database of interacting proteins. *Nucleic Acids Res*, 28(1):289–291, January 2000.
22. A. Zanzoni, L. Montecchi-Palazzi, M. Quondam, G. Ausiello, M. Helmer-Citterich, and G. Cesareni. Mint: a molecular interaction database. *FEBS Lett*, 513(1):135–140, February 2002.

CLUSTERING DATA IN CHEMOSYSTEMATICS USING A GRAPH-THEORETIC APPROACH: AN APPLICATION OF MINIMUM SPANNING TREE WITH PENALTY CONCEPT*

L. S. OLIVEIRA, V. C. SANTOS, L. SILVA,
L. MATOS AND S. CAVALCANTI

Computing Department, Federal University of Sergipe
Av. Marechal Rondon, São Cristóvão, SE, CEP 49100-000, BRAZIL
E-mail: liliane.stnt@gmail.com, valeriacssp@gmail.com, leila@ufs.br, lnmatos@ufs.br, socrates@ufs.br

Chemosystematics is the classification of plants according to their chemical composition. This paper describes a clustering algorithm, based on minimal spanning trees, that uses the penalty concept to determine the dissimilarity among objects. The algorithm is applied to a dataset of compounds from the essential oil of plants acquired for classification purpose. The results achieved are compared with the ones that consider the proposed algorithm without the penalty concept and the ones that consider a preprocessing of the original data using the principal component analysis method.

1. Introduction

Chemical data has been widely used to solve botanical problems. Chemotaxonomy is the classification of plants based on the presence and concentration of certain specific chemical compounds. It uses chemical markers, particularly secondary metabolites, such as, alkaloids, terpenoids, and flavonoids, from a group of organisms to classify species [1].

A detailed study of the chemical composition and the plant botanical identification is of the outmost importance to assist on industrial quality control and production processes of plant derived pharmaceutical products. Additionally, the development of the chemistry of natural products has showed that the phytochemical constituents may be used to characterize, describe and classify species. Correlations between traditional

*This work is supported by the Brazilian and Sergipean research agencies CNPq and FAPITEC.

systematic and chemical classification are found on the literature since 1699[3]. However, most of the relationship studies between botanical systematic and chemical composition has recently been published, driven by the emergence of new, faster, and precise analytical techniques.

Evidences revealed by the chemical composition of plants have implied on a number of reconsiderations on their botanical classification. For example, botanical classification has had no success on classifying some troublesome families. Analysis of the secondary metabolites of such families has assisted to proper classification. The *Bonnetiaceae*, which consists of two genders, *Bonnetia* and *Archytaeae* are better associated with the *Guttiferae* than with the *Theaeceae*, due to the presence of xanthones.

Many classes of metabolites have proved useful in establishing taxonomic relationships. For example, the distribution of indol, carbazol, 8-predinilated cumarins, and monoterpenes were combined in a way to substantiate the *Murraya* (*Rutaceae*) gender division in two diferent genders [2].

Several studies showed the significance of chemical data to the solution of taxonomical problems [4,5,6,7]. The furthermost difficulties found by researchers to solve taxonomical problems lay in the construction of a suitable database, utilization of statistical methods, and the development of software to validate the procedure.

Chemosystematics has been aided by two major statistical tools, Principal Component Analysis (PCA) [7,8] and Cluster Analysis [9,10,11]. PCA [12] is a powerful statistical method that has been used to determine variability of essential oils from plants in large multivariate data sets so that the data can be more easily interpreted. Essentially, PCA provides the means to reduce the often large number of independent (correlated) variables or principal components that are linear combinations of the original variables and explain most of the variation in the original data set. It can be used to highlight relationships (correlations) between plants as well as between essential oil composition and plants.

Cluster Analysis is an exploratory technique that classifies or groups objects based on similarities in the observational data set of variables selected for the analysis. That is, it defines the structure of a data set by grouping similar observations so that the objects in a cluster are more similar to one another than they are to objects in other clusters. Several approaches for clustering data have been proposed. The work of Jain *et al.* [13] is an interesting survey in the subject. These approaches have been used in several domains, including chemosystematics. Nevertheless, for the domain here addressed – clustering of plants based on the similarity of their oil essential composition– the authors are not aware of any other work that applies clustering techniques based on graphs, and in this sense, this work is a contribution in the investigation of this domain.

The graph-theoretical clustering approach here described is based on minimal spanning trees (MST). In fact, this work extends the approach proposed by Yujian [14] in two ways: the consideration of the penalty concept in the dissimilarity measure between plants and the establishment of a different metric to define the threshold cutting of the MST. The approach has been validate by developing 132 of experiments experiments, for each of the three scenarios: (1) the application of the adapted algorithm of Yujian with the new threshold cutting metric; (2) the extension of scenario (1) to include the penalty concept and (3) the use of pre-processed data, using the PCA technique, as the input of the algorithm in scenario (1). The aim of this analysis is to identify whether the consideration of penalty or PCA technique improves the quality of the generated groups in relation with the original approach.

This paper is organized as follows. Sections 2 and 3 presents the similarity criteria and the algorithm used, respectively. The results of the experiments performed using the data set of plants are summarized in Section 4. Finally, in Section 5 we conclude and present directions of future works.

2. The Similarity Criterion

Clustering algorithms are largely used to group individuals based on similarity criteria. The problem may be stated as follows.

Problem Let $P := \{p_1, p_2, \ldots, p_n\}$ be a data set of objects to analyse. The goal of the clustering process is to find a partition of non-empty subsets P_k of P, $1 \leq k \leq r$, such that $P = \bigcup_{k=1}^{r} P_k$, and $\forall i, j, i \neq j, 1 \leq i, j \leq r, P_i \cap P_j = \emptyset$. The subsets P_k are called *clusters* and elements belong to a same cluster according to a similarity criterion. In the domain of chemosystematics the objects are plants.

The similarity criterion is fundamental to the definition of a cluster. Each object p_i is represented as a list of attributes ($p_i := x_{i1}, x_{i2}, \ldots, x_{im}$) and objects with similar attributes should belong to a same cluster. For our domain of interest, $x_{is}, 1 \leq s \leq m$ are the concentration of essential oil of plant p_i.

To measure the degree of similarity it is common the use of a dissimilarity measure, based on the distance computation. The most popular metric for continuous attributes is the Euclidean distance, given by

$$d_E(p_i, p_j) = \sqrt{\sum_{s=1}^{m}(x_{is} - x_{js})^2} \tag{1}$$

In this paper we consider this metric and also the following criterion, that includes the penalty concept denoted as g_{ij}.

$$d_{EP}(p_i, p_j) = \sqrt{\sum_{s=1}^{m}(x_{is} - x_{js})^2 \times g_{ij}} \tag{2}$$

Definition 2.1. Let p_i and p_j be two objects with attributes $x_{s1}, x_{s2}, \ldots, x_{sm}, s = i, j$. The penalty g_{ij} is defined as the number of discrepant variables of p_i and p_j. A variable is said to be *discrepant* if either it is presented in only one of the objects (for a given $y, x_{yi} = 0$ and $x_{yj} \neq 0$, or vice-versa) and/or the difference of its value in these objects is greater than a user-established threshold t for the problem ($|x_{yi} - x_{yj}| > t$).

The reason of introducing penalty in the dissimilarity measure is to consider the knowledge of the domain here investigated. This concept captures the fact that, in the study of variability of plant essential oils, the lack of one attribute means the plant needs to be classified in a separate group.

3. The Algorithm Based on Minimal Spanning Trees

The algorithm described in what follows is an extension of the one presented by Yujian [14]; it basically differs on the equation to calculate the threshold cutting , denoted by θ, of the minimal spanning tree.

3.1. *Definitions and Representation of the Data Set*

Let $P := \{p_1, p_2, \ldots, p_n\}$ be the data set and d_{ij} be the distance measure between objects p_i and p_j of P; this distance may be calculated as in equations (1) and (2). Let $G(P) = (VG(P), EG(P))$ be a weighted (undirected) graph, where vertex set $VG(P)$ represents the elements of P. Each pair of vertices in $VG(P)$ is connected by an edge; thus $G(P)$ is a *complete* graph and $EG(P) := \{(p_i, p_j)|1 \leq i, j \leq n, i \neq j,\ n = |VG(P)|\}$. Each edge (p_i, p_j) has a weight that represents the distance d_{ij}. A *subgraph* of $G(P)$ is a graph $H(P)$ such that $VH(P) \subseteq VG(P)$ and $EH(P) \subseteq EG(P)$.

A *path* in $G(P)$ is a sequence of alternated distinct vertices and edges of P. If the first and the last vertices of this sequence are identical, the path is closed and it is called a *cycle*. A graph $G(P)$ is said to be *connected* if there is at least one path between each pair of vertices of $G(P)$.

A *tree* is a connected graph that does not contain any cycle. A *spanning tree* $ST(P)$ of $G(P)$ is a subgraph of $G(P)$ that is a tree, such that $VST(P) = VG(P)$. A tree T has the property that $|ET| = |VT| - 1$. A *minimal* spanning tree $MST(P)$ of $G(P)$ is a spanning tree ST of $G(P)$ such that

$$p(ST(P)) = \sum_{i=1}^{|VST|-1} p(e_i)$$

is minimal among all possible spanning trees $ST(P)$ of $G(P)$. The terms $p(e_i)$ and $p(ST(P))$ stands for the weight of the edge e_i and of the tree $ST(P)$, respectively.

Let $\emptyset$ represents the empty subset and define $\max\emptyset = 0$ and $\min\emptyset = \infty$. A maximal θ-distant subtree of $G(P)$ is defined as a subtree $T_\theta(P)$ of $G(P)$ which satisfies the following conditions:

(1) $T_\theta(P)$ is an $MST(P)$ of a subgraph $H(P)$ of $G(P)$;
(2) $p(e) \leq \theta$, $\forall e \in T_\theta$;
(3) $p(e) > \theta$, $\forall e = (x_i, x_j)$, such that $x_i \in VT_\theta(P)$ and $x_j \notin VT_\theta(P)$.

For example, for the tree shown in Fig. 1(a), if $\theta = a - \varepsilon$, where ε is a sufficiently small positive number, we have two maximal θ-distant subtrees (Fig. 1(b)), whereas if $\theta = b - \varepsilon$ we have three ones (Fig. 1(c)).

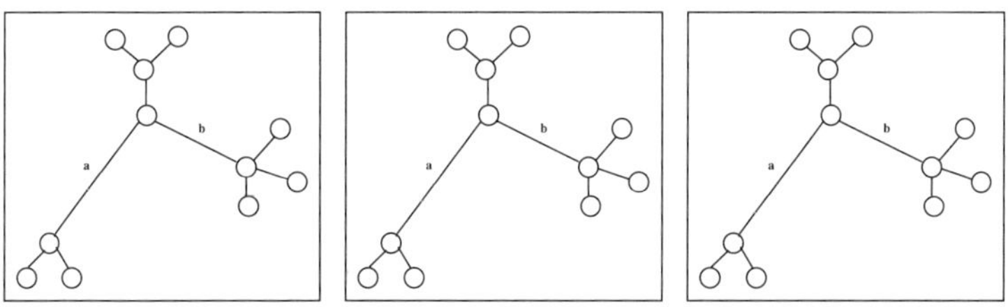

(a) case 1 (b) case 2 (c) case 3

Figure 1. A spanning tree and two levels of maximal θ-distant subtrees.

3.2. *The Clustering Algorithm Based on Maximal θ-distant Subtrees*

The classical clustering algorithm based on MST is given in what follows. The intuition is that two vertices with a short edge-distance (small weight value) should belong to the same cluster (subtree) and vertices with long edge-distance should belong to different clusters and hence long edge-distances should be cut.

algorithm MST-cluster

> *Let $G(P)$ be a graph.*
> *Use Kruskal algorithm* [15] *to calculate the $MST(P)$ of $G(P)$;*
> *Define a threshold cutting t;*
> *Use breadth-first search to collect the subtrees $T_k(P), k = 1, 2, \ldots \ell$ of $MST(P)$, such that $p(x_i, x_j) \leq t, \forall (x_i, x_j) \in ET_k(P)$;*
> *Let $C_k := T_k(P)$ be the clusters found.*

The algorithm based on maximal θ-distant subtrees, denoted as $\theta - MST$ *cluster*, is a variant of this algorithm that aims to compute the *MST-cluster* algorithm for several threshold t, with the aim to automatically establish the best one for cutting the tree. It can be described as follows. The threshold is denoted by θ, given by $\theta = Mean + Var$, where for a given tree T, with $|VT| = n$ and $e_i \in ET$,

$$Mean = \frac{1}{n-1} \sum_{i=1}^{n-1} p(e_i)$$

and

$$Var = \frac{1}{n-1} \sum_{i=1}^{n-1} (p(e_i) - Mean)^2.$$

The parameter N is empirically determined by $N = \sqrt{|VG(P)|}$ for large data sets and $N = 1, 2, 3$ for small ones. In the experiments of Section 4 we consider $N = 3$.

algorithm $\theta - MST$ *cluster*$(G(Y))$

{ *input: graph* $G(P)$; *output: collection of clusters* C (*initially* $C = \emptyset$)}
Use Kruskal algorithm [15] *to calculate the* $MST(Y)$ *of* $G(Y)$;
Define a threshold cutting $\theta = Mean + Var$;
Use breadth-first search to collect the subtrees $T_k(Y), k = 1, 2, \ldots \ell$ *of* $MST(Y)$, *such that* $p(x_i, x_j) \leq \theta, \forall (x_i, x_j) \in ET_k(Y)$;
Divide the subtrees $T_k(Y)$ *into two disjoint sets, where* $u + w = \ell$:
$F^1 = \{T_k(Y) | 1 \leq k \leq u, |T_k(Y)| \geq N\}$ *and*
$F^2 = \{T_k(Y) | 1 \leq k \leq w, |T_k(Y)| < N\}$;
Case
$u = 0$
{*overfragmentation of a cluster is stopped*}
$C := F^2$;
$u > 0$ *and* $w = 0$:
{*no small clusters are detected*}
for each $T_k(Y)$ *of* F^1 *do*
Let $G(Y)$ *be the complete graph*
such that $VT_k(Y) = VG(Y)$;
$C := C \cup \theta - MST$ *cluster*$(G(Y))$;
$u > 0$ *and* $w > 0$:
{*merge small clusters*}
Let $T^* := \bigcup_{k=1}^{w} T_k(Y)$;
if $|VT^*| \geq N$
then $F^{1*} := F^1 \cup \{T^*\}$
else { *mark each vertex in* T^* *as outliers* }
for each x_i *in* VT^* *do* $C := C \cup \{x_i\}$
$F^{1*} := F^1$;
for each $T_k(Y)$ *of* F^{1*} *do*
Let $G(Y)$ *be the complete graph*
such that $VT_k(Y) = VG(Y)$;
$C := C \cup \theta - MST$ *cluster*$(G(Y))$;
return C.

4. Results and Discussion

To evaluate the approach we have considered a data set comprising 87 plants, belong to 20 genera. Each plant is represented by a vector of the concentration of its essential oil composition. The number of compound in each plant is variable. The information used has been extracted from the literature.

Based on this data set we have considered 132 case studies for each scenario investigated. Each case study comprises two genera of plants; each one consisting of three to twenty plants. In this case the algorithm reaches a correct answer if it produces only two groups and, in each one, there would be elements of just one specie. However, in many practical situation it is not possible to achieve the desirable answer and some mismatches can arise. The algorithm may produce groups with elements of both species or may also produce three or more groups, some of them with just one element (outlier). We have developed our analysis based on the capability to avoid those mistakes.

We have investigated the performance of the $\theta - MST$ *cluster* algorithm in three different scenarios: (1) the application of the adapted algorithm of Yujian with the new threshold cutting metric; (2) the extension of scenario (1) to include the penalty concept and (3) the application of PCA technique. A scenario is, in fact, associated to a feature extraction procedure, that is, a pre-processing stage, prior to clustering, where feature vectors are formed. The result of each pairwise alignment, in each of those scenarios, was investigated.

Table 1 shows results of experiments with plants of Salvia genus in scenario (2). Column 1 refers to groups formed exclusively with plants of Salvia genus; Column 2 refers to groups formed exclusively with plants that do not belong to the Salvia genus; Column 3 refers to isolated plants (outliers) and Column 4 refers to mixed groups, comprising plants of both genera under investigation. We can see that some runs produced the desired result, whereas some others did not do it. It is crucial to investigate how bad or how good those results are. We have done it on assessing clustering performance in each scenario.

Based on Table 1 it has become suggestive that the $\theta - MST$ *cluster* algorithm with penalty dissimilarity measure are an effective approach for Salvia chemotaxonomy identification. However, to be able to fully understand adherence of method to data, it becomes necessary to investigate whether similar results occurs when testing the method with other genera. In Figure 2 we show the mean value (mean of columns 2 and 3 in Table 1), minimum and maximum of clusters per genus, for each genus and scenario. The adequacy of the feature extraction procedure can be visualized based on the spread (size of the vertical bar) and adherence of the mean value to the ideal number of cluster. A suitable method is the one that produces

Table 1. Results from Salvia genus in scenario (2)

Plants	Salvia groups	Not Salvia groups	Outliers	Mixed groups
Salvia e Sideritis	1	1	4	0
Salvia e Thymus	0	3	5	3
Salvia e Teucrium	0	3	4	1
Salvia e Satureja	0	0	2	1
Salvia e Mentha	0	0	1	1
Salvia e Origanum	0	0	1	1
Salvia e Lavandula	0	0	1	1
Salvia e Hesperozygis	**1**	**1**	**0**	**0**
Salvia e Melissa	0	0	1	1
Salvia e Micromeria	**1**	**1**	**0**	**0**
Salvia e Minthostachys	**1**	**1**	**0**	**0**
Salvia e Hedomea	**1**	**1**	**0**	**0**
Salvia e Hyptis	0	0	2	1
Salvia e Nepeta	0	1	6	2
Salvia e Phlomis	**1**	**1**	**0**	**0**
Salvia e Agastache	1	0	2	0
Salvia e Ocimun	1	1	1	0
Salvia e Stachys	**1**	**1**	**0**	**0**
Salvia e Cunila	2	1	1	0

small variance (small vertical bar) and mean value closed to one. We can see that there is a large variance in the number of cluster per genus and that there is no absolute feature extraction procedure, that is, a feature extraction procedure suitable for one genus may not be for another one. Nevertheless, we can show empirically, based on an analytical procedure, that penalty dissimilarity measure is, in average, the best one.

For each genus considered in the experiments we choose the best feature extraction method based on the absolute distance between the mean number of clusters per genus and the value one. For a particular genus, the feature extraction procedure with the smaller distance is the better one and we compute this as a success situation. Table 2 reveals, for each scenario, the mean number of cluster per genus (column 2), the standard deviation (column 3) and the total occurrence of success (column 4). Results of Table 2 help us on ranking the feature extraction procedure. We can see that, in average, PCA and penalty dissimilarity have the same mean number of clusters per genus but, since the standard deviation of penalty dissimilarity is smaller and since the total occurrence of success is higher, we can infer based on this criteria that the penalty dissimilarity (scenario (2)) is often better than the other ones.

We have also developed an analysis to measure how sensitive is the feature extraction procedure to the presence of outliers. In this case, we

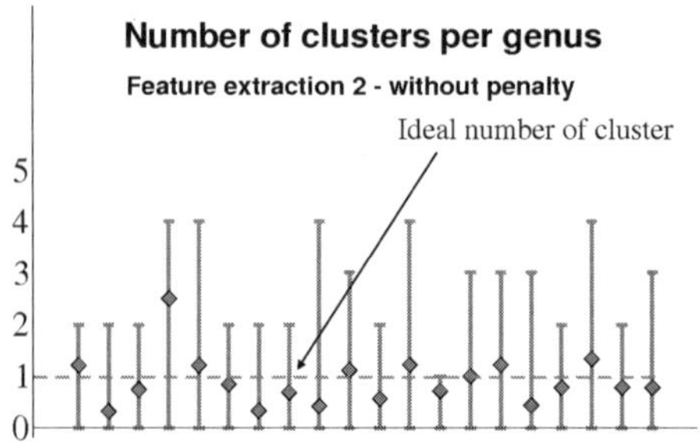

(a) Scenario 1

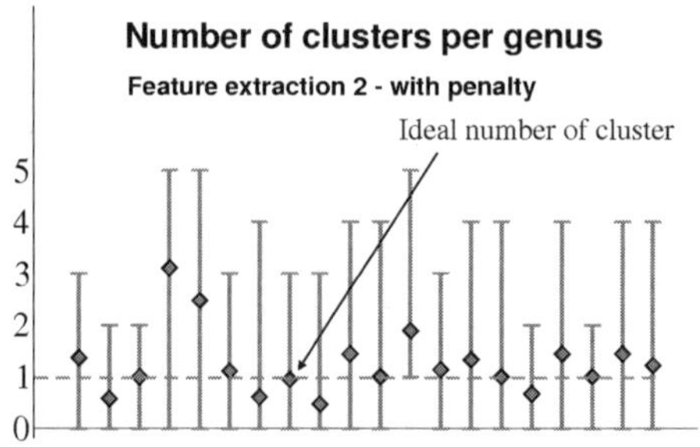

(b) Scenario 2

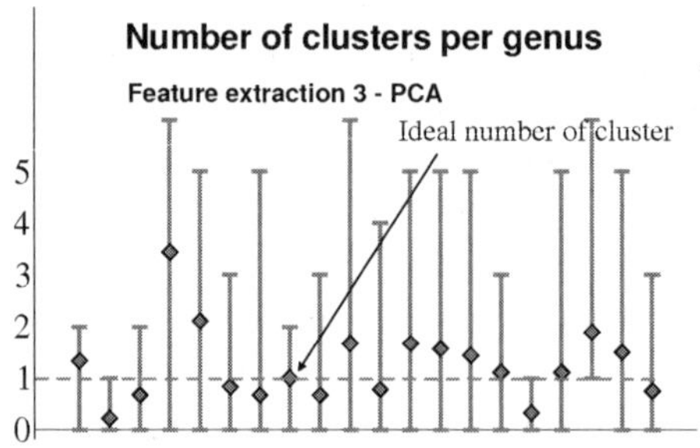

(c) Scenario 3

Figure 2. Dispersion of clusters per plant (minimum, average, maximum)

assign a success to a feature extraction procedure for a particular genus if, in average, the number of outliers is closer to zero. Table 3 reveals the mean number of outliers, standard deviation and number of success for each feature extraction method. We can see that when using PCA (scenario

Table 2. Analysis of cluster/genus

Feature extraction procedure (scenario)	Mean number of clusters/genus	Standard deviation	Number of success
(1)	0.64	0.82	5
(2)	0.89	1.00	10
(3)	0.89	1.14	5

Table 3. Analysis of outliers per genus

Feature extraction procedure (scenario)	Mean number of outliers/genus	Standard deviation	Number of success
(1)	3.37	2.05	0
(2)	2.51	1.76	6
(3)	1.74	1.24	14

(3)) the number of outliers is typically smaller then the ones we get when using other methods. Hence, whereas penalty dissimilarity furnishes more coherent groups, it is also more sensitive to the presence of outliers.

5. Conclusions

This work presents an extension of a graph-theoretical based approach to clustering data, proposed by Yujian [14]. The differential is to consider the penalty concept when computing the dissimilarity measure and to define a new threshold cutting expression. In the domain here addressed - clustering of plants based on the similarity of their oil essential composition– the authors are not aware of any other work that applies clustering techniques based on graphs, and in this sense, this work is a contribution in the investigation of this domain.

Moreover, we have exploited three scenarios using a data base of essential oil composition of plants: (1) the application of the adapted algorithm of Yujian with the new threshold cutting metric; (2) the extension of scenario (1) to include the penalty concept and (3) the use of pre-processed data, using the PCA technique, as the input of the algorithm in scenario (1).

We have developed 132 experiments for each scenario to validate the three situations. The results suggest that the scenarios (2) and (3) generate better results than scenario (1). Moreover, scenario (2) gives in average groups with better quality, although outliers are better avoided in scenario (3).

The next steps are to validate the algorithm with a larger data set and also to apply the algorithm to other domains, in order to reach a more conclusive result about the use of the three scenarios. In addition we would like to investigate the impact of the use of the penalty concept and PCA technique with other clustering algorithms.

References

1. R. D. Gibbs. *Chemotaxonomy of flowering plants*, McGill-Queen's University Press (1999).
2. P. G. Waterman, *Phytochemistry* **49** 1175 (1998).
3. D. E. Fairbrothers, *Annals of the Missouri Botanical Garden* **64** 147 (1977).
4. D. S. Rycroft, *J. Hattori Bot. Lab.* **93** 331 (2003).
5. M. I. Sampaio-Santos,M. A. C. Kaplan, *J. Braz. Chem. Soc.* **12** 144 (2001).
6. K. M. Valant-Vetschera, E. Wollenweber, *Biochem. Syst. Ecol.* **29** 149 (2001).
7. C. Zidorn, H. Stuppner, *Biochem. Syst. Ecol.* **29** 827 (2001).
8. Z. A. Rafii, E. Zavarin, Y. Pelleau, *Biochm Syst. Ecol. 19*, 249 (1991).
9. A. G. Figueredo, M Sim-sim, M. M. Costa, J. G. Barroso, L. G. Pedro, M. G. Esquive, F. Gutierres, C. Lobo, S. Fontinha, *Flavour Frag J.* **20** 703 (2005).
10. M. Dósea, L. Silva, M. A. Silva, S. Cavalcanti, *Chemometrics and Int. Lab. Syst.* **94** 1 (2008).
11. N. S. S. Magalhães, S. Cavalcanti, I. R. D. Menezes, A.A.Araújo, H. M. Oliveira, A. J.lves, *Eur. J. Med. Chem.* **34**, 83 (1999).
12. R. Kramer, *Chemometric Techniques for Quantitative Analysis*, Marcel-Dekker (1998).
13. A. K. Jain, M. N. Murty, P. J. Flynn, *ACM Computing Surveys* **31** 264 (1999).
14. L. Yujian, *Pattern Recognition* **40**, 1425 (2007)
15. R. Sedgewick, *Algorithms: Part 5 – Graph Algorithms*, Addison-Wesley (2002).

NATURAL CLUSTERING USING PYTHON

D. E. RAZERA

Federal Institute of São Paulo, Acesso Dr. João Batista Merlin, s/n - Jardim Itália - São João da Boa Vista - SP, Zip Code 13872-551, Brazil

C. D. MACIEL[‡]

J. C. PEREIRA

Electrical Engineering Department, University of São Paulo, Av. Trabalhador São-Carlense, 400, São Carlos, SP, Zip Code 13566-590, Brazil

[‡]*E-mail:* maciel@sc.usp.br

S. P. OLIVEIRA

Computer Science Department, University of Iowa, 101 MLH Iowa City, IA 52246, USA

Clustering involves the task of dividing data into homogeneous clusters so that items in the same cluster are as similar as possible and items in different clusters are dissimilar. The Fuzzy C-Means Clustering (FCM) algorithm is one of the most widely used fuzzy clustering algorithms. Using a combination of fuzzy clustering, resampling bootstrapping) and cluster stability analysis for all possible numbers of clusters of the dataset, it is possible to obtain the correct number of clusters. Real datasets present samples which may have some attribute values inconsistent within the same cluster. Using these samples can insert an error that interferes with the quality of classification. This can be solved by modifying the FCM algorithm to accept a degree of reliability for each attribute of each sample. Adapting this method to work with datasets with a large number of samples is computationally intensive. We use Python for the implementation of the proposed method. Python is a dynamic object-oriented programming language that offers strong support for integration with other languages and comes with extensive standard libraries. Because we use the MPI parallel routines with Python we developed a classification method based on FCM and resampling, which has excellent computing performance and greatly reduced implementation costs.

1. Introduction

Clustering algorithms can be applied in many fields like medicine, biology, exploratory data analysis, marketing, to generate potential clustering and hypotheses for subsequent studies[1].

Clustering involves the task of dividing data into homogeneous classes or clusters so that items in the same class are as similar as possible and items in different classes are as dissimilar as possible. Cluster analysis can be considered the most important unsupervised learning problem to investigate and interpret data. Clusters algorithms are expected to produce partitions that reflect the internal structure of the data and identify natural classes and hierarchies present in it[2,3,4]. Many different clustering algorithms[5,6] have been proposed for automatically uncovering the natural partition in the dataset. This research uses the Fuzzy C-Means algorithm similarly to Ref. [7].

However, most clustering algorithms suffer from the drawback of always generating a partition of the data, regardless of the correct number of classes. Thus, we need to validate the output of a clustering algorithm to verify if the classes classification represent a meaningful structure of the data, or are just an artifact of the algorithm[8]. The notion of the natural partition can be interpreted as the choice of the most consistent partitioning of the many elements in a dataset to maximize the stability. Resampling can be used to evaluate this stability[9,10]. The central idea of resampling is to compare a reference cluster with many clusters from sub-samples of the original dataset. The quality of classification is based on the traditional measures like F1, cross-classification, Hubert, and others. For a survey of metrics that can be used for measuring the quality of the clusterings, see Ref. [11]. The sub-sample is chosen randomly and is representative of the original dataset, and if there is discrepancy in the values of the metrics.

When data analysis methods are applied to real world applications, one often finds those datasets contain many inconsistent or missing data elements. Inconsistent datasets occur for many reasons, may arise as a result of human operators, misplaced observations or by failure of measurement equipment.

The method used in this work identifies the correct number of clusters in a dataset, by using metrics to evaluate the quality of a clustering. However the processing times of the initial implementation in Matlab are very high for databases with large numbers of samples and attributes. In many real applications the number of samples is extreme large, like genome databases, climate studies, analysis of market values, etc. Clustering such datasets could require excessive time, which might make a timely response impossible.

Adapting the method of Ref. [11] for datasets with a large number of samples and attributes requires the use of parallel programming techniques in order to significantly reduce the processing time. To achieve the parallel implementation quickly we turned to Python, which is as easy to use for writing programs as Matlab but allows parallelization through the use of MPI. Python is an interpreted scripting language, and now has sufficient resources for manipulation of arrays and files[12,13]. The advantage of using Python is the gain in time for design and implementation of the algorithm of the method[14].

The possibility of adding modules that expand the capabilities to allow manipulation of data makes Python an excellent choice to implement the algorithms under study. MPI allows high level parallel data structures. The module MPI4PY[15], used in this work, allows the program to utilize MPI and the facilities for numeric manipulation of data in Python.

We run our programs on a Beowulf cluster which is built with traditional microcomputers interconnected in a network, working like a supercomputer with multiple processors. This physical structure may use either dedicated or undedicated, homogeneous or heterogeneous processors, allowing variability in the number and type of hosts working on a problem. This enables any lab with personal computers in a network to provide a cluster computer and therefore run parallel programs.

Summarizing, the objective of this work is to implement a classification method based on modified FCM[7] and resampling[11], using MPI routines within the Python language. This will achieve greater computing performance while allowing our method to handle inconsistent datasets with large numbers of elements and attributes.

The structure of paper is follow: In section 2, we propose an algorithm that we implement; Section 3 presents a discussion of Python and MPI; Section 4 presents discussions about the implementation of the algorithms, datasets and computational clusters; Section 5 presents the results and discussions while Section 6 contains the conclusions.

2. Proposed algorithm

We used the modified fuzzy c-means (FCM) algorithm[11] that is similar to the original FCM[7] algorithm, except that the modified algorithm inserts a vector of uncertainties, Ij, where each attribute receives a degree of reliability with values between 0 and 1, with 1 indicating a completely reliable and consistent value, and 0 indicating values not collected or even completely divergent from the range of values. For a dataset matrix X (n x p), with n samples and p attributes in each sample, the c clusters centers are computed by

$$v_{ik} = \frac{\sum_{j=1}^{n} \mu_{ij}^{m} i_{jk} x_{jk}}{\sum_{j=1}^{n} \mu_{ij}^{m} i_{jk}} \tag{1}$$

where v_{ik} is the k_{th} attribute value of cluster center $\vec{v_i}$, x_{jk} is the k_{th} attribute value to j_{th} data in dataset matrix X(n x p), and i_{jk} is the k_{th} attribute of the uncertainty vector I_j .

The Euclidean distance in the FCM algorithm modified is calculated by again including the uncertainty vector:

$$d^2(\vec{x_j}, \vec{v_i}) = \frac{p}{\sum_{k=1}^{p} i_{jk}} \sum_{k=1}^{p} i_{jk}(x_{jk} - v_{ik})^2 \tag{2}$$

Cluster membership for incomplete feature vectors is determined by

$$\mu_{ij} = \frac{1}{\sum_{l-1}^{c} (\frac{\sum_{k=1}^{p} i_{jk}(x_{jk}-c_{ik})^2}{\sum_{k=1}^{p} i_{jk}(x_{jk}-c_{lk})^2})^{(m-1)}} \tag{3}$$

For the evaluation of clustering and stop parameter can be compared two partition matrices $U^{(1)}$ and $U^{(2)}$, where U(c x n).

Resampling[16,17] can be seen as a special case of the Monte Carlo method; that is, it is a method for finding solutions to statistical and mathematical problems using simulation. There are similarities and differences between resampling and Monte Carlo simulation. In resampling one could do all possible combinations, but it would be too time-consuming and computing-intensive.

To use resampling in the validation of clustering means to create a model of clusters for a dataset and with items that they have not been used, to create a validation cluster. These two clusters (model and validation) can be compared using itself measured as the described on Ref. [11]. Repeating these comparisons for different samples (constructed from the total dataset), the variability of the similarity measure for the partitions can be obtained to evaluate the variability of the structure of the clusters.

3. Python and MPI

Research in the sciences and engineering is increasingly demanding high computational performance. The complexity found in research is directly reflected in the difficulty of drawing up the codes in traditional compiled programming languages (like Fortran, C, and C + +), resulting in a large time for software design and construction, much greater than the time used in the analysis of data obtained in the search. Given this discrepancy of time for research and for programming, scripting languages suitable for computer science have attracted the attention of the scientific community for being more productive in processing the data. These tools provide a simple interface and a graphical visualization of results, without the inconvenience of worrying about the low level details associated with the traditional compiled programming languages.

However, scripting languages have been less efficient for processing when compared to a compiled programming language. So using scripting languages in more elaborate codes can be extremely painful in time for implementation of this code. An alternative that has attracted the attention of the scientific community is Python, which has the advantages of a scripting language and has a satisfactory performance in tasks that require massive computation.

Python is a free, flexible, high-level programming language composed of various elements and a configurable interface to each task: file and database management, data analysis, visualization, and automatic report generation. By the use of the Numeric, Scientific, SciPy, Visualization, and other modules, we can dramatically decrease the shortcomings of Python as a programming environment for scientific computing. This fact justifies the acceptance of the Python language by the scientific community. Interactive work[15], visualization[18], efficient multidimensional array processing[14], and scientific computing[12,19] are some example of what Python can do.

However, as this is a scripting language, some aspects of programming must be observed so that Python has satisfactory. For example, you should avoid excessive use of links and file access. Where possible, use specific modules for processing the data. As long as the core computationally-intensive components are compiled in a suitable language capable of achieving high performance, Python can provide the flexibility needed by researchers and users without compromising computational efficiency.

Over the last years, high-performance computing has attracted the attention of large segments of the scientific community, as can be seen in many publications [12,15,20,18]. This arose from the popularity of open source systems in, for example, Beowulf class clusters[21]. The union of quality open source software and commodity hardware has created a tool for search of several areas that need high-performance computing.

A model used in parallel programming is MPI[22]. MPI is standardized Message Passing Interface and is not a programming language but a model, and allows users to write portable programs in the main scientific programming languages. The standard defines the syntax and semantics of library routines in programming parallel computers. There are many implementations available from vendors of high-performance computers and open source projects like Open MPI[23]. Applications can run in clusters of (homogeneous and heterogeneous) workstations or dedicated nodes, (symmetric) multiprocessors machines, or even a mixture of both.

MPI works in a high-level environment, simplifying development, maintenance, and portability, without sacrificing performance. In the MPI model, the application is made up of N programs communicating with N1 others.

Following the trends, some researchers have tried to make available the benefits of massively parallel scientific applications for Python codes using MPI[15,20]. Python has enough networking capabilities as to develop an MPI implementation in pure Python, i.e., without using compiled languages or third-party MPI implementation. But this approach does not take advantage of Python. So the use of a module that allows the use of MPI directly in Python code is much more advantageous. In this work we use the MPI package for Python[15] providing the programming structures of the MPI standard.

4. Materials and Methods

4.1. *Assembly of the Cluster Computing*

Cluster is the name for the interconnection of more than one computer on a network with the objective of distributing the processing of any application from all computers on the network, thus working as a single supercomputer. This allows it to achieve performance with conventional computers (personal computers) that would only be possible in high-performance supercomputers.

Each computer is called a cluster node or node, and all must use the same operating system. The advantage of open source systems is the possibility to modify the code of the operating system to improve performance.

Among various possible configurations for the assembly of the cluster for this study, we chose a Beowulf cluster model, which allows the connection of the network through an Ethernet. Another factor in the possibility that the choice of using the MPI library as for the distribution of tasks.

We used five conventional computers completely different from each other, 4 computers with the processing architecture in 64 bits and 1 of 32 bits. This system is not dedicated, since computers are used in other daily tasks. All computers have the Linux operating system, MPI library, and the Python modules needed for the processing of the code of the method. This approach aims to demonstrate the feasibility of assembling a cluster using the computers of a conventional laboratory research, for applications in code using parallel routines.

4.2. *Datasets*

Six data sets had been analyzed applying the proposed classification method. In Table 1 we present a summary of the attributes for the six datasets used, including the number of clusters, the number of attributes (characteristics), the amount of data (total numbers of items in a dataset) and the number of items in each cluster.

Table 1. Summary of the attributes for the six datasets used in this work

Dataset Number of clusters, Number of attributes)	Number of data	Data by clusters
Artificial1 (4, 2)	400	100-100-100-100
Artificial2 (7, 2)	350	50-50-50-50-50-50-50
Artificial3 (4, 4)	2000	500-500-500-500
Artificial4 (5, 3)	1000	200-200-200-200-200
Wine[24] (3, 3)	178	59-71-48
Iris[24] (3, 4)	150	50-50-50

To obtain an incomplete dataset, we randomly select a specified percentage of its individual components to be labeled as an inconsistent or missing value. The method was tested using 2, 5, 10, 20 and 30% of the missing values.

The bootstrap were performed with 90% of the complete set dataset and 100 random samples. The incomplete dataset were clustered by FCM algorithm evaluated with 2 to 8 partitions. Each sample was clustered with the exactly the same number of clusters as used in the incomplete dataset. The centers obtained were reapplied in the incomplete dataset, then the second comparison matrix was generated (one for each sample).

The comparison between the first and the second clusterings is made using the measures described in the Introduction section. The final measures are obtained by computing the mean between all of the 100 samples.

When the analysis of final measures is obtained, it is possible to determine the probable number of clusters in the original dataset.

4.3. *Implementation in Python*

To have a basis for comparison between the parallel processing times and the one-machine times, it was necessary to build two codes. These codes are very similar and the main routines are the same. What differs in the parallel code are calls to MPI functions, so that the code would run in parallel. That is, with the inclusion of routines for running MPI code on all machines in the cluster at the same time. The code should take into account the process number and since each process knows who it is, they can perform different tasks for each other.

The following presents the parallel code in Python used in this work. Note that from # until the end of that line is comment; indenting indicates nesting of compound statements, and lines are continued by a backslash at the end of the line. MF is the Python package containing our routines, while MPI is the Python package containing the Python MPI interface.

```
Inconsistent  = [2 , 5 , 10 , 20 , 30];
tested_cluster =   [ 2 , 3 , 4 , 5 , 6 , 7 , 8]
subsample_number = 100
bootstrap = 90

# compute number of processes and myrank
numproc = MPI.COMM_WORLD.Get_size()
myrank = MPI.COMM_WORLD.Get_rank()
comm = MPI.COMM_WORLD
tag = 0

if (numproc < 2):     exit
if myrank==0 :
    # only node zero executes this part of code
    Data = numpy.loadtxt( data_file )
    tini = time.time()   # start count time
    # send data size
    MPI.COMM_WORLD.Bcast([Data.shape , MPI.FLOAT ],0)
    for (Node in range(1,numproc) ):
        # send data to nodes
        MPI.COMM_WORLD.Send([Data[:], MPI.FLOAT], Node, tag)
```

```
else:
    # all other processors execute this part of code
    status = MPI.Status()
    # get size of data
    MPI.COMM_WORLD.Bcast([data_size, MPI.FLOAT ],0)
    # reserve space in memory for Data
Data = numpy.empty(data_size, dtype=float)
# get data
MPI.COMM_WORLD.Recv([Data, MPI.FLOAT], 0, tag, status)
parte = (float)(porcentagem_faltantes.size)/(float)(numproc)
index = range( Inconsistent * myrank/numproc , \
  1 + Inconsistent * myrank/numproc )
indice_result = 0;

for p_falt in Inconsistent(index) do:
    data2 = insert data (Data.copy(),p_falt)
    I = MF.determinar_vetor_incompletos(data2.copy())
    for ngrupos in tested_cluster:
        Uref,Vref = MF.fuzzy_c_means_modificado \
          ( data2, I, 100, ngrupos , 2, 0.0001 )
        diff = 0
        acc  = 0
        f1   = 0
        parametros = numpy.zeros([1,5])
        for kk in [0, subsample_number]:
            data_redute = \
              MF.bootstrap(data2,bootstrap)
            Iboot = MF.determinar_vetor_incompletos \
                (data_redute.copy())
            Uboot,Vboot = MF.fuzzy_c_means_modificado \
                 ( data_redute, Iboot, 100, ngrupos, \
                   2, 0.0001 )
            Ucompara = MF.ajuste_centros_compara \
              (data2,I,Vboot,2)
            diff += MF.Diff(Uref,Ucompara)
            acc  += MF.Acc(Uref,Ucompara)
            f1   += MF.F1(Uref,Ucompara)
            parametros += \
              MF.parametros_mat_coincidencia \
```

```
                (Uref,Ucompara)
            diff = diff/float(subsample_number)
            acc  = acc/float(subsample_number)
            f1   = f1/float(subsample_number)
            parametros = parametros/float(subsample_number)
            if(myrank==0):
                print "proc= ", p_falt," grupo =  ",indice_result
            resultados[indice_result,:] = \
              [p_falt,ngrupos,diff,acc,f1, \
               parametros[0,0],parametros[0,1], \
               parametros[0,2],parametros[0,3],parametros[0,4]]
        indice_result=indice_result+1;
if myrank == 0:
    # get and print results
    status = MPI.Status()
    recvbuf = numpy.empty([(Inconsistent.size)* \
                                grupos.size,10],dtype=float)
    for No in range(1,numproc):
        index = range((int)(parte*No),(int)(parte*(No+1)))
        MPI.COMM_WORLD.Recv([recvbuf, MPI.FLOAT], No, tag, status)
        tfim = time.time()
    print "%5.2f s     " %(tfim-tini) ,
    try:
        file=open("arqtempos.txt","a")
    except IOError, message:
        print >> sys.stderr, \
          "Error when open file", message
        sys.exit(1)
    print >> file , arquivos_dados, " para ", numproc ," \
      processos levou ","%5.2f s \n" %(tfim-tini)
    file.close()
else:
    # send results to processor 0
    MPI.COMM_WORLD.Send([resultados[:], MPI.FLOAT], 0, tag)
```

A characteristic of our approach is that the data only has to be loaded on one machine that distributes it to the others. The advantage is that we can increase the number of computers in the cluster without any problems and this should naturally increase the parallelism.

The parallel code can use any computer on the network to access data and perform tasks. However, it may find problems when broadcasting data to the network if we use heterogeneous clusters with both 32-bit and 64-bit architectures. We dealt with this by transforming all integers to floating point.

The method was run on cluster with different combinations of machine and configurations, and also on single stand-alone computer for comparison. The comparison between timings indicates that the performance of our parallel method with Python is good.

5. Results

To show the performance of the presented method in parallel, first we compare the execution of the Python and Matlab implementations running in one processor. Table 2 shows the Matlab code running times presented in Ref. [25]. Table 3 shows the Python timings in a single 32-bit machine and then in a single 64-bit machine.

Table 2. Matlab timings

Dataset	Time processing (s)
Iris	660
Wine	786
Artificial1	2,364
Artificial2	2,058
Artificial3	58,680
Artificial4	14,610

Table 3. Python timings

Dataset	Time processing (s) 32 bits PC	Time processing (s) 64 bits PC
Iris	197.29	62.41
Wine	291.53	93.65
Artificial1	1,273.98	402.55
Artificial2	974.31	304.44
Artificial3	30,860.83	9,859.15
Artificial4	7,660.72	2,567.08

The improvement of performance just from changing the implementation language is clear. Codes in Python use manipulation of algebraic matrices and are more efficient than codes developed in Matlab for our applications. This is after the elimination of ties within the program and the use of appropriate modules for algebraic manipulation.

Finally, we show timings for the same six datasets on a homogeneous cluster of 64 machines (Figure 1) using 64-bit machines, and on a heterogeneous cluster when at least one of the computers was a 32-bit machine. Tables 4 and 5 show the timings on clusters with 2 through 5 computers.

Table 4. Time performance for Python in heterogeneous cluster

Dataset	Time processing for Heterogeneous Cluster (s)			
	2 nodes	3 nodes	4 nodes	5 nodes
Iris	79.67	42.42	38.69	39.15
Wine	116.92	62.31	58.32	58.20
Artificial1	504.40	256.25	250.29	263.38
Artificial2	386.86	200.77	199.83	210.53
Artificial3	12,518.24	6,622.78	6,219.13	6,067.73
Artificial4	3,078.39	1,627.52	1,538.20	1,634.14

The fact that the processing time with 5 processors is the same or even higher than with 4 processors is because the 32-bit computer is always being used in this case. This 32-bit computer is a bottleneck of the system. Despite this, the results show that the method can run in parallel on a non dedicated cluster since the time of execution are lower than the sequential codes.

Table 5. Time performance for Python in Homogeneous Cluster.

Dataset	Time processing for Homogeneous Cluster (s)		
	2 nodes	3 nodes	4 nodes
Iris	80.86	40.90	41.09
Wine	123.05	62.56	65.59
Artificial1	263.34	263.50	158.23
Artificial2	203.62	210.33	122.73
Artificial3	6,549.40	6,504.42	3,943.05
Artificial4	1,702.28	1,659.06	1,001.94

Comparing the values of tables 2 through 5 we observe that the timings represent a gain of approximately 90%. The chart in Figure 1 compares the values found for processing in stand alone, heterogeneous and homogeneous cluster.

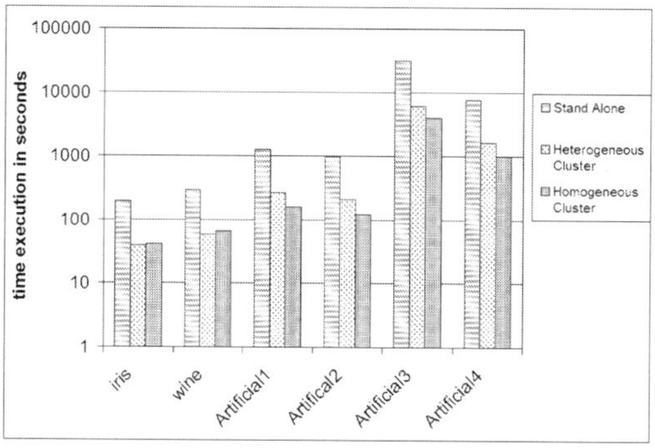

Figure 1. Time performance. Vertical axis is second.

6. Conclusions

With the Python-MPI implementation, the use of our method with a large amount of data is not prohibitive anymore. The ease of programming and manipulation of data provided by the Python should be highlighted. Its integration with MPI may be a determining factor in its future use in scientific applications. We want to point out that we may not need dedicated supercomputers. The improvement in performance can be accomplished with just basic clusters as available in any computer laboratory. An increase in the number of computers in the cluster will reduce the execution time even further with for our second approach (64 bits homogeneous).

References

1. E. Backer, Computer-Assisted Reasoning in Cluster Analysis, Prentice Hall, 1995.
2. C. Döring, M. J. Lesot and R. Kruse. Data Analysis with fuzzy clustering methods. Computational Statistics & Data Analysis, vol. 51, May 2006.
3. Wei-jen Hsu, Debojyoti Dutta, Ahmed Helmy, On the structure of user association patterns in wireless LANs, ACM SIGMOBILE Mobile Computing and Communications Review, v.11 n.2, April 2007
4. G. Kerr, H. J. Ruskin, M. Crane, P. Doolan, Techniques for clustering gene expression data, Computers in Biology and Medicine, v.38 n.3, p.283-293, March, 2008.
5. A. K. Jain and R. C. Dubes, Algorithms for Clustering Data. Prentice Hall, Englewood Cliffs, NJ, USA 1988.
6. A. K. Jain, M. N. Murty, P. J. Flynn, Data clustering: a review, ACM Computing Surveys (CSUR), v.31 n.3, p.264-323, Sept. 1999
7. R. J. Hathawhay and J. C. Bezdek, Fuzzy c-Means Clustering of Incom-

plete Data. IEEE Transactions on Systems, Man and Cybernetics Par B: Cybernetics, vol. 31, 2001.
8. E. Levine and E. Domany. Resampling Method For Unsupervised Estimation Of Cluster Validity. Neural Computation 13:25732593. MIT Press, Cambridge, MA, USA 2001.
9. M. H. C. Law and A. K. Jain. Cluster Validity by Bootstrapping Partitions. Technical Report MSU-CSE-03-5, Dept. of Computer Science and Engineering, Michigan State University, Michigan, USA 2003.
10. V. Roth, T. Lange, M. L. Braun and J. M. Buhmann. A Resampling Approach to Cluster Validation. In 15th Computational Statistics (COMPSTAT'02), Germany, 2002.
11. S.T. Milagre, C.D. Maciel, J.C. Pereira, A, A. Pereira, Fuzzy Cluster Stability With Missing Values Using Resampling. In: International Symposium on Mathematical and Computational Biology - BIOMAT, 2007, Búzios - RJ. BIOMAT 2007 - International Symposium on Mathematical and Computational Biology. Rio de Janeiro-Brazil : World Scientific Co. Pte. Ltd, 2007.
12. P.H. Borcherds, Python: a language for computational physics, Computer Physics Communications, no 177, 2007,pp. 199201
13. L. Anthony Drummond, Vicente Galiano, Violeta Migallón, Jose Penadés, Interfaces for parallel numerical linear algebra libraries in high level languages, Advances in Engineering Software, Vol 40, Issue 8, 2009, pp. 652-658
14. J.F. Lamarche, The Numerical Performance of Fast Bootstrap Procedures, Computational Economics,Vol 23, 2004, pp. 379389.
15. Lisandro Dalcín, Rodrigo Paz, Mario Storti, Jorge D'Elía. MPI for Python:Performance improvements and MPI-2 extensions, Journal of Parallel and Distributed Computing, Volume 68, Issue 5 (May 2008), pages 655-662, 2008
16. P. Good. Resampling Methods. Springer-Verlag, New York, NY, USA 1999.
17. C. Borgelt. Resampling for Fuzzy Clustering. Symposium on Fuzzy Systems in Computer Science 2006, Otto-von-Guericke-Universität Magdeburg, Germany.
18. J.H. Meinke, S. Mohanty, F. Eisenmenger, U.H.E. Hansmann, SMMP v. 3.0Simulating proteins and protein interactions in Python and Fortran, Computer Physics Communications, Vol. 178, Issue 6, 2008, pp.459-470.
19. S. Oliveira AND S.C. Seok, A Matrix-based Multilevel Approach to Identify Functional Protein Modules, Int. J. Bioinformatics Research and Applications, vol. 4, no. 1, pp 1127, 2008.
20. P.A. Kler, E.J. López, L.D. Dalcín, F.A. Guarnieri, M.A. Storti , High performance simulations of electrokinetic flow and transport in microfluidic chips, Computer Methods in Applied Mechanics and Engineering, Vol. 198, Issues 30-32, 2009, pp. 2360-2367
21. Beowulf.org, The Beowulf cluster site, http://www.beowulf.org
22. Message Passing Interface Forum, Message Passing Interface (MPI), Forum Home Page, http://www.mpi-forum.org, 1994.
23. Open MPI Team, Open MPI: open source high performance computing, http://www.open-mpi.org .

24. C. L. Balke and C. J. Merz. UCI Repository of Machine Learning Databases. University of California, Irvine, CA, USA 1998
25. S. T. Milagre. Análise do Número de Grupos em Bases de Dados Incompletas Utilizando Agrupamentos Nebulosos e Reamostragem Bootstrap. 2008. Doctoral thesis in Electrical Engineering, University of São Paulo, São Carlos, SP, Brazil.

ON THE DYNAMICS OF REINFECTION: THE CASE OF TUBERCULOSIS

C. CASTILLO-CHAVEZ[1,2,3,4]

X. WANG[1]

1 Mathematical, Computational Modeling Sciences Center,
PO Box 871904, Arizona State University, Tempe, AZ 85287

2 School of Human Evolution and Social Changes,
Arizona State University, Tempe, 85287

3 School of Mathematics & Statistics
Arizona State University, Tempe, 85287

4 Santa Fe Institute, 1399 Hyde Park Road Santa Fe, NM, 87501

J. P. APARICIO

Instituto de Investigación en Energías no Convencionales,
Universidad Nacional de Salta, 4400 Salta, Argentina

Z. FENG

Department of Mathematics, Purdue University, West Lafayette, IN 47907

Endogenous infection and exogenous reinfection are two mechanisms responsible for the reactivation or regeneration of active tuberculosis (TB) in individuals who have experienced prior active TB infections. We provide a brief review of a classical reinfection model, introduce a more general model, and include some new results. We conclude with a snapshot on the use of reinfection models in the study of the evolution of TB.

1. Introduction

Tuberculosis (TB) is an ancient infectious disease caused by *Mycobacterium* tuberculosis. Worldwide, TB accounts for more deaths among adults than all other infectious diseases combined[26] . The tubercle bacillus was identified by Robert Koch in 1882. In 1931, Calmett and Guerin discovered a "tamed" living bacterium named Bacillus Calmett-Guerin (BCG) which was then used to develop a "vaccine" that enhances the ability of the immune system to either prevent TB colonization (infection) or to possibly reduce the bacterium's ability to become re-activated. Britain adopted a BCG vaccination program in the 1950's but BCG has not been used in the US due to a lack of satisfactory studies that certify its effectiveness. The first effective anti-TB antibiotic Para-Amino-Salicylic (PAS) was discovered in 1946 by Lehmann. The wonder drug "Isoniazid" was used in New York to treat TB in 1952. The ability of TB to generate rather quickly resistant strains eventually led to a multi-drug protocol treatment that has been quite effective in reducing the number of active-TB cases in countries where long-treatment procedures can be effectively implemented. Today, a combination of several drugs is used to treat patients with active TB. Those infected with drug resistant TB (a group that often involves individuals who did not complete treatment) present huge challenges to the medical establishment. The lack of effective compliance with TB treatment can lead to catastrophic increases in the prevalence of DR-TB (drug resistant TB) including the generation of XDR-TB (extensively drug-resistant TB). It is surprising to see that there was no use of mathematical models in the study of TB dynamics despite its high mortality and morbidity rates (prior to the 20th century) before the 1960's.

Mathematical Epidemiology, a "subfield" of theoretical epidemiology, uses dynamical systems to characterize and understand epidemic patterns. The birth of mathematical epidemiology goes back to the work of Ross(1911) on malaria [33] and Kermack and McKendrick (1927) on childhood diseases [25] . The first mathematical model specifically linked to TB that we are aware of is in the form of a linear discrete system. It was introduced by Waaler in 1962, that is, nearly five decades after the work of Ross. Waaler's model includes three epidemiological classes: susceptible, latent (TB), and infectious (active TB) [43]. The goal of Waaler's model is to capture primarily the patterns of reported TB data. For a "recent" review of models and modeling approaches at the population level, we refer the readers to Ref. [12].

TB progression is not uniform, that is, some infected individuals are more likely to develop active TB. Models that incorporate a long and variable rate of progression have been formulated and studied [6,7,17]. Feng *et al.* [18] modeled the impact of variability in latency using arbitrary distributions and found out that such a generalization only resulted in quantitative (dynamical) differences [18]. Models that have included multiple strains of TB have also been developed [7,9]. Feng *et al.* [19] introduced a model with multiple strains and variable latency periods. These researchers found that their generalized two-strain model led to similar conclusions as as those in Ref. [9]. That is, antibiotic induced resistant (just as pesticide resistance) enhances, often substantially the likelihood that sensitive TB and drug-resistant TB will coexist. The use of variable periods of latency, the result of the inclusion of arbitrary latency period distributions, did not drive qualitative dynamical differences.

Some individuals with prior TB infections experience TB recurrence, that is, symptoms may come back after the effect of treatment is over (active TB). Endogenous infection and exogenous reinfection are well-documented routes towards TB progression, that is, paths that lead to active tuberculosis in individuals who have experienced prior active-TB infections. Whether or not the major mechanisms behind the recurrence of TB include endogenous reactivation (exacerbation of an old infection) or exogenous reinfection (recurrent infection by a different strain) has been debated for decades [27]. Endogenous reactivation is the leading suspect in the search for a major mechanism responsible for tuberculosis progression since the 1960s [38]. Earlier mathematical TB modeling has incorporated endogenous reactivation (relapse) by allowing the generation of secondary cases from contacts between treated individuals and those with active infections [3,4,6,17,30,44]. However, there is strong evidence that support the view that exogenous reinfection is also an important source of TB recurrence among the "cured" [3,4,13,17,35,44]. Recent studies show that exogenous reinfection is common in areas with a relatively high incidence of TB. In fact, a recent Shanghai study found that 61.5% of TB recurrent cases from 1999 through 2004could be attributed to exogenous reinfection [35].

Mathematical models have been formulated to study the impact of exogenous reinfection on the long-term (asymptotic) dynamics of TB [3,4,17,44]. Here, we review the essence of this research, introduce an extension of the model in Ref. [17] and establish new mathematical results.

This paper is organized as follows: Section 2 reviews the dynamics associated with exogenous reinfection using the classical model [17], introduces an extended framework [44], provides some preliminary mathematical analysis and contrast the differences between the results generated by both models; a brief overview on the use of reinfection models in the study of the evolution of TB progression is the focus of Section 3; final thoughts and future directions based on earlier work are discussed in Section 4.

2. Mathematical Models of Tuberculosis with Exogenous Reinfection

Exogenous reinfection is believed to play a significant role on TB dynamics in environments where tuberculosis infectivity and prevalence are relatively high. Mathematical models have been developed and analyzed in studies of the impact of exogenous reinfection on TB dynamics [17,44]. It is the goal of this section to review briefly the results in Ref. [17] and publish some of the new results in Ref. [44].

2.1. *Model I*

The exogenous reinfection model introduced in Ref. [17] is given by the following system of nonlinear differential equations:

$$\begin{aligned}
\frac{dS}{dt} &= \Lambda - \rho c S \tfrac{I}{N} - \mu S, \\
\frac{dE}{dt} &= \rho c S \tfrac{I}{N} - p\rho c E \tfrac{I}{N} - (\mu + k)E + \sigma \rho c T \tfrac{I}{N}, \\
\frac{dI}{dt} &= p\rho c E \tfrac{I}{N} + kE - (\mu + r + d)I, \\
\frac{dT}{dt} &= rI - \sigma \rho c T \tfrac{I}{N} - \mu T \\
N &= S + E + I + T,
\end{aligned} \tag{1}$$

where $S(t)$ denote the susceptible population at time t; $E(t)$ is the latently infected (assumed not infectious) class at time t; $I(t)$ denote the actively infected (assumed infectious) class at time t; and $T(t)$ is the size of the effectively treated class at time t. Here Λ is the constant recruitment rate; c is the contact rate per individual; k is the per capita rate of departure of individuals from the latent class to the infectious class; d is the per-capita TB disease induced death rate; r is the per-capita TB treatment rate; σ denotes the reduction in susceptibility due to a prior endogenous infection; p represents the reduction in susceptibility of a latently-infected individual; and ρ denotes the probability that a contact is effective for TB transmission given that a susceptible had contacts with an actively infected individual. $\frac{I}{N}$ denotes the likelihood that the encounter is with an actively infectious individual (random or uniform mixing).

We set $\beta = \rho c$, which is interpreted as the average number of effective contacts a susceptible has per unit of time. The basic reproductive number, that is, the number of secondary infections generated by an actively infectious individual in a population of susceptibles at a demographic equilibrium, is given by

$$\Re_0 = \frac{\beta}{\mu + k + r + d} \tag{2}$$

Feng *et al.* [17] showed that, in the absence of reinfection ($p = 0$) TB will die out provided that the reproductive number $\Re_0 \leq 1$ but that TB will persist if $\Re_0 > 1$. In Ref. [17], it was shown that exogenous reinfection can lead to a subcritical or backward bifurcation (with p positive and small). In other words, reinfection can lead to a situation where $\Re_0 = 1$ is no longer the threshold for disease eradication. In fact, if we let

$$p_0 = \frac{k(\mu + r + k)}{(\mu + r)\mu}$$

the following results hold:

(1) $\Re_0 > 1$ implies that System (1) has exactly one positive equilibrium, that is, the model supports an endemic state.

(2) $\Re_0 < 1$ and $p > p_0$ mean that for each such p, a positive constant $\Re_p < 1$ can be found, such that Model (1) supports exactly two positive equilibria as long as $\Re_0 > \Re_p$. However, only one positive equilibrium is possible if $\Re_0 = \Re_p$. No positive equilibrium is supported if $\Re_0 < \Re_p$.

(3) $\Re_0 < 1$ and $p < p_0$ imply that Model (1) supports no positive equilibrium. $\Re_0 < 1$ and $p = p_0$ imply that Model (1) supports exactly one positive equilibrium.

The possibility of multiple positive equilibria and bi-stability is collected in a (backward) bifurcation diagram (Figure 1).

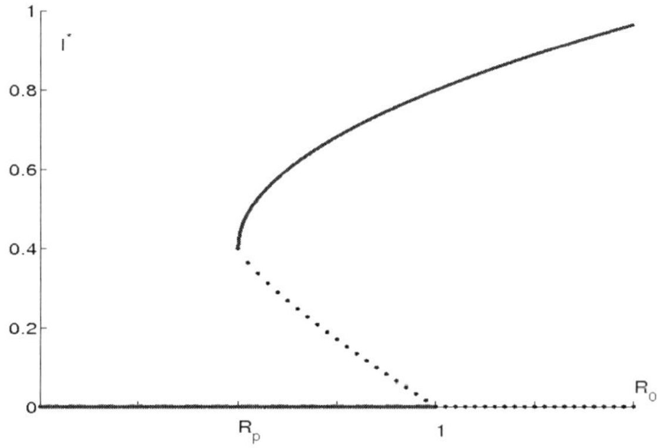

Figure 1. Backward bifurcation diagram when exogenous reinfection is included in model (1). When $\mathcal{R}_0 < \mathcal{R}_p$, the disease-free equilibrium is globally asymptotically stable. However, when $\mathcal{R}_p < \mathcal{R}_0 < 1$, there are two endemic equilibria. The upper ones are stable, and the lower ones are unstable.

The results in Feng *et al.* [17] attracted the attention of theoretical epidemiologists. In a letter to the editor of the *Journal of Theoretical Biology*, M. Lipsitch *et al.* [28] claimed that the existence of two positive equilibria depends on unrealistic assumptions about the epidemiology of TB. Although his back of the envelop calculation is fine, there seems to be a misunderstanding on the premises behind their calculation. It seems that the computed "probabilities" are not comparable. Specifically, Lipsitch *et al.* note that the rate (P_1) that a contact between an infectious individual and a susceptible individual will lead to disease is $\beta\frac{k}{\mu+k}$, while the rate (P_2) that exogenous reinfection (the contact between an infectious individual and a latently infected individual will lead to disease) is $p\beta$. They conclude, from their calculation, that the existence of multiple positive equilibria is unrealistic as it would require that $P_1 > P_2$ which in their opinion, would contradict the biological belief that latent infections provide some immunity to reinfection. Their possible misinterpretation derives from the fact that both transition rates are not directly comparable. Both expressions give the rate of movement to the active-TB class but not from the same epidemiological state. That is, each computation starts from a different epidemiological class and therefore the comparison must be carefully assessed. Furthermore, even if true, there is no data that has tested the potentially sensible biological belief. If we assume that the development of active TB is the result of "natural" progression from the latent state then the probability of developing active TB from such a latent TB state is just $\frac{k}{k+\mu}$. If on the other hand, we assume that exogenous reinfection is the only route towards active TB then the probability of developing active TB in a population at an endemic state is $\frac{p\beta I_\infty/N}{p\beta I_\infty/N+\mu}$. In order to get an idea of the magnitude of this last number, let's recall that the total number of yearly cases of active TB in the world is approximately 8×10^6. If all these cases were to occur within a single nation with high prevalence, lets say India, then $\frac{I_\infty}{N} \approx \frac{8\times10^6}{10^9} = 0.008$. Hence, the probability of developing active TB via natural progression ($\frac{k}{k+\mu}$) is much bigger than the probability of developing active TB through exogenous reinfection. The precise rate condition that must be met is that $k > \beta p \frac{I_\infty}{N} = 0.008p\beta$. This result does not contradict the conditions for the existence of multiple positive equilibria given in Ref. [17] and, the above analysis seems to respond to the concerns in Ref. [28].

2.2. *Model II*

An important feature of TB, as Arthur M. Dannenberg, Jr. points out, is that "....after the healing of a primary infection of TB, the number of T lymphocytes [cells] with specific receptors for the bacillary antigens in the blood and tissues decreases with time, and the tuberculin reaction may even disappear. These events may occur if the tubercle bacilli and their antigens are eliminated"[15] . In other words, recovery from infection only offers temporary immunity.

Lee B. Reichman and Earl S. Hershfield[31] report between 49-56 per 100,000 new cases in Singapore per year. In fact, these researchers reported a total of 1712 new cases in 1997. Specifically, they reported that 55 cases per 100,000 developed TB for the first time and 265 cases (also per 100,000 TB cases) were directly attributed to relapse. If we let γ denote the ratio of the infection rate per recovered person to the infection rate per never infected person, we conclude that γ is quite large in Singapore (1997). If we can't distinguish the precise cause of relapse (reinfection or reactivation) then γ is about 6.25. The immunity that individuals gain from recovery will decline and consequently, the chance of becoming infected again must be higher than that for "Joe the plumber" [42] . In general, the probability of becoming reinfected by TB depends on multiple (not independent) factors that include "genetics", compliance to chemotherapy, age, sex, race, socioeconomic status, and history of prior infections [14] .

While mathematical models have been used to study the role of partial immunity(reduced susceptibility) in TB dynamics (see, for example, Ref. [6, 7, 10, 30]), the impact of temporary immunity has not been specifically addressed. In order to avoid confusion, let's revisit the meaning of partial immunity here, it means that treated (recovered) individuals can become infected again (but probably at a reduced rate) while the phrase temporary immunity means that a recovered individual (here) will remain immune for a period of time before becoming susceptible to TB again.

The assumption that recovered TB individuals gain temporary immunity from prior infections and that the infection rate of those who have lost their temporary immunity is higher than that of susceptible individuals leads to a modeling framework that divides the uninfected population into three sub-groups: naive susceptibles, that is, S individuals with no prior exposure; modified susceptibles, that is, S_F individuals who have recovered from prior infections but lost their immunity; and R: immune individuals. It is assumed that the infection rate of modified susceptibles is γ times that of naive susceptibles. Data suggest that $\gamma > 1$. In fact, γ may be much larger than 1 (Singapore data). However, if we do not or cannot differentiate between the precise causes of relapse, reinfection or reactivation, then γ should be estimated as just the ratio of new cases of infection from S_F to S. Using the Singapore data we see that

$$\gamma = \frac{265/S_F}{1712/S} \approx \frac{265/(1712 * 45)}{55/100000} \approx 6.25,$$

under the assumptions that the average life span of individuals after their loss of immunity is 45 years when TB prevalence has reached an endemic state. In this new model [44] , the population is divided into five epidemiological classes: S, S_F, latently infected (non-infectious) individuals L, active TB infectious individuals I, and recovered/immune individuals R with $N = S + S_F + L + I + R$ denoting the total population.

The transfer diagram in Fig. 2 highlights the "motion" between classes via disease progression. Additional notation is still required to formulate the new model. We let ν denote the per capita transition rate from L to I; η and σ are the per capita treatment rates for L and I individuals, respectively; α is the per capita rate at which a recovered individual loses immunity (becoming susceptible again); μ_L and μ_I are the per capita extra death rates due to TB disease in the L and I classes, respectively; $\lambda(I) = c\frac{I}{N}$ where c is the infection rate for an S individual, and γ represents the increased infection rate for an S_F individual; $\beta(N)$ is the recruitment/birth rate, a function of the total population N, satisfying $\beta(N) > 0, \beta'(N) \leq 0$. All parameters are naturally assumed to be positive. Using these definitions we can describe TB dynamics in this extended sense via the following system of nonlinear ODEs [44] :

$$\frac{d}{dt}S = \beta(N)N - \mu S - \lambda(I)S,$$

$$\frac{d}{dt}S_F = \alpha R - (\mu + \gamma\lambda(I))S_F,$$

$$\frac{d}{dt}L = \lambda(I)S + \gamma\lambda(I)S_F - (\mu_L + \mu + \eta + \nu)L,$$

$$\frac{d}{dt}I = \nu L - (\mu_I + \mu + \sigma)I,$$

$$\frac{d}{dt}R = \eta L + \sigma I - (\alpha + \mu)R,$$

where as it was noted before, the infection rate for S_F individuals is assumed to be higher than that for S individuals, that is, $\gamma > 1$.

Existence and uniqueness of solutions of (3) for all $t \in (0, \infty)$ are established in the "usual" way. Hence, the model is well-posed, that is, all solutions with nonnegative initial data will stay nonnegative for $t > 0$.

We proceed to re-scale some of the variables through the following change of variables $x = \frac{S}{N}$, $u = \frac{S_F}{N}$, $y = \frac{L}{N}$, $z = \frac{I}{N}$ and $w = \frac{R}{N}$. Letting $\beta(N) = \beta$, that is, a constant, leads to the following particular re-scaled version of Model 3:

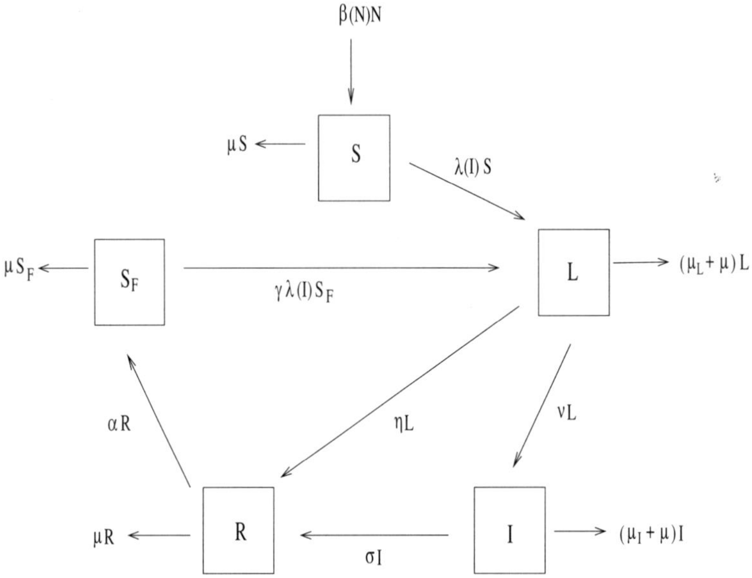

Figure 2. System diagram.

$$\frac{d}{dt}x = \beta(1-x) - (c-\mu_I)zx + \mu_L xy,$$

$$\frac{d}{dt}u = \alpha w - \beta u - (\gamma ck - \mu_I)zu + \mu_L uy,$$

$$\frac{d}{dt}y = czx + \gamma czu - (\beta + \mu_L + \eta + \nu)y + \mu_I zy + \mu_L y^2, \tag{3}$$

$$\frac{d}{dt}z = \nu y - (\beta + \mu_I + \sigma)z + \mu_I z^2 + \mu_L zy,$$

$$\frac{d}{dt}w = \eta y + \sigma z - (\alpha + \beta)w + \mu_I zw + \mu_L yw,$$

where $x + u + y + z + w = 1$.

The trivial steady-state of System(3) $E_0 = (1, 0, 0, 0, 0)$ is stable if and only if

$$R_0 = \frac{c\nu}{(\beta + \mu_L + \eta + \nu)(\beta + \mu_I + \sigma)} < 1. \quad (4)$$

Letting $\phi = \mu_L + \beta + \eta + \nu$, $\zeta = \mu_I + \beta + \sigma$ and $\theta = \frac{\eta\zeta}{\nu} + \sigma$, allow us to rewrite the reproductive number as $\Re_0 = \frac{c\nu}{\zeta\phi}$ where c denotes the number of secondary infections produced by one infectious individual per susceptible per unit of time, ν/ϕ is the probability that a latent individual that survived enters the infectious stage, and $1/\zeta$ is the mean sojourn time an infectious individual stays in the I stage. $\Re_0$ naturally gives the number of secondary infections produced by one infectious individual during the entire infectious period in a population at demographic steady state.

System (3) not only always supports the disease-free equilibrium (DFE) $E_0 = (x_0, u_0, y_0, z_0, w_0) = (1, 0, 0, 0, 0)$ but from the Jacobian at E_0, we see that all eigenvalues have negative real parts if $\Re_0 < 1$ and that at least one eigenvalue will have a positive real part if $\Re_0 > 1$. Hence the following standard result holds.

Theorem 2.1. *For system (3), the disease-free equilibrium E_0 is locally asymptotically stable if $\Re_0 < 1$ and unstable if $\Re_0 > 1$.*

Next, we use a similar method as the one used in Ref. [12] and the center manifold theorem to show that a backward bifurcation occurs at E_0 when $\Re_0 = 1$.

To compute the center manifold of the system at $E_0 = (x_0, u_0, y_0, z_0, w_0) = (1, 0, 0, 0, 0,)$ and $\Re_0 = 1$, we first translate the equilibrium E_0 to the origin $\hat{E}_0 = (0, 0, 0, 0, 0)$ via the change of variables:

$$p_1 = x - 1, \quad p_2 = u, \quad p_3 = y, \quad p_4 = z, \quad p_5 = w.$$

Since $\Re_0 = 1$ is equivalent to $c = \zeta\phi/\nu$, we can rewrite the Jacobian matrix associated with $E_0 = (0, 0, 0, 0, 0)$ at $\Re_0 = 1$ as follows:

$$\begin{bmatrix} -\beta & 0 & -\mu_L & \frac{\zeta\phi}{\gamma} - \mu_I & 0 \\ 0 & -\beta & 0 & 0 & \alpha \\ 0 & 0 & -\phi & \frac{\zeta\phi}{\gamma} & 0 \\ 0 & 0 & \gamma & -\zeta & 0 \\ 0 & 0 & \eta & \sigma & -\alpha \end{bmatrix},$$

and the corresponding Jordan Form as

$$\begin{bmatrix} 0 & 0 & 0 & 0 & 0 \\ 0 & -\alpha-\beta & 0 & 0 & 0 \\ 0 & 0 & -\beta & 0 & 0 \\ 0 & 0 & 0 & -\beta & 0 \\ 0 & 0 & 0 & 0 & -\zeta-\phi \end{bmatrix}.$$

Thus, after the following transformation

$$\begin{pmatrix} p_1 \\ p_2 \\ p_3 \\ p_4 \\ p_5 \end{pmatrix} = \begin{bmatrix} -\frac{(\mu_I\nu+\zeta(\mu_L-\phi))(\alpha+\beta)}{\beta(\eta\zeta+\nu\sigma)} & 0 & 0 & 1 & \frac{(\alpha+\beta-\zeta-\phi)((\mu_L+\zeta)\phi-\mu_I\nu)}{(\beta-\phi-\zeta)(\nu\sigma-\eta\phi)} \\ \frac{\alpha}{\beta} & -1 & 1 & 0 & \frac{\alpha}{\beta-\zeta-\phi} \\ \frac{(\alpha+\beta)\zeta}{\eta\zeta+\nu\sigma} & 0 & 0 & 0 & -\frac{\phi(-\alpha-\beta+\phi+\zeta)}{\eta\phi-\nu\sigma} \\ \frac{(\alpha+\beta)\nu}{\eta\zeta+\nu\sigma} & 0 & 0 & 0 & \frac{\nu(-\alpha-\beta+\phi+\zeta)}{\eta\phi-\nu\sigma} \\ 1 & 1 & 0 & 0 & 1 \end{bmatrix} \begin{pmatrix} x_1 \\ x_2 \\ x_3 \\ x_4 \\ x_5 \end{pmatrix} \tag{5}$$

we can rewrite System(3) into a system of ODEs for the state variables x_i, $i = 1, 2, 3, 4, 5$ where $x_1 \in R^c$ (center manifold) and $(x_2, x_3, x_4, x_5) \in R^s$ (stable manifold). We use the parameter $s \equiv c - \frac{\zeta\phi}{\nu}$ ($\Re_0 = (>, <)1$ is equivalent to $s = (>, <)0$) and consider the the following parameterized functional forms for the state variables

$$x_{i+1} = h_i(x_1, s) = a_i x_1^2 + b_i x_1 s + c_i s^2 + O(3), \text{ for } i = 1, 2, 3, 4. \tag{6}$$

Plugging the above expressions into the system of x_i's and comparing the coefficients of x_1, x_1^2, $x_1 s$ we proceed to solve for a_i, b_i, and c_i, for $i = 1, 2, 3, 4$. The vector field restricted to the center manifold is given by

$$\dot{x_1} = \frac{\nu}{\zeta+\phi} s x_1 + \frac{(\nu\mu_I+\zeta\mu_L)(\alpha+\beta)(\zeta\phi+\beta(\zeta+\phi))+\phi\gamma\alpha(\eta\zeta+\nu\sigma)-\zeta\phi^2(\alpha+\beta)}{\beta(\eta\zeta+\nu\sigma)(\phi+\zeta)} x_1^2 + O(3). \tag{7}$$

System (7) has two equilibria, $\tilde{x}_1 = 0$ and

$$x_1^* = \frac{\beta(\eta\zeta + \nu\sigma)\nu s}{(\nu\mu_I + \zeta\mu_L)(\alpha+\beta)(\zeta\phi + \beta(\zeta+\phi)) + \phi\gamma\alpha(\eta\zeta+\nu\sigma) - \zeta\phi^2(\alpha+\beta)}. \tag{8}$$

If we let

$$\gamma_c \equiv \frac{\nu\beta}{\phi\alpha} + \frac{\zeta\phi(\alpha+\beta)}{\alpha(\eta\zeta+\nu\sigma)} - \frac{(\nu\mu_I + \zeta\mu_L)(\alpha+\beta)(\zeta\phi + \beta\zeta + \beta\phi)}{\phi\alpha(\eta\zeta+\nu\sigma)}, \tag{9}$$

then from the above definitions and computations we arrive at the following result.

Theorem 2.2. *If $\gamma > \gamma_c$ then System (3) supports a backward bifurcation at $\Re_0 = 1$. That is, the trivial equilibrium E_0 is l.a.s. for $\Re_0 < 1$ and unstable for $\Re_0 > 1$ while a nontrivial equilibrium E^* exists that it is l.a.s. for $\Re_0 < 1$ (but close to 1).*

Proof. Since $\Re_0 < 1$ then $s < 0$ and therefore $\tilde{x}_1 = 0$ is stable. For $\gamma > \gamma_c$, x_1^* given in (8) is positive. Rewriting (7) as

$$\dot{x_1} = Ax_1 + Bx_1^2 + O(3),$$

implies that the stability of x_1^* is determined by the sign of $A + 2Bx_1^*$. Since $x_1^* = \frac{-A}{B}$ and $A = \frac{\nu}{\zeta+\phi}s < 0$ then

$$A + 2Bx_1^* = -Ax_1^* > 0.$$

Therefore, x_1^* is unstable. Notice that $x_1 > 0$ and $x_i \approx 0$ for $i = 2, 3, 4, 5$. From the transformation matrix (5), we see that $p_1 < 0$, $p_i > 0$ for $i = 2, 3, 4, 5$. Thus, $u > 0$, $y > 0$, $z > 0$ and $w > 0$ if $x_1 > 0$.

Numerical studies of System (3) verify that the backward bifurcation is "there". Fig.(3) is a bifurcation diagram that uses $\Re_0$ as its bifurcation parameter. The graph is generated with XPPAUT. For the selected set of parameters used the lower bound for $\Re_0$ for multiple endemic equilibria is $\Re_p = 0.232513$. The simulation results show that for a backward bifurcation to occur γ may be as small as 5. The parameter values used for Fig. 3 are $\beta = 0.012$, $\mu_I = 0.14$, $\mu_L = 0.0014$, $\sigma = 10$, $\nu = 0.04$, $\eta = 0.002$, $c = 10.0$ and $\gamma = 20$.

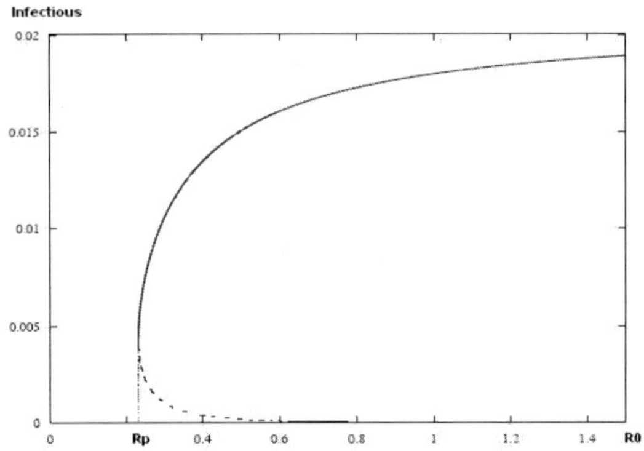

Figure 3. Backward bifurcation diagram. $\Re_p = 0.232513$, dotted line indicates unstable branch and solid line indicates stable branch.

It is worth observing that if we rewrite Condition (9) as $\gamma > \gamma_c$ where γ_c is the constant defined by the right hand side of (9) then we can re-state our results via the critical ratio of infection rate of the modified susceptibles to that of the naive susceptibles: a backward bifurcation is possible only if γ exceeds this critical value γ_c. Condition (9) can also be reformulated in terms of the relapse parameter α and a threshold condition $\alpha > \alpha_c$ (where α_c is the critical relapse rate determined form (9)). When the relapse rate is high ($\alpha > \alpha_c$) a stable endemic equilibrium is possible for $\Re_0 < 1$, and consequently, disease eradication may not be possible even if the reproductive number is brought below 1. Many existing TB models have been shown to have the property that the disease will die out if $\Re_0 < 1$ (see for example, Ref. [1, 8, 24]). These results show that the condition $\Re_0 < 1$ may not be enough to eradicate a disease when the disease relapse rate is high enough. The new threshold condition should be $\Re_0 < \Re_p < 1$ or that the rate of TB relapse must be low. Similar conclusions have been observed in the study of drinking dynamics[34] or ecstasy use[37] or even the spread of extreme ideologies[11] . In some instances, explicit computations of these exact thresholds have been carried out [12] .

For the case of $\Re_0 > 1$, it is difficult to prove the existence of a unique equilibrium and to determine its stability. Numerical studies supports the

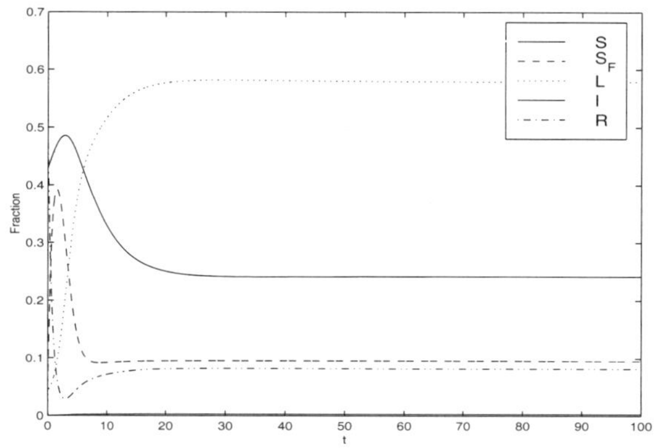

Figure 4. Simulation of System (3) when $\Re_0 > 1$.

view that a unique endemic equilibrium exists and is stable (see Fig. 4).

3. Reinfection Models of TB Dynamics: A Perspective on TB Evolution

Historical data from TB trends from various nations show that TB is a disease with high prevalence and changing progression rates. The identification of the mechanisms behind the observed extreme reduction of active TB cases over that last century may be teased out via the use of models that focus on TB dynamics over very long-time scales. Such studies require the incorporation of demographic changes. Naturally, we cannot assume constant contact, progression, treatment, or mortality rates over long-time scales. The quantitative dynamics of TB has changed dramatically over time. In fact, the mortality rate dropped dramatically between 1880 and 1960 Fig. 5 (Left) (U.S. Bureau of the Census, 1908, 1975 Ref. [41]). These dramatic shifts took place prior to the discovery (or massive introduction) of antibiotics. Models have been used to study the dynamics of tuberculosis over long-time scales [2,4] . The emphasis in Ref. [2, 4] has been on the use of demographic and epidemiological data to parameterize "rather" simple compartmental models. That is, models whose flexibility is enhanced through the use of time-dependent parameters that must be estimated from data. These non-autonomous models are used not only to capture the patterns of TB over long time horizons but also to identify evolutionary changes on the epidemiology of TB.

Recent work (Ref. 4) has incorporated second order effects such as exogenous reinfection in order to capture a detailed epidemiological picture of the long-term dynamics of active TB. A recent model developed to fit US Massachusetts data incorporates mechanisms for new infections and reinfections [4] . The population is therefore divided into seven classes (Fig. 6). Uninfected individuals, U class; First-infected individuals, members of the high-risk latent class E, individuals assumed to be asymptomatic and non-infectious but capable of "quickly" progressing to the clinical disease (or active-*TB*) at the per-capita rate k; Individuals who do not progress to the active *TB* class quickly enough from E are moved to the low-risk latent class L at the per-capita rate α. The introduction of high- and low-risk latent classes capture in an organizational way the observed patterns of fast TB progression of a small but significant fraction of recently infected individuals. L individuals progress to active-TB at the per-capita rate k_L and are assumed to be susceptible to reinfection. Re-infected individuals move to the second high-risk class E^* from where they can progress to the active-TB class at the per-capita rate k^*. Individuals that escape progression (returning to the class L) do so at the per-capita rate α. Prior infections with *M. tuberculosis* afford some immunological protection that is modeled through the introduction of reduced susceptibility to reinfection and/or reduced risk to progression to E^* ($k^* < k$), the second high-risk latent class. New cases of active-TB are classified as pulmonary with probability q and collected in the A_p class. Extrapulmonary cases which are assumed to be not infectious are moved to the class A_e. Recovered individuals (class R) may develop active TB again (TB-relapse) at the per-capita rate k_{Rp}.

In order to formulate the full model, additional definitions are needed. The definition of contact number used here and incorporated in the model facilitates the evaluation of various evolutionary strategies and connections

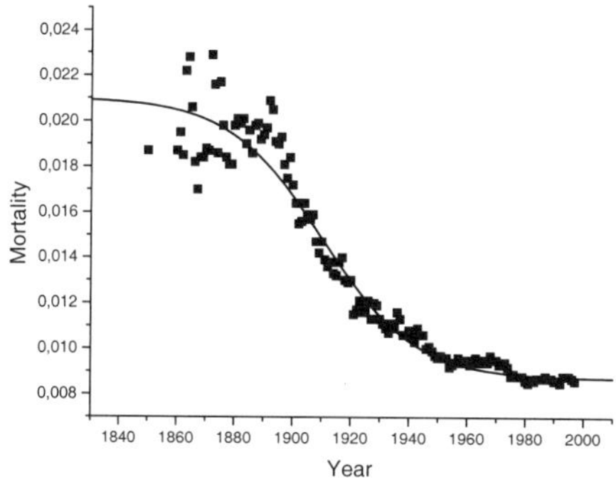

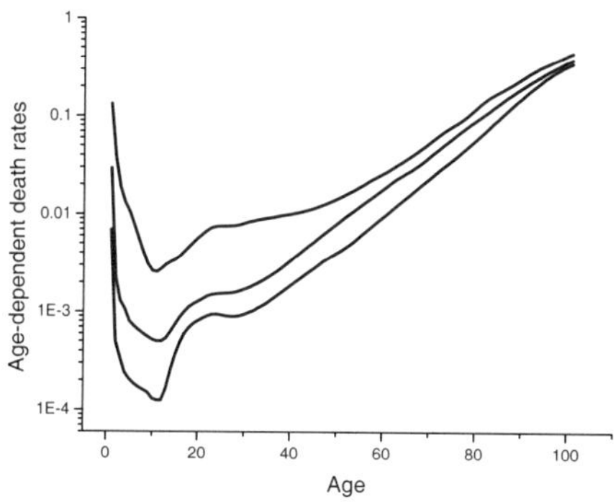

Figure 5. Left: Observed total mortality rates (squares) and their approximations (continuous line). Right: Age-dependent probabilities of dying within one year (q_x) from period life tables for the United States based in the decennial census data from 1900, 1950, and 2000 taken from Ref. [4].

to data. The contact number definition derives from the observation that an average infectious individual is capable of producing on average, Q_0 total secondary infections when placed in a fully susceptible population. If the fraction U/N of his/her contacts is with first-infected individuals then the number of infections produced is $Q_0 U/N$. The infection rate (per unit of time) is obtained by dividing this number by the infectious period $1/\gamma$. It

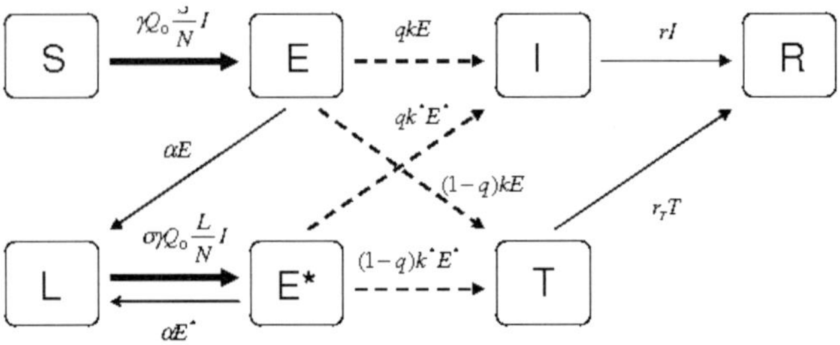

Figure 6. Transfer Diagram. Individuals are recruited into the uninfected class U; moved into the high-risk latent class E after the infection from where they progress to active TB. A fraction q develops pulmonary TB (A_p) while the complementary fraction 1-q develops extrapulmonary TB (A_e). E individuals move to the low-risk latent class L or to A_p or A_e. L individuals may become reinfected and move to the high-risk class E^* from there they may develop active-TB, or may return to the low-risk class L. Active cases if they recover move to the R class. Transmission is represented in the above transfer diagram using thick arrows and progression with dashed arrows. Progression from low-risk latent class L is not shown. Transfer rates per unit of time are also shown.

is therefore given by

$$\gamma Q_0 \frac{U}{N} A_p. \tag{10}$$

The re-infection rate is therefore

$$\sigma\gamma Q_0 \frac{L}{N} A_p \tag{11}$$

where σ accounts for the possible protection conferred by previous infections ($\sigma \leq 1$).

Using the above definitions and notation, we arrive at the following TB transmission model (see also Fig. 6):

$$\frac{dU}{dt} = B - \mu U - \gamma Q_0 \frac{U}{N} A_p,$$

$$\frac{dE}{dt} = \gamma Q_0 \frac{U}{N} A_p - (k + \mu + \alpha)E,$$

$$\frac{dL}{dt} = \alpha(E + E^*) - (\mu + k_L)L - \sigma\gamma Q_0 \frac{L}{N} A_p,$$

$$\frac{dE^*}{dt} = \sigma\gamma Q_0 \frac{L}{N} A_p - (k^* + \mu + \alpha)E^*, \tag{12}$$

$$\frac{dA_p}{dt} = q(kE + k^*E^* + k_L L + k_{Rp}R) - \gamma A_p,$$

$$\frac{dA_e}{dt} = (1 - q)(kE + k^*E^* + k_L L + k_{Rp}R) - \gamma_e A_e,$$

$$\frac{dR}{dt} = rA_p + r_e A_e - (\mu + k_{Rp})R,$$

where $\gamma = \mu + d + r$ and $\gamma_e = \mu + d_e + r_e$. Recruitment brings individuals only to the uninfected class, that is, a limitation of these models comes from the fact that immigration does not include the recruitment of infected individuals. In general, parameters like μ, k_L, and k^* must be time-dependent functions as well as B (see Ref. [4]).

The above model easily accounts for the variation in time of parameters since demography and changes in contact and epidemiological parameters over time. In Ref. [4], US census data[41] have been used[5] as well as estimated time tables for the period 1850-1900[23] to compute the probability of death within one year per year. Figure 5 highlights the type of "fits" used in the estimating functions such as age-dependent probabilities from available data [4] . Age-dependent and age-independent models are fitted to data in Ref. [4]. Here a graph of the non-agestructured model (12) fitted to the Massachusetts TB incidence data is highlighted in (Fig. 7). The fact that an cxccllcnt fit is obtained with the use of real recruitment and mortality data is not surprising given the number of free-parameters. However, the story is more complicated (see Ref. [4]) because the goal is not just to fit the data but to identify mechanisms capable of slowing down TB progression (while fitting incidence and prevalence TB data).

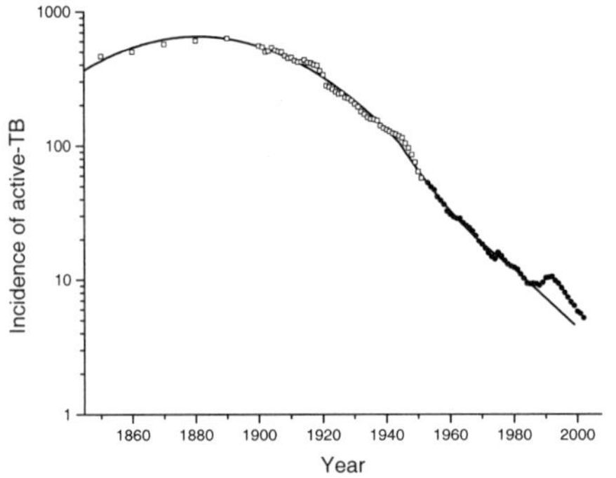

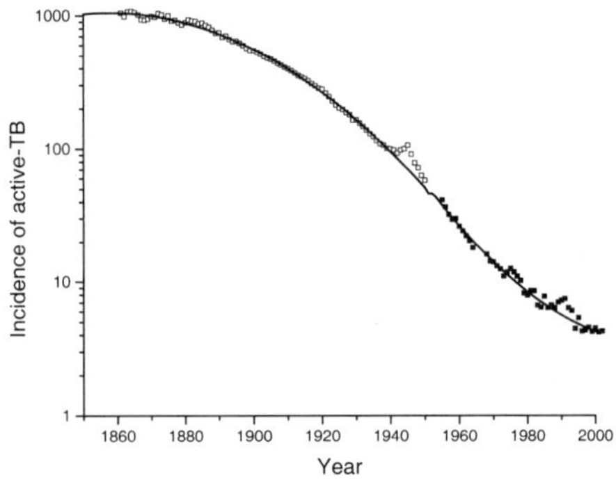

Figure 7. Observed incidence of active-TB (all forms, solid squares, per year and per 100,000 population), estimated incidence of active-TB (estimated as 2.875 times TB mortality rates, open squares), for the United States (left) and Massachusetts state (right). The simulated incidences obtained from model (12) are shown in continuous lines.

The causes that have been identified responsible for the observed decrease in active TB incidence over the past century and a half include: (i) purely dynamical [6,20,21,22] ; (ii) a reduction in transmission, due to public health measures that include the isolation of active cases in sanatoria, use of antibiotic treatment and the improvement of the living and working conditions (ventilation and reduction in overcrowding); (iii) a reduction in progression tied in to the theory of improving of living conditions [29] and a known hypothesis of host-parasite co-evolution [39,40] .

In Ref. [4], we show that vital-dynamics driven epidemics cannot account for the timing or the rate of observed active-TB decline even when reinfection is excluded. The results in Ref. [4] strongly support the view that the most likely mechanism is the one that reduces transmission and/or progression parameters over in time. The hypothesis of reduction in transmission as the cause behind the decline of tuberculosis rates is by far the most popular. In Model (12) reductions in transmission correspond to reductions in the contact number Q_0. However, Q_0 is not an independent factor, Q_0 is a function of the mean length of infectious period $(1/\gamma)$, the average risk of infection per case per susceptible (β), and the size of the network of close contacts of a typical active-TB case (specified in a more general model in Ref. [4]). Hence, reductions in Q_0 may come from:(a) a decrease in the mean infectious period $(1/\gamma)$, which can be achieved by the 'removal' of infectious individual soon after diagnosis (isolation or treatment); (b) a decrease in the environmental risk of infection (β), the result, for example of improving ventilation in workplaces; or (c) a decrease in the number of contacts per individual, the result of reductions in mean household size (reductions in crowdedness).

Public health measures like the sanatorium movement[16] or the widespread use of antibiotics after the fifties, have reduced the effective mean infectious period but these reductions do not appear to be strong enough to explain the data[4] . The fact that over the 20th century most US individuals move from the country into cities to join crowded living and working environments suggest that the level of crowdedness experienced by individuals has probably not gone down. Hence, the strongest plausible explanation must come from a shift on the ability of TB to progress from the latent to the active-stage. The fact that about one third of the world population lives with TB and that only 8×10^6 cases of active-TB are documented each year in a world with high contact rates that are the results of demographic and organizational changes that have given rise to thousands of urban centers. Changes that are driven by the birth of massive systems of public transportation, cities capable of supporting tremendous increases in population density, the existence of large and crowded public school systems, massive rates of international travel, increases in the migration and more sub-populations with high life expectancy. These changes support the view that progression must have slowed down given the reductions in active TB incidence. In fact, we see from the detailed analysis in [4] and references there in, that high reductions in progression rates must be the leading hypothesis.

4. Final Thoughts and Discussion

Tuberculosis (TB) is a lightly "virulent" disease (based on the relative low per capita disease induced deaths) that has colonized 2 billion humans. Despite dramatic increases in population density, the growth of mega-cities, intense population mobility and higher contact rates (from massive use of public transportation and other factors), the likelihood of developing active-TB, which if untreated will kill one third of its victims, has gone down dramatically.

Today we "only" see about 8 million of new cases of active TB per year which end in about 3 million deaths per year around the world. Despite the successes in reducing the likelihood of TB progression, partly due to reductions in malnutrition, improved living conditions and "enhanced" immunological host responses, the fact that prevalence is so high means that TB is nothing but a "ticking" bomb. The fact that one out of every three human is the host of a strain of tuberculosis means that the potential for massive re-activation is there. Hence, efforts to identify and understand re-infection or re- activation mechanisms are of critical importance today. The appearance of HIV and its dramatic growth in some parts of the world have resulted in large increases rates of TB re-activation (see Ref. [32]). The lack of effective compliance with multi-drug TB treatment protocols has also contributed to the observed increases in the number of cases of DR- TB and XDR-TB. Increases in drug resistance pose a global health threat that cannot be minimized.

In this article, we have shown that re-activation or re-infection mechanisms bring are capable of supporting qualitative changes in TB dynamics. Specifically, these mechanisms could be responsible for supporting multiple endemic states and bi-stability. The impact of these mechanisms is likely to increase due to factors that include increases in HIV prevalence and deteriorating economic conditions (famine and the like). Hence, the eradication or effective control of TB becomes a challenging and critical enterprise.

Looking at policies that bring $\Re_0$ to a value less than one may no longer sufficient if reactivation or re-infection are strong. Here, we have looked at rather simple models deliberately ignoring a large number of epidemiological factors, social and socioeconomic, host's heterogeneity, population structure and "movement" patterns. We have also neglected new and critically important activation mechanisms that include co-infections (HIV) or DR-TB or XDR-TB. However, we believe that the addition of these additional levels of complexity will strengthen our results. That is, that re-activation or re-infection mechanisms are more likely to play an important role in the context of complex social environments and highly heterogeneous communities.

Acknowledgements

The research of Wang and Castillo-Chavez was supported by the National Science Foundation (DMS-0502349), the National Security Agency (DOD-H982300710096), the Sloan Foundation, and Arizona State University. The research of Feng was supported by NSF grant DMS-0719697. JPA is a member of the CONICET.

References

1. Anderson, R.M. and May, R.M. *Infectious Diseases of Humans*, Oxford Science Publications, Oxford, 1991.
2. Aparicio, J.P., Capurro, A.F., Castillo-Chavez, C. Markers of disease evolution: The case of tuberculosis. *J. theor. Biol.*, 2002, **215**: 227–237.
3. Aparicio, J.P., Capurro, A. F. and Castillo-Chavez, C. Long-term dynamics and re-emergence of tuberculosis, in *Mathematical Approaches for Emerging*

and Reemerging Infectious Diseases: An Introduction, Blower, S., Castillo-Chavez, C., Kirschner, D., van den Driessche, P. and Yakubu, A.A. (eds.) , Springer-Verlag. 2002, 351-360.

4. Aparicio, J.P. and Castillo-Chavez, C. Mathematical Modeling of Tuberculosis Epidemics. *to appear in MBE.*
5. Bell, F.C. and Miller, M.L. Life Tables for the United States Social Security Area 1900-2100, *Actuarial Study No. 120*, Social Security Administration, Pub. No. 11-11536, 2005.
6. Blower, S. M., McLean , A. R., Porco, T. C., Small, P. M., Hopwell, P. C., Sanchez, M. A. and Moss, A. R. The intrinsic transmission dynamics of tuberculosis epidemics. *Nature Medicine*, 1995, **1** (8): 815-821.
7. Blower, S. M., Small, P. M. and Hopwell, P. C. Control strategies for tuberculosis epidemics: new models for old problems. *Science*, 1996, **273**: 497-500.
8. Brauer, F. and Castillo-Chavez, C. *Mathematical Models in Population Biology and Epidemiology.* Springer, 2001.
9. Castillo-Chavez, C. and Feng, Z. To treat or not to treat: the case of tuberculosis. *J. Math. Biol.*, 1997, **35**: 629-656.
10. Castillo-Chavez, C. and Feng, Z. Global stability of an age-structure model for TB and its applications to optimal vaccination strategies. *Mathematical Biosciences*, 1998, **151**: 135-154.
11. Castillo-Chavez, C. and B. Song, Models for the Transmission Dynamics of Fanatic Behaviors, in *Bioterrorism: Mathematical Modeling Applications to Homeland Security*, Banks, T. and Castillo-Chavez, C. (eds.) SIAM Series Frontiers in Applied Mathematics, 240 pp. Volume 28, 2003.
12. Castillo-Chavez, C. and Song, B. Dynamical Models of Tuberculosis and applications. *Journal of Mathematical Biosciences and Engineering*, 2004, **1**, (2): 361-404.
13. Chiang, C.Y. and Riley, L.W. Exogenous reinfection in tuberculosis. *Lancet Infect Dis.*, 2005, **5**: 629-636.
14. Comstock, George W. In *Tuberculosis*, 2000, **144**: 129-156, Marcel Dekker, Inc..
15. Dannenberg Jr., Arthur M. Chapter 2 in *Tuberculosis*, Marcel Dekker, Inc., 2000.
16. Davis, A.L. History of the sanatorium movement, in *Tuberculosis* Rom, W.N. and Garay, S.M. (eds.), Little, Brown and co, 1996.
17. Feng, Z., Castillo-Chavez, C. and Capurro, A.F. A Model for Tuberculosis with exogenous reinfection. *Theor. Pop. Biol.*, 2000, **57**: 235-247.
18. Feng, Z., Huang, W. and Castillo-Chavez, C. On the role of variable latent

periods in mathematical models for tuberculosis. *Journal of Dynamics and Differential Equations*, 2001, **13**, (2): 425-452.

19. Feng, Z., Iannelli, M. and , Milner F. A two-strain TB model with age structures. *SIAM J. Appl. Math.* **62** (5): 1634-1656.
20. Grigg, E.R.N. The arcana of Tuberculosis. *Am. Rev. Tuberculosis and Pulmonary Diseases*, 1958, **78**: 151-172.
21. Grigg, E.R.N. The arcana of Tuberculosis III. *Am. Rev. Tuberculosis and Pulmonary Diseases*,1958, **78**: 426-453.
22. Grigg, E.R.N. The arcana of Tuberculosis IV. *Am. Rev. Tuberculosis and Pulmonary Diseases*, 1958, **78**: 583-608.
23. Haines, M.R. Estimated Life Tables for the United States, 1850-1900. Historical Paper n° 59. *National Bureau of Economic Research*, 1994.
24. Hethcote, H.W., Qualitative analysis for communicable disease models. *Math. Biosc.*, **28**: 335-356.
25. Kermack, W.O. and McKendrick, A.G. A contribution to the mathematical theory of epidemics. *Proc. Royal Soc.* London, **115**: 700-721.
26. Kochi, A. The global tuberculosis situation and the new control strategy of the World Health Organization. *Tubercle*, 1991, **72**: 1-6.
27. Lambert, M.L., Hasker, E., van Deun, A., Roberfroid, D., Boelaert, M., van der Stuyft, P. Recurrence in tuberculosis: relapse or reinfection? *Lancet Infect Dis.* 2003, **3**: 282-287.
28. Lipsitch, M., Murray, M.B. Multiple equilibria: Tuberculosis transmission requires unrealistic assumptions. *Theoretical Population Biology*, 2003, **63**: 169-170.
29. McKewon, T. and Record, R.G. Reasons for the decline of mortality in England and Wales during the nineteenth century. *Population Studies*, 1962, **16**: 94-122 .
30. Porco, Travis C., and Blower, Sally M. Quantifying the intrinsic transmission dynamics of tuberculosis. *Theoretical Population Biology*, 1998, **54**: 117-132.
31. Reichman, Lee B. and Hershfield, Earl S. *Tuberculosis.* Marcel Dekker,INC., 2000.
32. Roeger, Lih-Ing W., Feng, Z. and Castillo-Chavez, C. The Impact Of Hiv Infection On Tuberculosis, *to appear in MBE.*
33. Ross, R. *The prevention of Malaria*, 2nd ed. (with Addendum), John Murray, London, 1991.
34. Sanchez, F., Wang, X., Castillo-Chavez, C., Gruenewald, P. and Gorman,D. Drinking as an epidemic–a simple mathematical model with recovery and relapse, in *Therapist's Guide to Evidence Based Relapse Prevention,*

Witkiewitz, Katie and Marlatt, G. A. (eds.), 2007: 353-368.

35. Shen, G., Xue, Z., Shen, X., Sun, B., Gui, X., Shen, M., Mei, J., Gao, Q. Recurrent tuberculosis and exogenous reinfection, Shanghai, China. *Emerging Infectious Disease*, 2006, **12**:1176-1178.
36. Song, B., Castillo-Chavez, C. and Aparicio, J.P. Tuberculosis Models with Fast and Slow Dynamics: The Role of Close and Casual Contacts. *Mathematical Biosciences*, 2002, **180**: 187-205.
37. Song, B., Garsow-Castillo, M., Rios-Soto, K., Mejran, M., Henso, L., and Castillo-Chavez, C. Raves Clubs, and Ecstasy: The Impact of Peer Pressure. *Journal of Mathematical Biosciences and Engineering*, 2006, **3**, (1): 1-18.
38. Stead, W. W. The pathogenesis of pulmonary tuberculosis among older persons. *Am Rev Respir Dis*, 1965, **91**: 811-22.
39. Stead, W. W. Genetics and resistance to Tuberculosis. *Annals of Internal Medicine*, 1992, **116**: 937-941.
40. Stead, W. W., and Bates, J.H. Geographic and evolutionary epidemiology of tuberculosis, in *Tuberculosis. Little, Brown and co.* Rom, W.N. and Garay, S.M. (eds.), 1996: 77-84.
41. U. S. Bureau of the Census. *Historical statistics of the United States: colonial times to 1970.* Washington DC: Government Printing Office, 1975.
42. Verver, S., Warren, R.M., Beyers, N., Richardson, M., van der Spuy, G.D., Borgdorff, M.W., *et al.* Rate of reinfection tuberculosis after successful treatment is higher than rate of new tuberculosis. *AM J Respir Crit Care Med*, 2005, **171**: 1430-1435.
43. Waaler, H.T., Gese, A., and Anderson, M. The use of mathematical models in the study of the epidemiology of tuberculosis. *AM. J. Publ. Health.*, 1962, **52**: 1002-1013.
44. Wang, X. Backward Bifurcation in a Mathematical Model for Tuberculosis with Loss of Immunity. *Ph.D. Thesis*, Purdue University, 2005.

THE SPREAD OF THE HIV INFECTION ON IMMUNE SYSTEM: IMPLICATIONS ON CELL POPULATIONS AND R_0 EPIDEMIC ESTIMATE

M. ROSSI

Department of Forensic Medicine, Medical Ethics, Social and Occupational Medicine, School of Medicine, São Paulo University,
Av. Dr. Arnaldo, 455,
01246903 São Paulo SP Brazil
E-mail: mrossi@dim.fm.usp.br

L. F. LOPEZ

Department of Forensic Medicine, Medical Ethics, Social and Occupational Medicine, School of Medicine, São Paulo University,
Av. Dr. Arnaldo, 455,
01246903 São Paulo SP Brazil
E-mail: lopez@dim.fm.usp.br

Understanding the spread of HIV virus in patient HIV-positive and that immune response involvement has motivated a great number of works in Mathematical Immunology. The models representing healthy and infected cell populations show variations of the dynamics of viral infection in some scenes, trying to demonstrate the ways and mechanisms that leading HIV virus invades the target-host cells, unbalancing the immune system. The use of the Basic Reproduction Number concept becomes an important parameter to quantify the viral proliferation and disease evolution, requiring a precise definition in the model of viral transmission for threshold condition calculation and in the continuous viral infection. In this work, the study was performed with the "Next Generation Matrix", methodology used to calculate the ?0 expression from a mathematical model that considers the viral infection in target cells populations that consist of macrophages, dendritic cells, lymphocytes TCD4 and CTL evidencing the most important parameters to obtain the infection threshold. As conclusion, was obtained an expression that relates parameters of HIV infection, correlated with the population of infected macrophages and dendritic cells to lymphocytes TCD4 cells, with control exercised by CTL population, demonstrating the sufficient condition to the establishment of viral proliferation and evolution of the patient of AIDS.

1. Introduction

Infectious diseases and epidemics have been studied for years, considering the type of vector, coverage of the illness, people involved and especially the force of infection characteristic of the disease. Different approaches related to the control and eradication of infection require knowledge of the kinetics of the epidemic process and the involvement of human immunity in containing with the infecting agent and increase of new cases of sick people. Different vaccination schemes may yield different levels of immunization of the population; these do not prevent completely infection but reduces the probability of occurrence of new cases and the metabolic consequences of infected host, reducing the burden of disease. The models that represent these phenomena, usually in the form of dynamic systems (classical models formed mainly by EDOs, difference equations and stochastic models) are developed according to the level of complexity, due the host-pathogen relationship and people involved, victims of the disease. Under the most varied ways of mathematical representation, these studies deal with the kinetics and control of infectious process, considering strategies for reduction of new cases and increasing the immunization of susceptible individuals, with the goal of total eradication of the disease in question. A conception of great interest in epidemiology is the R_0, the Basic Reproduction Number, which value is the expected number of secondary cases per primary cases of infection in a completely susceptible population[16,11,15] . The conception was initially proposed by the MacDonald in the 1950s where in the case of one vector and one host showing that for cases where $R_0 > 1$ the endemic state is established, and where $R_0 < 1$ the progress of the disease is not observed. But for more complex systems, the deduction of simple expressions of R_0 is complicated because the persistence of infection is composed by several quantities closely related on expression of classical R_0. For these more complex cases, the definition developed by Diekmann and Heesterbeek[4] , defined as Basic Reproduction Ratio, has been used to study the establishment of the criterion of threshold in heterogeneous systems. Therefore, the R_0 is an important indicator of efforts to eliminate the infection. We introduced the concept of R_0 into a mathematical model representing the phenomena of the human infection by HIV-1 and the dynamics of the immune system sent to the control and elimination into innate immune response. The equations set, composed of a SI model with nonlinear and delay terms, describe the cellular immune mechanisms and which are part of efforts to control the viral infection, caused by HIV-1, and co-adaptation to the host. Population of dendritic cells and macrophages, with their infected sub-populations, are considered for proliferation and maintenance of the infectious process within the individual, the lymphocyte TCD4 cells (and their chronically- and productively-infected) and CTL, showing the appearance of heterogeneous population. Due to the physiological character of the cells in this model, it is utilized the concept of "generations" developed by Roberts and Heesterbeek[17] , where the word generation refers to different hosts that can influence in different ways, the course of infection in different locations or it may has some other characteristic that can differentiate them from the epidemiological point of view. Based on the work of Diekmann and Heesterbeek[3,4] , it was developed an matrix operator of new generation to establish the threshold of viral infection based on immune mechanisms involved in the control and eradication of HIV-1.

2. The Threshold Calculation

The mechanisms of HIV infection are still unclear and therefore the range of maximum efficiency of a particular vaccine is not reached even by the number of resulting mutant strains found during the treatment scheme with antiretroviral drugs or by the formation of reservoirs cellular compounds, mainly by macrophages and lymphocytes TCD4 cells chronically infected. Since viral secondary infections can occur both in the antigens presentation and in the lymphocytes activation or by macrophages activation on inflammation sites, the result of the expression of R_0 should consider these factors so that efforts to control and eradicate the disease are more effective. Therefore, the need to consider other cellular reservoirs and not just approach on a host, show the deduction and evaluation importance of information threshold from a tool to control the generation of infectious HIV-1. The SI model equations set, Eq. (1), representing the biochemical phenomena found in viral infection, proliferation and the immune response, are shown below. The considerations related to the dynamics of the model are in ref [18].

$$\frac{d(Mp)}{dt} = s_m + l_1 M_p - \beta_1 Mp\frac{v^n}{v^n + K_v} - \delta_f Mp$$

$$\frac{d(Mpi)}{dt} = \beta_1 Mp\frac{v^n}{v^n + K_v} - (\delta_f + \alpha`)Mpi$$

$$-\beta_{mf} MpiT - k_1 MpiCTL$$

$$\frac{d(iDC)}{dt} = s_{id} + l_2 iDC - K_1 iDC \frac{v}{v + K_v} - \delta_{id} iDC$$

$$\frac{d(mDC)}{dt} = K_1 iDC \frac{v}{v + K_v} - \delta_{dc} mDC - \beta_{dc} mDC.T - k_2 mDC.CTL$$

$$\frac{dT}{dt} = s_4 + (\beta_{mf} Mpi + \beta_{dc} mDC) T$$

$$-a \left(\frac{1 - e^{-ct}}{c} \right) T (\xi Mpi + \phi mDC) - \delta T$$

$$\frac{dI}{dt} = f * a \left(\frac{1 - e^{-ct}}{c} \right) T(t - \tau)(\xi Mpi(t - \tau) + \phi mDc(t - \tau))$$

$$e^{-(\delta + \alpha^*)} - (\delta + \alpha^*) I + 0,0005L - k_3 I.CTL$$

$$\frac{dL}{dt} = (1 - f) * a \left(\frac{1 - e^{-ct}}{c} \right) T(t - \tau_{lat})(\xi Mpi(t - \tau_{lat})$$

$$+\phi mDC(t - \tau_{lat}))e^{-(\delta + \alpha^*_{lat})} - (\delta + \alpha^*_{lat}) L - 0,0005L$$

$$\frac{d(CTL)}{dt} = s_8 - \delta_{ctl} CTL - (\xi' Mpi + \phi' mDC) CTL$$

$$+ \left(k_1^{ctl} Mpi + k_2^{ctl} mDC + k_3^{ctl} I \right) CTL$$

$$\frac{dv}{dt} = Q_1(\delta + \alpha^*) I + Q_2(\delta + \alpha^*_{lat}) L + Q_3 Mpi - K_2 iDC.v$$

$$-\delta Mp. \frac{v^n}{v^n + K_v} - cv \qquad (1)$$

In this equation, it is studied a mathematical model representing the viral infection, with penetration and viral proliferation mechanism of HIV-1 and immune systems components representing the immune response, which come in contact with the virus and the physiologic depletion that virus offers. Mp and Mpi represent the population of active and inactive and infected macrophages; iDC and mDC, the immature dendritic cells and those that have matured after viral contact; T, I and L, the lymphocyte TCD4 cells nave, infected and chronically infected patients, respectively; CTL, the population of cytotoxic active lymphocytes and v, the viral load in the infected individual. The macrophages could be considered as reservoirs of viral proliferation of HIV-1 and also perform the function of presenting antigens to the lymph organs together with the dendritic cells[18] . Thus, the macrophages subpopulation infected is considered on feature to calculus methodology, presented by Roberts and Heesterbeek[17] , the dendritic cells plus virus subpopulation and lymphocytes TCD4 infected. The parameters for the activation of CTL lymphocytes, the number of virus per cell and the intracellular delay at the beginning of the viral cycle are shown when necessary. Therefore, the matrices with the entries in terms of death, by Heesterbeek and Diekmann, are $(M_2 + D)$ and $(M_1 + \Sigma)$ and contains the values of death rates of subpopulations of infected macrophages, complexes formed by dendritic cells and viral particles and lymphocytes productively and chronically infected. The likelihood of macrophages and dendritic cells activation into sites of infection and level of immune response are demonstrated by the population of CTL lymphocytes. The matrix $(M_2 + D)$ (Eq. 2) carries the death rates of infected cell populations, such as infected macrophages, mature dendritic cells with viruses and lymphocytes TCD4 cells productively infected. The last two vectors are zero because they are representatives of populations of CTL lymphocytes and viral load.

$$(M_2 + D) = \begin{pmatrix} -(\mu_f + \alpha') & 0 & \cdots & \cdots & \cdots \\ 0 & -\mu_{dc} & 0 & \cdots & \cdots \\ \vdots & 0 & -(\mu + \alpha^*) + c & 0 & \cdots \\ \vdots & \vdots & 0 & -(\mu + \alpha^*_{lat}) - c & 0 \\ \vdots & \vdots & \vdots & 0 & 0 \\ \vdots & \vdots & \vdots & \vdots & 0 \end{pmatrix} \quad (2)$$

where c is equal to 0,0005.

Since HIV-1 penetrates the tissue barrier, it starts a cascade of cellular and biochemical mechanisms responsible by recognition and anti-invasive pathogen. The extinction rates the representing the CTL action over infected cells population, exercised by entry and exit flux, can be seen in the matrix $(M1 + \Sigma)$ (Eq. (3). The first two vectors, represent the likelihood of infection that occur by the cells due to viral encounter[18] , and this encounter occurs on a nonlinear feature, under the law mass action form hypothesis. The fifth vector describes the involvement of the immune response against the intruder, including in each column the difference between the likelihood of action and cytotoxic cell death induced by activation. The effect of activation and death induced by macrophages infected (ξ), and the dendritic cells (ϕ) are presented in the first two columns, with viral particles trapped on SIGN complexes through the presentation of antigens during the infection history, showing the equilibrium living between control of infection and escape by HIV-1. The probability of the human body to eliminate the virus is related in the last vector in the form of the clearance rate C.

$$(M_1 + \Sigma) = \begin{pmatrix} 0 & \dots & \dots & \dots & \dots & \beta_1 Mp \\ \dots & 0 & \dots & \dots & \dots & K_1 iDC \\ \dots & \dots & 0 & \dots & \dots & \dots \\ \dots & \dots & \dots & 0 & \dots & \dots \\ (k_1^{ctl} - \xi') & (k_2^{ctl} - \phi') & k_3^{ctl} & 0 & -\mu_{ctl} & 0 \\ \dots & \dots & \dots & \dots & 0 & -C \end{pmatrix} \tag{3}$$

The matrix T(s) Eq. (4) show the parameters and the likelihood that the dynamics of the evolution of infectious HIV-1. The beginning of the infection and the immune response occurs during the contact and subsequent activation of the cells of the innate response, such as macrophages and dendritic cells that has characteristics of Antigen-Presenting Cells (APC). These activate the lymphocytes naïve TCD4 cells through the presentation of these antigens to lymphocytes set that, once it was activated, modulate the shape and the control intensity of proliferation of the virus[8,13]. The level of activation is shown in the first two vectors. As the "infectious synapse" between lymphocytes and APC is successful, there is the formation of subpopulations of infected lymphocytes, represented by the third and fourth vectors, where each input is related to kinetics of viral infection and proliferation. The fact that the intracellular infectious process is not instantaneous is admitted, and occurs after a small delay due to virus found into intracellular mechanisms for the synthesis of your genetic material, and such information is included in the model through a delay (τ). The last two vectors represent the probability of activation of lymphocytes CTL during the antigens presentation by macrophages and dendritic cells infected and the cytolytic activity of macrophages and dendritic cells on the viral load of HIV-1. We define the matrices $V := M_2 + D + M_1 + \Sigma$ and $F := T(S)$. The matrix $K = -FV^{-1}$, eq. (5), is shown below. The matrix

V, obtained here by analysis of in and out compartments flows of infected and infectious cells, is a Metzler matrix because the off-diagonal entries represent such flows or, possibly, the probability of state transition to appear into cellular infected subpopulations[9] . Interested readers can read about the definition and application examples of Metzler matrix and M-matrix on papers by Mitkowski[14] , Esteva-Peralta[6] and van den Driessche[19] . Due the nature of V and F matrices, the Next Generation Matrix K will always be a nonnegative matrix, and the Matrix Theory guarantees the existence of a positive eigenvalue whose modulus is at least as large as all others eigenvalues founded. The offdiagonal entry in structure of K matrix comes from the fact that infected macrophages and dendritic cells plus HIV subpopulations produce new infected individuals almost on contacts between TCD4 naïve lymphocytes and are controlled by CTL lymphocyte population. Moreover, these new infected lymphocytes are used by virus to growth and proliferate the infection into the human body. Diekmann and Heesterbeek conceptualize the Basic Reproduction Number, R_0, as the eigenvalue of greatest value of the matrix $-FV^{-1}$, ie, the spectral radius of this material is $R_0 = \rho(-F.V^{-1})$. To calculate the largest eigenvalue of the matrix K (therefore, R_0), the matrix results were obtained by transforming the principal matrix into four sub-matrices and forming a block matrix, where A and D are Metzler matrices[5,10] and defined as the determinant of the blocks, det $(A_{22} - A_{12}A_{11}^{-1}A_{21})$.

$$\mathbb{K} = [\mathbf{A_{11}}\ \mathbf{A_{21}};\ \mathbf{A_{21}}\ \mathbf{A_{22}}]$$

The conditions of stability of this product matrix with Metzler matrices can be seen in Ref. [10] and threshold conditions of the system can be calculated in the study of the unique endemic equilibrium. In the last matrix, each element show the particular contribution of each population to disease proliferation and cell's death. It is a kind of "who infects whom" point-of-view and Next Generation Operator methodology is applied on then.

The final mathematical expression of R_0 Eq. (6) is a mixture of two others R_0 Eq. (7, 8), because each represents the populations involved on installation and proliferation of HIV-1 infection in humans. Diekmann and Heesterbeek[4] consider that for this case, the R_0 predominant is the largest eigenvalue, which coincides with the expression given in ref. [18].

$$T(S) = \begin{pmatrix} -\beta_{mf}T_0 & 0 & \cdots & \cdots & \cdots & 0 \\ 0 & -\beta_{dc}T_0 & 0 & \cdots & \cdots & 0 \\ f\xi T(\tau) & f\phi T(\tau) & f\xi Mpi(\tau) & f\xi mDC(\tau) & 0 & \cdots \\ a & b & (1-f)\xi Mpi(\tau_{lat}) & (1-f)\xi mDC(\tau_{lat}) & 0 & \cdots \\ \cdots & \cdots & \cdots & 0 & -(\xi' Mpi + \phi' mDC) & 0 \\ \cdots & \cdots & \cdots & \cdots & 0 & -(\delta.Mp + K_2 iDC) \end{pmatrix} \quad (4)$$

$$K = \begin{pmatrix} -\frac{\beta_{mf}T_0}{\mu_f+\alpha'} & 0 & \cdots & \cdots & 0 & -\frac{\beta_{mf}T_0}{\mu_f+\alpha'}\frac{\beta_1 Mp}{c} \\ 0 & -\frac{\beta_{dc}T_0}{\mu_{dc}} & 0 & 0 & 0 & -\frac{\beta_{dc}T_0}{\mu_{dc}}\frac{K_1 iDC}{c} \\ \frac{f\xi T(\tau)}{\mu_f+\alpha'} & \frac{f\phi T(\tau)}{\mu_{dc}} & \frac{f\xi Mpi(\tau)}{(\mu+\alpha^*)+0{,}0005} & 0 & 0 & \frac{f\xi T(\tau)}{\mu_f+\alpha'}\frac{\beta_1 Mp}{c} + \frac{f\phi T(\tau)}{\mu_{dc}}\frac{K_1 iDC}{c} \\ \frac{a}{\mu_f+\alpha'} & \frac{b}{\mu_{dc}} & 0 & \frac{(1-f)\phi mDC(\tau)}{(\mu+\alpha^*_{lat})-0{,}0005} & 0 & \frac{a}{\mu_f+\alpha'}\frac{\beta_1 Mp}{c} + \frac{b}{\mu_{dc}}\frac{K_1 iDC}{c} \\ \frac{\xi' CTL_0}{\mu_f+\alpha'} & \frac{\phi' CTL_0}{\mu_{dc}} & 0 & 0 & \frac{-(\xi' Mpi+\phi' mDC)}{\mu_{ctl}} & \frac{\xi' CTL_0}{\mu_f+\alpha'}\frac{\beta_1 Mp}{c} + \frac{\phi' CTL_0}{\mu_{dc}}\frac{K_1 iDC}{c} \\ 0 & \cdots & \cdots & \cdots & 0 & \frac{-(\delta.Mp+K_2 iDC)}{c} \end{pmatrix} \quad (5)$$

where $a = (1-f)\xi T(\tau_{lat})$ and $b = (1-f)\phi T(\tau_{lat})$.

$$R_0 = \sqrt{R_0^{mf} + R_0^{dc}} \tag{6}$$

where

$$R_0^{mf} = f * a\left(\frac{1 - e^{-ct}}{c}\right)\frac{\xi T(t-\tau)e^{-(\mu+\alpha^*)*\tau}}{(\mu_f + \alpha')}\frac{\beta_1}{c}Mp_0 \tag{7}$$

$$R_0^{dc} = f * a\left(\frac{1 - e^{-ct}}{c}\right)\frac{\phi T(t-\tau)e^{-(\mu+\alpha^*)\tau}}{(\mu_{dc})}\frac{K_1}{c}iDC_0 \tag{8}$$

The terms T_0, Mp_0, iDC_0 and CTL_0 correspond to the equilibrium value of these variables in the disease-free equilibrium (DFE). The R_0 first mathematical expression Eq. (7), show that, to be progress in infection, the average value of effective contact between infected macrophages and lymphocytes TCD4 cells must be greater than the life expectancy of these macrophages, multiplied by the efficiency of infection of HIV-1 on the same cell type. If the viral infection efficiency on the phagocytic cells is higher than HIV-1 clearance rate, then the subpopulation of infectious cells will increase, carrying the virus to lymphoid organs and germinal centers, contributing to the constant activation of nave lymphocytes, driving to depletion of lymphocytes TCD4 cells level and total exhaustion of the immune system, leading the individual to AIDS. The same thought can be used for the second term of R_0 Eq. (8) on the subpopulation of dendritic cells containing viable viral particles attached in their complex DC-SIGN and transfering to lymphocytes at the "infectious synapse".

3. Discussion

The (i, j) entry of the Next Generation Matrix K is a expected number of secondary infections produced in compartment (i) by an case initially in compartment (j) and are related to the balance between production of new infections in the group of macrophages and dendritic cells and lymphocytes and the control of viral proliferation (with their removal) generated by HIV-specific CTL through the action of apoptosis mechanism on the infected. The performance of all macrophages infected with viral particles impact new generations of infection, remaining as a major reservoir (if not more important) in the continuity of the infection process of lymphocytes TCD4+ cells because of the constant antigen presentation and immune response activation on infection or inflammatory process triggered, leading to increased activation of the population of TCD4, depleting the immune system to total exhaustion. Some works cites[1,2,7] that the viral population existing within the infected macrophages have a different mutant profile that is found in the subpopulation of infected lymphocytes, probably due to different selective pressures to which HIV-1 is exposed, whether in an individual in treatment or not, what would be the rapid dominance of wild virus after cessation of use of therapeutics drugs. In addition, the specificity of dendritic cells in relation to the fragments epitopes or complete HIV particles is represented by gp120 viral glycoprotein high affinity to the receptor lecithin SIGN, and carried to lymphoid organs due to the recognition of antigens and immune response triggered by germinative centers in lymph organs. If the binding rate (dependent of affinity between the viral ligand and receptor-specific) between dendritic cell and viral particle is above the clearance rate, there will be accumulation of viral load in patients and increase of new infections cell, leading to continuous infection in the host, contributing to the decline of the concentration of lymphocytes naïve TCD4+ cells and diminished the ability of the immune system to combat this and other diseases. The two equations, 7 and 8, shows the mechanism of immune response that may be related to the targets of new drugs or new forms of treatment. The impact of new approaches to treatment which can lead to the partial eradication of the disease, is the formation of strains resistant to antiviral drugs, either by action of the cell tropism viral subtype presents. Then, the model shows that populations of macrophages and dendritic cells form a set of possible therapeutic targets to combat HIV/AIDS and targets for vaccination schemes.

References

1. Aquaro, S et al. Macrophage and HIV infection: Therapeutical approach toward this strategic virus reservoir. *Antiviral. Res* (2002), **55**: 209-25.
2. Collman, R G et al. HIV and cells of Macrophage/Dendritic lineage and others non-Tcell reservoirs: new answers yield new questions. *J. Leuko. Biol.* (2003), **74**: 631-634.
3. Diekmann, O; Heesterbeek, J A P; Metz, J A J. On the definition and the computation of the basic reproduction ratio R0 in models for infectious diseases in heterogeneous populations. *J. Math. Biol.* (1990), **28**: 365-382.
4. Diekmann, O; Heesterbeek, J A P. Mathematical epidemiology of infectious diseases: Model building, analysis, and interpretation. Chichester, New York. John Wiley, 2000.
5. Dumont, Y; Chiroleu, F; Domerg, C. On a temporal model for the Chikungunya disease: Modeling, theory and numerics. *Math. Biosc.* (2008), **213**: 80-91.

6. Esteva-Peralta, L; Velasco-Hernandez, J X. M-matrices and local stability in epidemic models. *Math. Comp. Model.* (2002), **36**: 491-501.
7. Gorry, P R et al. Pathogenesis of macrophage tropic HIV-1. *Curr. HIV Res.* (2005),**3**, 53-60.
8. Haase, A T. Population biology of HIV infection: Viral and CD4+ Tcell demographic and dynamics in lymphatic tissues. *Annu. Rev. Immunol.* (1999), **17**:625-656.
9. Jacques, J A; Simon, C P. Qualitative theory of compartmental systems. *SIAM Rew.* (1993), **35**:43-79.
10. Kamgang, J C; Sallet, G. Computation of threshold conditions for epidemiological models and global stability of the disease-free equilibrium (DFE). *Math. Biosc.* (2008): **213**: 1-12.
11. Longini, I M Jr. et al. Containing pandemic influenza at the source. *Science*, (2005), **309**: 1083-1087.
12. Lopez, L F; Coutinho, F A B; Burattini, M N; Massad E. Threshold conditions for infection persistence in complex host-vectors interactions. *C. R. Biologies.* (2002), **325**: 1073-1084.
13. McCune, J M. The dynamics of CD4+ T cell depletion on HIV disease. *Nature* (2001), **410**: 974-979.
14. Mitkowski, W. Dynamical properties of Metzler systems. *Bull. Pol. Ac. Tech.* (2008), **56**, 309-312.
15. Riley, S et al. Transmission dynamics of the etiological agent of SARS in Hong Kong: Impact of public health interventions. *Science* (2003), **300**:1961-1966.
16. Roberts, M G. The pluses and minuses of R0. *J. R. Soc. Interface* (2007), **4**: 949-961.
17. Roberts, M G; Heesterbeek, J A P. A new method for estimating the effort required to control an infectious disease. *Proc. R. Soc. Lond. B* (2003), **270**: 1359-1364.
18. Rossi, M; Lopez, L F. Mathematical Model of HIV infection: Dendritic cells and Macrophage presence on infection proliferation. *PloS Comput. Biol.* ***(in press)***.
19. van den Driessche, P; Watmough, J. Reproductive numbers and subthreshold endemic equilibria for compartmental models of disease transmission. *Math. Biosc.* (2002), **180**: 29-48.

REAL-TIME FORECASTING FOR AN INFLUENZA PANDEMIC IN THE UK FROM PRIOR INFORMATION AND MULTIPLE SURVEILLANCE DATASETS*

G. KETSETZIS, B. COOPER, D. DEANGELIS AND N. GAY

Health Protection Agency,
Centre for Infections,
National Health Service,
61 Colindale Avenue, NW9 5EQ, London, USA
E-mail: Georgios.Ketsetzis@hpa.org.uk

The recent emergence of swine flu outbreak with worldwide confirmed cases indicates that we may be at the brink of an influenza pandemic. Government preparedness policies require the real-time modelling of flu outbreaks and knowledge of possible scenarios of what may happen, to guide decision making. There are various challenges in real-time modelling and forecasting due to the uncertainty in the epidemiological characteristics of the virus, behavioural changes of the population during a pandemic and the difficulties in national surveillance. In this work we develop an appropriate pandemic flu model and integrate it within an adaptive Bayesian framework to predict the outcome of a historic epidemic in the UK(1969) in order to test the performance of this proposed framework. Our results show that with appropriate choices of priors for some of the model parameters it is possible to predict accurately the possible outcome of an ongoing pandemic, implying that good prior information which is based on initial surveillance data analysis is essential to enable reliable predictions.

1. Introduction

The swine flu outbreak in Mexico in late April 2009 and the subsequent reporting of cases around the world have raised the WHO alertness level to one step before declaring an influenza pandemic. There are already some studies investigating the pandemic potential of this strain of influenza A [1] and initial estimates for R_0 for the local outbreak in La Gloria provide a value greater than 1. The Case Fatality Ratio (CFR) which is a clear indication of the severity of the outbreak is higher in Mexico than any other current local outbreak and the burden of disease also seems to vary by

*This work is supported by the uk department of health

country. This uncertainty in key epidemiological parameters of this H1N1 strain (recently identified as **H1N1swl**) make it difficult to give a global prediction of the spread of the virus and rather a more localized prediction is sought, ideally by country. This is because there is great variability in contact patterns between different age groups between countries, different surveillance systems-therefore different available data- and different policies in the availability and distribution of antivirals. As a result, many current up to date models of flu are restricted to at most a national level.

Real time modelling aims to use available surveillance data and appropriate models of flu transmission to predict the outcome of an outbreak (by fitting to data and subsequent extrapolation) and to update these predictions as new data are obtained, in a daily or weekly basis. During the early stages of such an outbreak, because of sparse data, data contamination with noise and a range of uncertainties in several parameters (for example the contact population patterns) there is large uncertainty about how the outbreak will progress, but as new and more data are obtained, parameters can be estimated with improved confidence and useful predictions of the possible outcome of the outbreak can be obtained. These predictions can then be used by decision makers to decide the application and time of interventions (school closure, transport restrictions, etc) and see, in real time, the result of these interventions and the subsequent forecasts. It appears therefore that real time modelling of pandemic influenza is particularly useful for national pandemic preparedness.

Despite some early current estimates of key epidemiological parameters [1], there has not been much work on real time modelling of pandemic influenza. Hall *et al.* [2] successfully fitted a mass action model to historic surveillance data by standard regression analysis to predict the outcome of the outbreak. Their model did not account for heterogeneous mixing, did not included delays in surveillance data or changes in reporting behaviour, and only allows for a constant basic reproduction number. Their results indicated that very tight bounds should be placed on some key parameters to achieve good predictions for pandemic influenza incidence but it may not be possible to obtain such accurate information from early surveillance data. Other work by White and Pagano [3] focuses on real-time estimates of serial interval and reproductive number while Bettencourt and Ribeiro [4] also employ a Bayesian method for estimation of the reproductive number. An individual based model was recently employed by Atti *et al.* [5] to investigate the real time impact of interventions in Italy but based on scenarios rather than data as it would be difficult in general to fit individual based models of whole country populations to real data.

In this work we develop a new model of flu and integrate it within a Bayesian inference framework to fit to real surveillance data from the 1969 UK pandemic and predict the outcome of this historic epidemic. In doing so, we provide real time estimates of more parameters than those estimated by Refs [3] and [4]. Further and extensive details of the present work will appear elsewhere [8].

2. Model

The model we employ is a deterministic and discrete time consisting of three parts. First a population transmission part which is a generalized age specific SEIR model with two latent and two infectious compartments. The transmission model is specified by a system of ordinary differential equations:

$$\frac{dS(t,j)}{dt} = -\lambda S(t,j) \tag{1}$$

$$\frac{dE_1(t,j)}{dt} = \lambda S(t,j) - (2/\mathrm{LP})E_1(t,j) \tag{2}$$

$$\frac{dE_2(t,j)}{dt} = (2/\mathrm{LP})E_1(t,j) - (2/\mathrm{LP})E_2(t,j) \tag{3}$$

$$\frac{dI_1(t,j)}{dt} = (2/\mathrm{LP})E_2(t,j) - (2/\mathrm{AIP})I_1(t,j) \tag{4}$$

$$\frac{dI_2(t,j)}{dt} = (2/\mathrm{AIP})I_1(t,j) - (2/\mathrm{AIP})I_2(t,j) \tag{5}$$

$$\frac{dR(t,j)}{dt} = (2/\mathrm{AIP})I_2(t,j) \tag{6}$$

where $S(t,j)$ is the number of susceptibles at time t and j denotes the age group. $E_{1,2}(t,j)$ are the number of infected at time t at stages 1 and 2 respectively, $I_{1,2}(t,j)$ is the number of infectious at stages 1 and 2 respectively and $R(t,j)$ is the number of recoveries from flu. λ denotes the force of infection given by the Reed-Frost formula [7], LP is the latent period and AIP is the average infectious period.

The second part of the model describes the effects of the virus on the health system as number of clinical cases, GP consultations, hospitalizations, deaths and antivirals used; in any particular health system these measurable quantities can be substituted appropriately depending on the corresponding available surveillance data. A certain proportion of infections will develop symptoms and from those clinical cases a proportion will visit their GP, Hospital, will die or use Antivirals, with some delay from the day of onset of symptoms.

The third part of the model considers the delays in surveillance which can be considerable as a country's health system may be compromised by the extend of an outbreak and there may be delays due to lengthy coroner investigations pertaining to deaths of young people. Such delays in surveillance need to be taken into account in modelling as they imply that the outbreak is always ahead of the observations as is reported and recorded by surveillance systems (see Ref. [8] for details). This is the principal difference of this model from other flu models in the literature.

Because the transmission model is age specific, it requires the population mixing matrix which we choose as the one published by the recent POLYMOD study [6]. The model is initialized by its initial approximately exponential growth phase with an initial number of infections in the region of interest.

The second part of the model requires the distribution of delays from onset of symptoms to GP consultation, hospitalization, death and use of antivirals, which are all chosen to be gamma-distributed. The third part of the model requires the distribution of delays from GP consultation, hospitalization, death and use of Antivirals to reporting of such an event by the surveillance system.

In addition, the model explicitly accounts for reported background influenza like illnesses (ILIs) that may be also circulating during a particular flu outbreak, such as, but not limited to, seasonal flu or/and pneumonia. Early in a pandemic outbreak where data are scarce, it may be difficult to distinguish between the pandemic flu signal and other circulating ILIs and there will be an inherent contamination of surveillance data with ILI "noise" which necessitates appropriate modelling. In addition, when the outbreak enters an exponentially growth phase and the clinical infection numbers become large, further serological tests that can distinguish outbreak strain from other circulated strains may be time consuming and therefore impractical in a real-time framework. Hence it may not be possible to remove background ILI noise from the flu data during an ongoing pandemic. For that reason we decided to model background ILI model as a Poisson approximation of a stochastic ILI background rate for each age group and estimate it in real time by fitting to the available surveillance data.

3. Model parametrization

Depending on the available data, the model parameters that we aim to estimate in this work are the basic reproduction number R_0, the average infectious period, which we denote by **AIP** henceforth, the proportion of clinical cases who visit their GP (p), the initial force of GP consultations $p\lambda(t = 0, j)$ and the background ILI rate per year per 100000. For multi-compartmental SEIR models like the one we employ here, the basic reproduction number R_0, the exponential growth λ_e and **AIP** are related according to Ref. [10]:

$$R_0 = \lambda_e \text{AIP} \frac{\left(\frac{\lambda_e \text{LP}}{m} + 1\right)^m}{\left(1 - \frac{1}{\left(\frac{\lambda_e \text{AIP}}{m} + 1\right)^n}\right)} \qquad (7)$$

m and n are the number of latent and infectious sub-compartments of the SEIR model. There is a number of remaining age specific parameters that assumes fixed values as detailed in Table 1. If hospital, death and antiviral data are present the model parameters increase accordingly.

Table 1. Age specific model parameters that are kept constant when fitting only to reported GP data

Parameter	Under 1	1-4	5-14	15-24	25-44	45-64	65+
% of Susceptibles	1	1	1	1	1	1	1
Latent Period (days)	2	2	2	2	2	2	2
proportion of symptomatic infections	0.5	0.5	0.5	0.5	0.5	0.5	0.5
mean incubation period	3	3	3	3	3	3	3
std incubation period	1	1	1	1	1	1	1
mean delay from onset to GP	1	1	1	1	1	1	1
standard deviation from onset to GP	0.5	0.5	0.5	0.5	0.5	0.5	0.5
mean delay from GP to report	1	1	1	1	1	1	1
std delay from GP to report	0.5	0.5	0.5	0.5	0.5	0.5	0.5

4. Bayesian Framework

For each of the parameters of interest (section 3) there is prior information encoded in the form of probability distributions. In general, some priors arise from applying elicitation methods to quantitative data (see Ref. [9]), or from information coming from overseas or from analysis of the first few hundred datasets of clinical cases and their contacts. We then use a Poisson likelihood to model the available data and within a Bayesian framework we can then calculate the posterior probability distribution of all parameters of interest via Markov Chain Monte Carlo (MCMC). We implement the standard Metropolis-Hastings algorithm with a symmetric (normal) proposal distribution with proposal variance that is adapted so that the rate of acceptance of the proposed moves is between 10% and 40%. The prior of the current and the previous sample are evaluated, as well as the corresponding likelihoods and we obtain a probability of accepting the proposed move by the Metropolis-Hastings formula:

$$a = \min\left(1.0, \frac{\pi(x_n)\mathcal{L}(x_n)}{\pi(x_o)\mathcal{L}(x_o)}\right), \tag{8}$$

where π denotes the prior distribution and $\mathcal{L}$ denotes the Poisson likelihood. n stands for the proposed and o for the current sample. The acceptance-rejection criterion is implemented by sampling from a random sample within the unit interval and if that sample is smaller than the probability of acceptance a, the proposal is accepted, otherwise it is rejected.

Once the MCMC chain has mixed satisfactorily, the remaining samples are stored and considered to be from the full posterior. A random subsample of the full posterior is used to generate model runs that are extrapolated beyond the currently available data to provide predictions about the possible outcome of the outbreak, based on the current prior information and surveillance data.

5. Available data and preliminary analysis

The purpose of this work is to validate the proposed framework on real data from the 1969 pandemic in the UK. This is done by selecting to fit to a particular subset of the available data and then predict what will happen and compare it with the actual outcome. The data available to us are age specific regionally aggregated UK GP consultations reported weekly (Figure 1). The model requires the 1969 UK mid-population estimates which are also available.

The starting day in the model is the 25/10/1969 which coincides with the first day of the available data. In the first subsequent weeks after that date there is no signal from the flu pandemic in the data. At this stage, the pandemic flu cases cannot be easily distinguished from different background ILI GP data, which, as appear in Figure 1, are age dependent and we account for that by a constant, age independent background GP ILI rate. The data also exhibit different rates of increase of GP consultations for the different age groups, but to avoid over-fitting and/or identifiability problems between the parameters we have chosen not to fit for an age

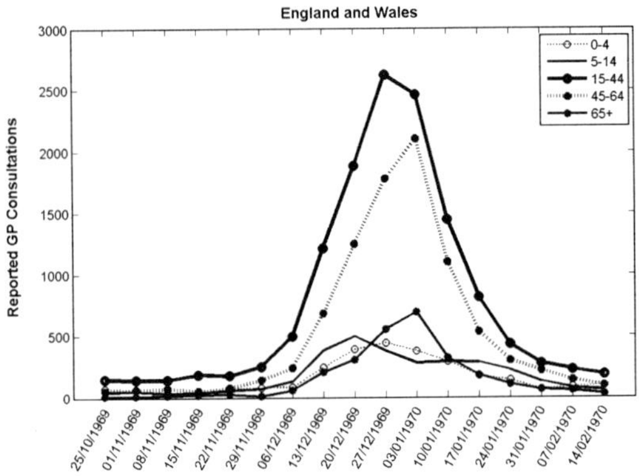

Figure 1. Age distributed regionally aggregated weekly reported GP data from the 1969 flu pandemic in the UK.

specific proportion of symptomatic who visit their GP in the present work. The reason for this initial approach is that at early stages of the outbreak, a higher than expected background ILI rate combined with a lower proportion of observable GP consultations gives equivalent results as those by a lower than expected background ILI rate combined with a higher proportion of observable GP consultations. If sufficient background ILI data were available this uncertainty would be resolved but there is the additional consideration that during an outbreak there may be reduced reporting of background ILIs because of the dominance of the outbreak virus. Hence the estimation of background ILIs could still be important during an ongoing outbreak and this is what the present work emphasizes. In subsequent work we will attempt to capture and predict predominantly the effects of reporting rather than background ILI. In addition, we anticipate a change in the rate of reported GP consultations as the epidemic progresses but this will be further investigated in subsequent work.

6. MCMC for fitting to data and predictions

We initially fit to data up to week 5 (35) days and we initialize the MCMC chain and other model parameters using existing estimates consistent with Ref. [11]. The prior distributions used here for 35 days are summarized in table 2:

The prior distributions of MCMC parameters were chosen to be uninformative to test the performance of the framework and to reflect uncertainty. To ensure that satisfactory mixing of the chain is achieved, the length of

Table 2. Prior Probability distributions for model parameters fit

Parameter	Distribution	left bound	right bound
λ_e	Uniform	0.1	0.3
proportion GP consultations	Uniform	0.1	0.3
p foi(t=0)	Uniform	310^{-11}	810^{-3}
Background GP ILI ≤ 1	Uniform	1000	3000
Background GP ILI 1-4	Uniform	1500	4000
Background GP ILI 5-14	Uniform	2000	4500
Background GP ILI 15-24	Uniform	1500	4000
Background GP ILI 25-44	Uniform	2000	4500
Background GP ILI 45-64	Uniform	1500	3000
Background GP ILI 65+	Uniform	500	2000
Average Infectious Period	Uniform	2	4

the MCMC chain is taken to be 200.000 with the first 50.000 being the adaptive period. The choice of adaptive period is made so that any burn in period in the MCMC chain is discarded. We adapt the proposal variance of every parameter of interest every 1000 iterations and the final output chain is also thinned, selecting every 50, to give us the posterior parameter estimates, from which we sample and extrapolate the model to predict the possible outcomes of the outbreak (Figure 2).

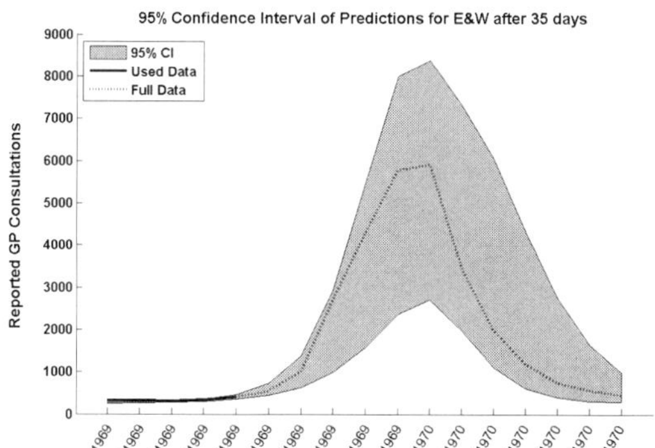

Figure 2. Predicting the 1969 Pandemic in England and Wales from fitting to early data: Model fits the data up to 35 days (solid line) and confidence intervals of the predictions obtained by extrapolating samples from the joint MCMC posterior. The 95% confidence interval of the predictions contains the dotted line which is the true subsequent data that are yet unobserved. We have aggregated the different age groups for convenience in display.

The parameter posteriors for 35 days are then used to inform the priors for fitting to 42 days of data and the above process is repeated for subsequent weeks into the outbreak. The results obtained by fitting to the 42 days worth of data by using the posteriors from fitting to 35 days data are displayed in Figure 3.

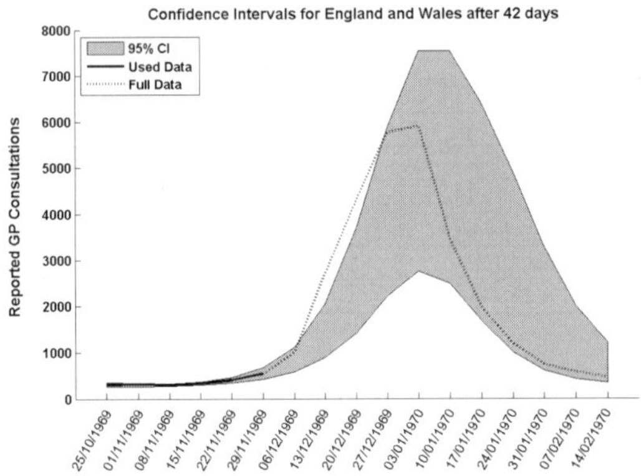

Figure 3. The model fits well to the available data of 42 days (solid line) when the posterior of the previous fit are used as the priors of the next, but it is not designed to capture the sudden increase in the data that follows week starting 06/12/1969. The 95% confidence interval of the predictions contains some part of the dotted line which is the true subsequent data that are yet unobserved.

The results obtained by predicting the outcome using only 42 days worth of data are due to a data feature that the current model is not designed to address: on the week starting 6/12/1969 the data display a sudden rapid increase and further data analysis shows that at this point the exponential growth of the observed epidemic increases suddenly. Therefore the data seem to deviate from the SEIR dynamics (adopted here) which only allow for a single exponential growth rate during the initial exponential growth of the epidemic. In this work, we show that this difficulty can be overcome by incorporating on the subsequent priors this sudden change in the incoming data, whilst keeping the priors uniform, therefore allowing for uncertainty.

Fitting to 49 days worth of data we obtain a better prediction of the course towards the peak of the outbreak (Figure 4) but not the intermediate data points due to the previous explanation.

When fitting to 56 days worth of data we observe the sudden increase in the slope of the available data curve which is due to the sudden change in reporting and to compensate for this we increase the lower bound in our prior for p. The fitting to the data we are able to predict better the exponential growth of the data up to the peak (Figure 5) but our predictions past the peak are not so good. This may be explained by spatial effects

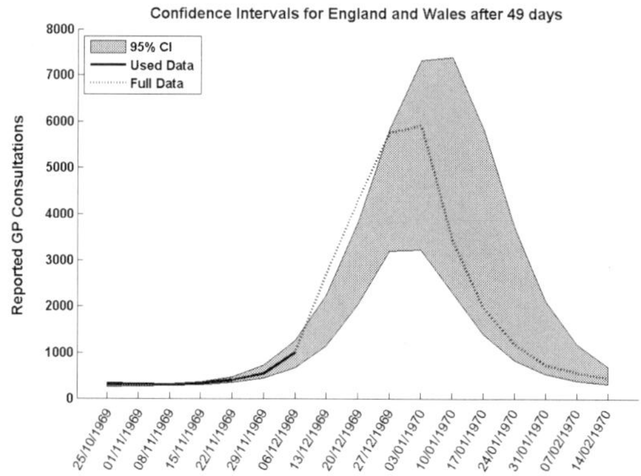

Figure 4. A further improvement in predictions by fitting to 49 days worth of data.

that slow down the end of the epidemic or change in the contact patterns or other seasonal effects that our model does not account for.

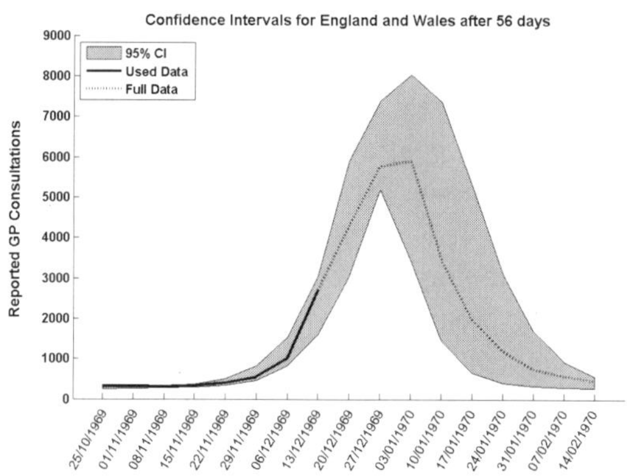

Figure 5. After 56 days of available data we can predict better than before the stage of increase of the outbreak and the approximate peak size, but not the decline of the epidemic.

By using the posterior of the fit to 56 days to fit data up to 70 days, which is very close to the actual time of the peak of this outbreak in the UK, our predictions become very accurate (Figure 6). Overall, this approach enables the largest or most relevant part of the outbreak progression to be consistently contained in the 95% confidence intervals of the predictions of the proposed framework.

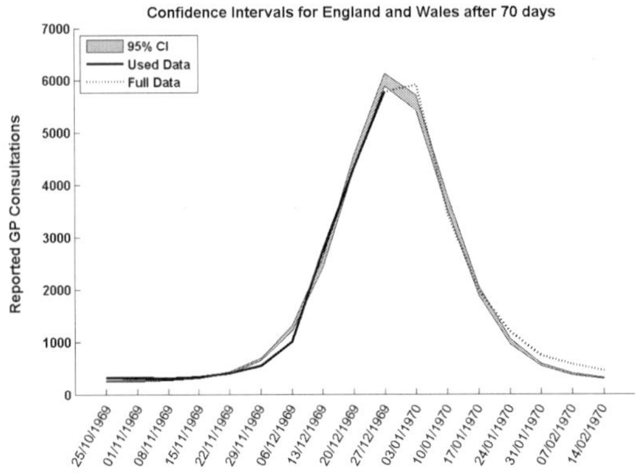

Figure 6. After the time of the peak the epidemic outcome can be predicted very accurately.

7. Discussion and Conclusions

Real data are contaminated by noise and exhibit characteristics that are difficult to model, such as spatial effects, changing mixing patterns and population behaviours that can dynamically influence the progression of an outbreak, to name a few. It is often difficult to model and understand the impact of such events but simpler models can still be used to provide a good picture about the likely progression of an outbreak from available data and prior knowledge of key epidemiological characteristics of the disease.

In this work we have presented a model of pandemic influenza within a Bayesian MCMC framework in order to predict the possible outcome of the 1969 pandemic in the UK. We have shown that it is possible to predict the outcome of such an outbreak even if the adopted model cannot fully describe the available data and we demonstrated that one can compensate for this by appropriate choice of priors that is motivated by the available incoming data in real time.

In future work we aim to improve the current model to age specific reporting of GP consultations and to include random effects to account with sudden changes in the data, such as those observed in the 1969 pandemic in the UK. Furthermore, because our results indicated the importance of prior information, prior elicitation is potentially a key component for better predictions.

References

1. C. Fraser, C. A. Donnelly *et al.*, *Sciencexpress* **DOI: 10.1126/Science.1176062**, 11 May (2009).
2. I.M Hall and R. Gani and H.E. Hughes and S. Leach, *Epidemiology and Infection* **135**(3), 372–385 (2007).
3. L.F. White and M. Pagano, *Stat. Med.* **v27**(16), 2999–3016 (2008).
4. L. M. A. Bettencourt and R. M. Rebeiro, *PLoS ONE* **3**(5), e2185 (2008).
5. M. L. C. D. Atti and S. Merler and C. Rizzo and M. Ajelli and M. Massari and P. Manfredi and C. Furlanello and G. S. Tomba and m. Iannelli, *PLos ONE* **3**(3), e1790 (2008).
6. J. Mossong *et al.*, *PLoS Med.* **5**, e74 (2008).
7. F. Ball. *Journal of Applied Probability* **20**(1), 153–157 (1983).
8. G. Ketsetzis and B. Cooper and D.Deangelis and N. Gay. *in preparation.*
9. I. H. J. Van Der Fels-Klerx and L. H. J . Goossens and H. W. Saatkamp and S. H.S. Horst *Risk Anal.*, **22**(1), 67–81 (2002).
10. H. J. Wearing and P. Rohani and M. J. Keeling, *PLoS Medicine*, **2**(7), e174 (2005).
11. I. M. Longini and M. E. Halloran and A. Nizam and Y. Yang, *American Journal of Epidemiology*, **159**(7), 623–633 (2004).

A PROBABILISTIC CELLULAR AUTOMATA TO STUDYING THE SPREADING OF PNEUMONIA IN A POPULATION

Y. SAITO

Departamento de Matemática, Universidade Federal de São Carlos,
Rod. Washington Luis, km 235,
São Carlos, SP, Brasil
E-mail: `otiasiruy@gmail.com`

M. A. A. DA SILVA

Departamento de Física e Química, FCFRP Universidade de São Paulo,
14040-903
Ribeirão Preto, SP, Brasil
E-mail: `maasilva@fcfrp.usp.br`

D. ALVES

Departamento de Medicina Social, FMRP Universidade de São Paulo,
Av. Bandeirantes, 3.900,
Ribeirão Preto, SP, Brasil
E-mail: `doquiron@gmail.com`

A simple analytical and simulation framework to study the diffusion of pneumococci in finite population of humans is proposed, using probabilistic cellular automata. Furthermore, this epidemic spatial model permits to reproduce explicitly the interaction of two types of transmission mechanisms in terms of global and local variables, which in turn can be adjusted to simulate respectively the populational mobility and geographical neighborhood contacts. It was possible with this alternative model:
1) to observe the pneumococcal spreading through a population,
2) the study of antibiotics effects in disease's control in relation with effectiveness and percentage of individuals covered by the responsibility analyses of CCC (child care center) in dissemination process, and
3) the understanding of the relationship of hours spent in CCC and pneumococcal transmission.

1. Introduction

Among the infectious diseases, pneumonia is highlighted because of its extensive dispersion region and high number of infections[1]. Thus considering a comprehensive strategy for understanding the spread of pneumonia in a population is essential. This strategy in turn must be based on theoretical quantitative method, derived from a firm understanding of how this disease spreads in a community, including the emergence of resistant strains. There are few studies related to mathematical modeling and simulation of the diffusion of pneumococci in the community, especially if we consider the study in small populations. A recent work in this area proposes a transmission model described by EDOs (Ordinary Differential Equations)[2], which have relevant restrictions on its applicability to account for the large proportion of the variability in the prevalence of pneumococcal carriage across communities. The main idea of this work is to show the advances in mathematical and computational modeling of the dynamics of pneumonia transmission in a population. Indeed, the simulation of biological systems, using computational tools, allows the realization of experiments in virtual laboratories where we can test hypotheses about the process epidemic. It is done through the development of a probabilistic cellular automaton, whose laws mimic the interactions between individuals of a population and its mobility[1].

2. Modeling Pneumonia Transmission

Development of cellular automata model based on probabilistic required a detailed study of the spread of pneumococcal in communities from a model based on ordinary differential equations[2] which does not consider the problem of small populations, heterogeneity of interactions and spatial population distribution. The motivation for this is that we can prototype and validate a more robust model for spread of pneumonia as an extension of that original study. Indeed, the study on which we rely discusses a differential equation that describes the behavior of the dispersion of pneumococcal in small communities. Transmission results from the interaction child-to-child and contacts occur in a period of eighty-four hours per week. The number eighty-four is due to twelve hours in a week. The parameter *(1-f)* is the fraction of the CCC not attend model. Model separates the population into two subsets, all the visitors and non visitors of CCC. The functions $Xc(t)$ and $Yc(t)$ represent the group attending the CCC, the population of non-carriers and carriers of the population density of pneumococcal, respectively.

Table 1. Variables used in the differential equations model.

Description of parameter	*Variable*
Transmission rate of pneumococcal outside the CCC	β_1
Transmission rate of pneumococcal within the CCC	β_2
Transmission rate from infected state to susceptible state	μ
This is the proportion of children attending the CCC	f
Average number of hours spent in the CCC	g
Proportion of children using antibiotics for a week	r
Proportion of effectiveness of the antibiotic	c

The functions $Xn(t)$ and $Yn(t)$ represent the group that did not attend the CCC, the population of non-carriers and carriers of the population, respectively. Thus, this model is used to obtain evidence about the importance of CCC in the spread of pneumococcal community[2].

$$\frac{dY_N}{dt} = \beta_1 X_N Y_N (1-f) + \beta_1 X_N Y_C (1-g/84) - \mu Y_N - crY_N \,, \tag{1}$$

$$\begin{aligned}\frac{dY_C}{dt} &= \beta_2 X_C Y_C (g/84) + \beta_1 X_C Y_C f (1-g/84) \\ &\quad + \beta_1 X_C Y_C (1-f)(1-g/84) - \mu Y_C - crY_C \,.\end{aligned} \tag{2}$$

Through changes of variables we obtain the following equations:

$$\frac{dX_N}{dt} = (\mu + cr)Y - \beta XY \,, \tag{3}$$

$$\frac{dY_N}{dt} = \beta XY - (\mu + cr)Y \,. \tag{4}$$

Where $Y(t)$ and $X(t)$ are functions that represent the densities of carriers of pneumococcal and non-carriers, respectively[2]. It is assumed that the carriers have only one type of host.

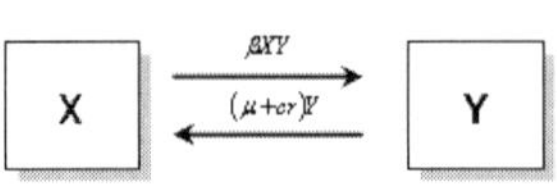

Figure 1. Diagram of the model of clearance and acquire of pneumococcal in CCCs.

As already commented, on basis of this model, we have developed and implemented a probabilistic cellular automaton model, in which every single cell of the grid represents a single individual address by a position *(i,j)*, initially testing and comparing with its results. Particularly in this study we use the state diagram of the original model based on differential equations for this purpose. Note that cellular automata are discrete dynamical systems (discrete time, discrete space and discrete number of states) that follow local rules[4]. The cellular automaton can be defined as graphs with a discrete variable at each vertex. The variables change according to pre-established interactions with neighbors at each step of time, i.e. the dynamics is defined according to the updates done in the previous time step for an arbitrary number of individuals modified simultaneously. In cellular automata the individuals that make up the population to be studied are represented by individuals that keep the states made possible by them. Since we are focusing our study in epidemics spreading dynamics, a set of two possible states was adopted, representing the health/disease status of every individual at a given time: susceptible (an individual can become infected), infective (can transmit the disease to the susceptible individuals)[3]. For implementation of the cellular automaton we adopted a two-dimensional network and a neighborhood of Moore interaction, where each individual network has eight neighbors. Changes of state (health of each individual in the network) are defined by transition rules that determine the states of individuals in each step of time and depend on the state of the individual and the current state of its neighboring individuals. As already mentioned this system has two states which refer the status of each individual at a given moment, susceptible or infected. Let us consider a discrete dynamical system (discrete space and discrete time) in which a population of N individuals is distributed on the sites of a two-dimensional toroidal lattice $M = m_{i,j}$, with i and j varying from 1 to L $(N = L\ x\ L)$[4]. Each individual site $m_{i,j}$ is assigned to receive three personal attributes: a spatial lattice position *(i,j)*, a set of two possible status, namely, s and i, specifying a clinical disease stage of each particular individual and an infection period τ, specifying how many units of time an infected individual can propagate the contagious agent. In particular, such infectious period (in the model measured by network updates, which can represent days, weeks, months or even years if up to) will define how much time a single element will stay contaminating its contacts[4]. Note that in each instant of time

$$\sum S + \sum I = N\,, \tag{5}$$

with N constant.

The most important feature of this alternative model, compared to the traditionally cellular automata used in epidemiology, is the use of two types of interactions between its individuals: the local interactions, due the influence that the neighborhood of every susceptible individual exerts over him, and the global influences, in which all individuals have equal probability to establish contact with everyone of the grid and thus making possible the dissemination to occur in further regions. Here, the probability of every individual to become infected is modeled as the superposition of these two kinds of interactions through the following equation:

$$p_S = \Gamma p_G + \Lambda p_L \,. \tag{6}$$

We assume also the normalization condition: $\Gamma + \Lambda = 1$, for the parameters used to adjust the interactions of short range (responsible for the formation of clusters) and long range (mean-field type), respectively, between individuals of the same population, such that the condition $p_S \in [0, 1]$ is true. The global influence in this equation is modeled considering the total number of infected individuals that are on the network, i.e., due to the presence or movement of any individual in the population infected, and can be described by the equation:

$$p_G = \frac{\alpha}{N} \sum_s \sigma_i \,. \tag{7}$$

We must emphasize that global probability is responsible to increases the simulations speed, what can approximate the probabilistic cellular automaton model and ODE model. The parameter α is set between zero and one and the index s to the total number of infected individuals in the cellular automata network. Actually, the sum in the Eq. (7) just counts the instantaneous number of infectious individuals *I(t)* in the population. The local probability is given by:

$$p_L = 1 - (1 - \lambda)^n \,. \tag{8}$$

Thus, the probability p_L deals with interactions between a individual and its n infected nearest neighbors. The parameter λ represents the probability of disease transmission between individuals. The probability p_L as in Eq. (8), and in a number of alternative forms[2],3 has been employed as a typical term for epidemic spread using cellular automata modeling. Moreover, we model the probability p_C of an individual become healthy due antibiotics and the probability of an individual receiving treatment with antibiotics, p_r. The algorithm of the cellular automaton, checks for each susceptible point. Through Eq. (6), is calculated the probability p_S of a susceptible individuals to be infected by any infected individual. If the individual becomes infected a time counter to cure that individual is started, because of its immune system. For each time "step", is recorded each infected individual, and treated with antibiotics; the individual will recover due to the treatment if probability p_C is less or equal to the parameter of antibiotic healing[5]. For the non-treated infected individuals, the recovery is reached when the time of infection is τ.

3. Results

The ordinary differential equations model presents a balance dependent on the transmission rate of the pneumonia, the average time in which the individual is infected in the state, the effectiveness of antibiotics and the percentage of users of the product. The authors found evidences that relate directly the proportion of individuals that attend the CCC f to determine the equilibrium state, where as greater is the percentage of children that do not attend the CCC *(1-f)* it will be lower the area of pneumococcal carrier density and the moment of time in which the equilibrium state is reached, because as greater the proportion of CCC are, the visitors will be less time required for the equilibrium state is reached, which enabled conclude that the CCC are important areas of pneumococcal dissemination and measures that should be studied to verify the reason for this phenomenon. Still, the results obtained by the ordinary differential equation model, the balance state between carriers people and non-carriers of pneumococcal introduced antibiotics in the proportion of dependent children attended, the average number of hours per week of attendance in CCC and coverage of vaccination which can be verified by the model of probabilistic cellular automaton[3]. It was observed a growth in the equilibrium value when CCC attends a bigger number of children; however this result is independent of the initial conditions of the problem. The article defines the parameter β_1 as a constant, however the simulations of the model could not confirm that the balance depends on the initial condition and therefore does not alter the relationship of people and prevalence β_1. No details about the origin of the value of β_1, but one can see that there is no risk of outbreaks of pneumonia where the rate of transmission of pneumococcal is less than 0.15. Evidently the average number of hours attended in the CCC is related to the carrier density of the population of pneumococcal in the growth of average weekly number of hours in the CCC increases the percentage of individuals, but this fact is also between the group of non-attendees, who despite not using the CCC, are indirectly affected due to contact with the group of attendees[6]. They should discriminate between different users types of the CCC, as if the average weekly use is reasonably small, the implications in the community will be minimal in relation to the spread of pneumococcal community. The model has limitations on the scope of risk factors, considering only the characteristics associated with age and does not consider the important variables: resistance to antibiotics and rates of mutation[7]. Another point not assessed to the transmission of pneumonia among other possible family members of individuals, which often occurs in real situations.

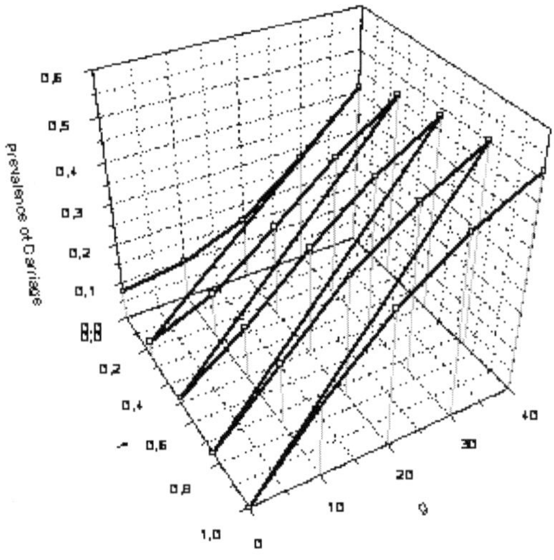

Figure 2. Relation between prevalence of carriage, proportion of children attending the CCCs and the average number hours spent in CCCs, concerning the ODE model.

The Figure 2 shows, results from the ODE model, of the direct relation between prevalence of carriage, proportion of children attending the CCCs and the average number hours spent in CCCs, where the parameter g presents greater weight than the parameter f in relation to prevalence of carriage. This result was used to building the probabilistic cellular automata model, because the heterogeneity is the higher peculiarity in this model.

Figure 3. Simulation of an epidemic of pneumonia using cellular automaton, in the initial moment and in the equilibrium moment respectively, for a network with 2500 individuals, $\tau = 5$, $p_C = 0{,}22$, $p_r = 0{,}046$ and $p_G = 0$.

The probabilistic cellular automaton model presented a behavior similar to the type of ordinary differential equations, but some differences should be highlighted. Particularly Figure 3 shows a simulation using the probabilistic cellular automaton model developed, showing the spatial characteristics of this type of modeling. This figure show the situation of the population in a few moments of time from an initial configuration in which we have 10% infected individuals in the center of the network. The equilibrium in the cellular automaton is reached in a larger period of time in relation to the ODE model and this conclusion is directly related with size of the network and the global probability, due the specificity of this case.

The simulation with probabilistic cellular automaton model had the network with 2500 individuals, the time of infection τ equal 5 days and the number of days of experiment was 400. The global probability of scattering is zero for simplicity in Part A. However, in Part B, we must highlight the importance of global probability p_G, because it is through her that we can approximate the probabilistic cellular automaton model and ODE model, where p_G can increases the simulations speed. An infectious disease grows easier and faster with the small-world phenomena, here characterized by adjusting the variable Γ of global influence of interactions among network individuals. In this case one can observe the formation of new clusters of infectious individuals over the time and this formation becomes more intense as Γ increases. Also one can see that aggressiveness of the pneumococcal carries a reasonable influence on the behavior of the pneumococcal spread in

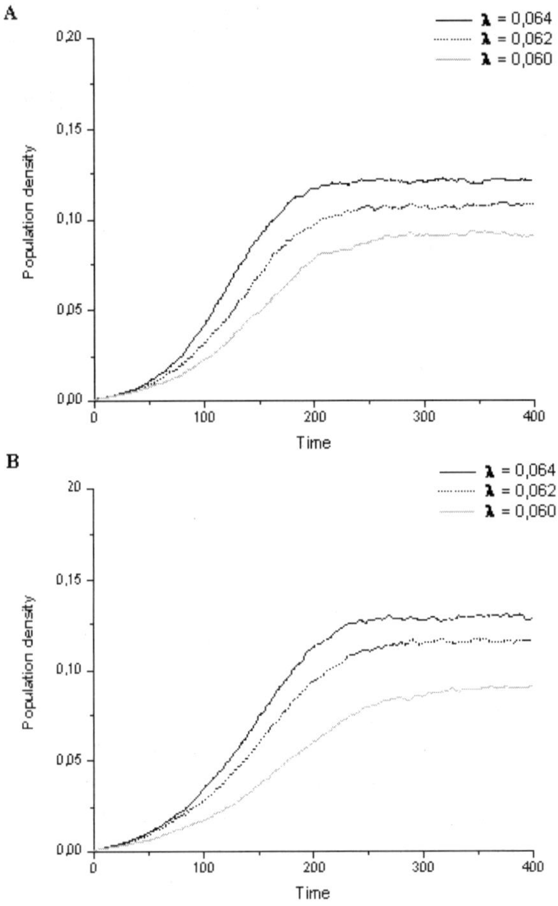

Figure 4. Part A: graph with the population density of pneumococcal carriers to the probabilistic cellular automaton model to different populations from the average of 100 samples, N = 2500, $\tau = 5$, $p_C = 0{,}22$, $p_r = 0{,}046$ and global probability $\Gamma = 0{,}001$. Part B: the graph has the same parameters configuration, with except the global probability $\Gamma = 0$. Transmission parameter λ is related to aggressiveness of the pneumococcal.

the population under study, with transmission parameter less than or equal to 0.062 a value capable of obtaining good results in behavior the point of equilibrium for the differential equations model and for the probabilistic cellular automaton model.

The simulation in Figure 5 used λ with following values 0.060, 0.062 and 0.064, which are reasonable numbers to compare results from the ODE

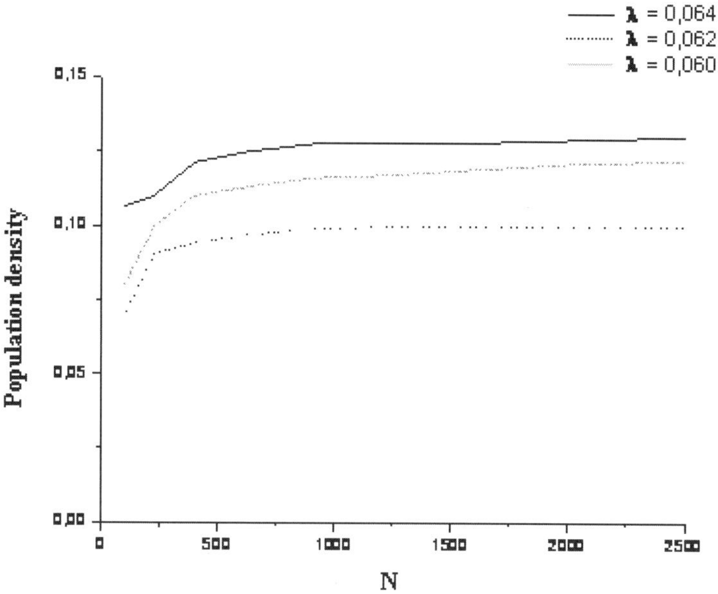

Figure 5. A phase diagram with the population density of pneumococcal carriers to the probabilistic cellular automaton model to different populations from the average of 100 samples, N being the size of the network, $\tau = 5$, $p_C = 0{,}22$ and $p_r = 0{,}046$.

model. The time of infection τ equal 5 days, global probability of scattering is zero for simplicity and the time of the experiment was 400 days, as is the time required to visualize the point of equilibrium of the system. The initial condition considered a single element in the center of the infected network. The probability of an individual receiving treatment with antibiotics p_r was 0.046 and the probability of cure by the antibiotics p_C used was 0.22. One can check the dependence of the moment when the equilibrium point is reached on the size of the population considered in the simulation, but cannot be said that the value of balance is dependent on that parameter (Figure 5).

A more complete phase diagram above show the conditions at which distinct phases can occur at equilibrium. One can see that space where the experiments were realized belongs to zone endemic. Even working in an extreme scenario ($\Gamma = 0$) with the developed model, we have succeeded in obtain similar results as those presented by the ODE model.

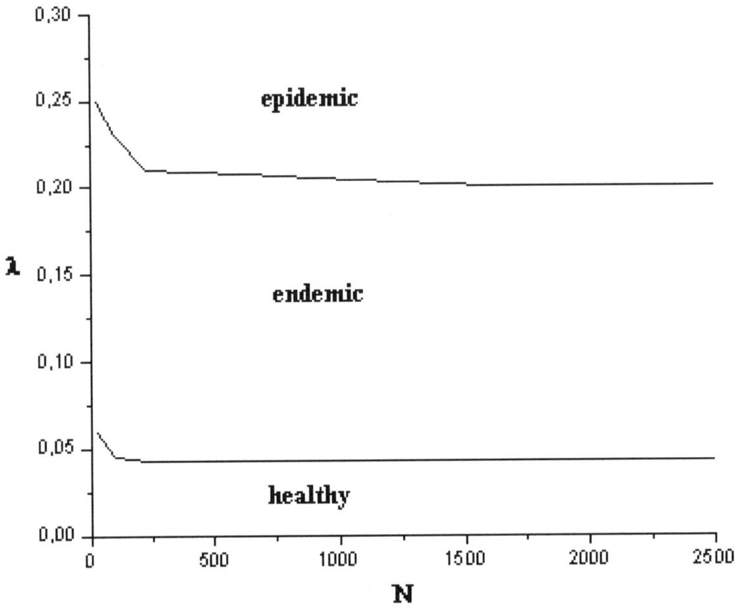

Figure 6. A phase diagram with the relationship between N and λ to the probabilistic cellular automaton model to different populations from the average of 100 samples, $\tau = 5$, $p_C = 0{,}22$ and $p_r = 0{,}046$.

4. Conclusion

Through mathematical tools we can get information about how a disease spreads in a population, and in essence, determine actions to prevent or contain such spread. The main points raised by the study were replicated and the results suggest that the model is really valid to study the pneumococcal dispersion and that there is evidence on the relationship pneumonia and CCC. The probabilistic cellular automaton model presented behavior similar to the type of ordinary differential equations. It was considered that comparisons between the results obtained by the model of cellular automaton and the differential equations model were carried out only in relation to their behavior, because one refers to the number of individuals and considers the spatiality of the problem, and the other the density population, respectively. The article states that although use of the data set of communities of children studied, the number of samples proved to be insufficient for testing in communities using the findings derived from the ordinary differential equations model, however it is possible the study of small sample simulations using the probabilistic cellular automaton model. Understanding the transmission dynamics of the spread of pneumococcal is the goal that promises to yield extraordinary benefits for both planning and control measures and implementation of surveillance. Rather than just a caricature of the original formulation of pneumonia transmission ODE model, the approach presented in this paper may be viewed as a simpler and generic alternative for investigating the spread of this disease in a population, which may greatly facilitate the analysis of a number of distinct epidemic scenarios. Particularly, a system with increasing topological complexity can be easily tackled, varying the full parameter space of the model. Finally we would like to highlight that the probabilistic cellular automata model will be used to study the heterogeneities in the location and size of hospitals and its influence on the transmission of multiply resistant organisms between locates and hospitals. However, more detailed considerations on the investigation of this avenue of research is left for a future contribution.

Acknowledgments

We thank FAPESP (Grant No. 2007/04220-4) and (Grant No. 2009/09118-9) for funding.

References

1. S. S. Huang, J. A. Finkelstein and M. Lipsitch, *Clin. Infect. Dis.* **v.40**, 1215 (2005).
2. D. J. Austin and R. M. Anderson, *Am. J. Infect. Dis.* **179**, 1883 (1999).
3. D. Alves, V. J. Haas and A. Caliri, *J. Biol. Phys.* **v.29**, (2003).
4. N. Boccara , K. Cheong, *J. Phys. A: Math and Gen.* **v.25**, 2447 (2000).
5. R. Dagan and K. L. O'Brien, *Clin. Infect. Dis.* **v.40**, 1223 (2005).
6. M. Lipsitch, *Emerg. Infect. Dis.* **v.5**, 336 (1999).
7. C. L. Byingtong, M. H. Samore, G. J. Stoddard, S. Barlow, J. Daly, K. Korgenski, S. Firth, D. Glover, J. Jensen, E. O. Mason, C. K. Shutt, A. T. Pavia, *Clin. Infect. Dis.* **v.41**, 21 (2005).

CONTRIBUTION OF WATERBORNE TRANSPORT IN THE SPREAD OF INFECTION WITH *TOXOPLASMA GONDII*

DECCY Y. TREJOS A.

Grupo de Investigación Ciencias Matemáticas y Tecnologías. Universidad Distrital Francisco José de Caldas. Cra. 3 No. 26A-40. Bogotá, Cundinamarca, Colombia

IRENE DUARTE G.

Escuela de Investigación en Biomatemática,. Universidad del Quindío. Cra. 15 Calle 12 Norte. Armenia, Quindío, Colombia

Toxoplasmosis is a parasitic zoonosis worldwide distributed, infecting a large proportion of human and animal populations, produced by the parasite *Toxoplasma gondii*. Some individuals are at high risk of serious or fatal illness due to this parasite, including fetuses and newborns with congenital infection and immune-impaired people. Some epidemiological studies have shown that in most of the world the presence of cats is critical for transmitting the parasite to various intermediary hosts (humans, pets). In addition, an outbreak in Vancouver, Canada, was related to the contamination of reservoir water from the city for a wild cat, and in Brazil, an epidemiological survey also linked the consumption of unfiltered water with infection in disadvantaged socio-economic strata. This paper evaluates the impact of transport by water, through rain, rivers, streams, etc., in the spread of *T. gondii*. A mathematical model proposed by Duarte and Trejos (2005) for the dispersion of the concentration of parasites from *T. gondii* in a host population of cats was used. This model combines the transmission model of SIR type with an epidemic spread of the parasite in a rectangular area; the coefficient of dispersion of the parasite includes all factors that influence the transport of the parasite (birds, rodents, insects, etc.). The assessment of the contribution of waterborne transport in the spread of *T. gondii*, is revealing in the model mentioned a term that describes it, independently with other transmission mechanisms. To do this, a velocity vectorial field on the hydrological map of the Department of Quindío, Colombia, was constructed and was subsequently incorporated into the model. The system simulation was made and the results with and without waterborne transport were compared; concluding that it is fundamental to the spread of the parasite.

1. Introduction

With the advance of science and technology, some infectious illnesses have disappeared, others have reduced and many others can be controlled; as it has occurred with illnesses produced by bacteria or virus. This is not the case of those produced by protozoan parasites or those that before were proper of animals and now are being transmitted to humans, some of which, have increased notoriously in recent years.

Toxoplasmosis is a parasitic zoonosis worldwide distributed, infecting a large proportion of human and animal populations, produced by the parasite *Toxoplasma gondii.* Some individuals are at high risk of serious or fatal illness due to this parasite, including fetuses and newborns with congenital infection and immune-impaired people.

In Colombia, numerous seroepidemiologic studies of the toxoplasmosis acquired during the pregnancy have been carried out in recent years. According to the health national study in 1980, the rate of positiveness is 47.1%, with high titles, mainly, in the Atlantic zone. In the same study 1.8% of the pregnant women presented high titles of antibodies. This is probably a sign of recent infection. In the Department of Quindío, rates between 0.7 and 1.6% of expectant mothers with serologic marker of acute infection have been reported. In the municipality of Armenia, the prevalence of the toxoplasmosis is 60%. This constitutes a priority in the public health of the locality[1].

Some epidemiological studies have shown that in most of the world the presence of cats is critical for transmitting the parasite to various intermediary hosts (humans, pets).

The *Toxoplasma gondii* parasite is a coccidian of felines and human beings. Its life cycle comprises an intestinal phase that takes place in the epithelium of the intestine of the felines and an extra intestinal phase that occurs as much in the final hosts as in the intermediary guests[2,3]. There are three infectious stages of the parasite: oocysts, tachyzoites and bradyzoites (tissue cyst). Felines shed oocysts ingesting any of the three infectious forms of the parasite; moment when an enteroepithelial cycle begins.

Once the parasite is on ground, it can be transported by means of the rain, the wind, the worms, the snails, the slugs, the insects, the birds, the mice, etc. That is why, wild and domestic species, fresh vegetables, pasturelands, irrigation water and water for human consumption can be contaminated. The rate of infection prevalence depends on the number of cats by unit of area and of the number of hosts that participate in the prey-predatory cycle[3].

The works carried out by Refs. [4, 5] demonstrates the importance of water as mechanism of transmission of the infection by *T. gondii*, since in spite of the process of purification of the water, the parasite survives and can infect through consumption. That is to say, a hypothetical inoculum could be contributing to the transmission of the infection by this parasite. This occurs in some place along the way of the water that finally supplies a population of intermediary or final hosts.

On the other hand, an outbreak in Vancouver, Canada, was related to the contamination of the city's water reservoir for a wild cat, and in Brazil, an epidemiological survey also linked the consumption of unfiltered water with infection in disadvantaged socio-economic strata.

Besides, a work carried out in the North of Rio de Janeiro in Brazil related the consumption of unfiltered water to a risk factor for the prevalence of the toxoplasmosis in disadvantaged socio-economic strata.

Trejos and Duarte[6] designed a mathematical model for dispersion of the concentration of parasites of *T. gondii* in a host population of cats. It combines the transmission model of SIR type with an epidemic spread of the parasite in a rectangular area. This model confirmed the importance of cats in the transmission of the infection by *T. gondii* showing that a single infected cat is capable of carrying the infection to adjoining areas, where initially, did not exist. The coefficient of dispersion of the parasite included all of the factors that have influence on the transport of the parasite (birds, rodents, insects, etc.) and on the transport by water (rain, rivers, streams, etc.).

This paper evaluates the impact of transport by water in the spread of *T. gondii* putting in evidence, in the model of Trejos and Duarte, a term that describes this transportation in an independent way from the other mechanisms of transmission. To obtain this term, a vectorial velocity field was built on the hydrologic map of the Department of Quindío, Colombia. Results with and without transportation by water were compared.

2. Mathematical Model

The model proposed to describe the dynamics of dispersion of the parasite *Toxoplasma gondii* among the host cats takes into account the following assumptions:

(1) Natural birth and mortality rates for the population of cats.

(2) Cat's migration and immigration rates are not considered.
(3) Transmission is indirect[7], It results from acquired contact (water, prey or meat infected consumption) between the concentration of the parasite *T. gondii* and the susceptible cats.
(4) The probability that a cat gets infected is proportional to the concentration of parasites that it consumes. The infecting form of the parasite is not taken into account.
(5) The time elapsed among the consumption of the parasite and the expulsion of the oocysts to the floor it is not taken into account.
(6) The infection does not induce death in the hosts.
(7) The concentration of the parasite and the host population are variables in space and time.
(8) Initially, acquired immunity and vaccination of the susceptible are not considered.
(9) The cats infected through their excreta contribute to the parasite rate increase in the environment.
(10) Natural decline rate of the pathogen.
(11) Cat's natural immunity is taken into account.

Cat's population is considered to be divided into susceptible $S(x,y,t)$, infected $I(x,y,t)$ and immune $R(x,y,t)$, and that of the parasites $P(x,y,t)$. The parameters γ and μ are the cat's birth and mortality rates, respectively, β is the number of parasites that an infectious cat excretes to the environment during its period of infectiousness $1/\eta$, θ is the natural decline of the parasite; D is the coefficient of dispersion of the parasite (longitude^2/time); $\lambda(P)$ is the measure of the efficient transmission of the parasite to the susceptible cat, so that $\lambda(P)S$ is the proportion of susceptible cats that get infected when consuming parasites, $P(x,y,t)$. As the infecting form of the parasite is not taken into account (tissue cyst, oocysts or tachyzoites), a uniform distribution is considered, that is to say, the probability that a cat gets infected is given by the uniform function:

$$
\begin{array}{ll}
0, & if \;\; P < P_{min} \\
\lambda(P) = (P - P_{min}/(P_{max} - P_{min}), & if \;\; P_{min} < P < P_{max} \\
1, & if \;\; P > P_{max}
\end{array}
$$

Where, P_{min} is the minimum quantity that a susceptible cat should consume to be infected and P_{max} indicates that above this quantity, there is infection for sure. This function is an approximation of the logistics with the algorithmic advantage that it is lineal or constant (0 or 1). Thus, $1/\lambda(P)$ is the average period of the susceptible state of the cat, $1/\mu$ is the life average of the susceptible and infected cats, $1/\beta$ is the oocysts excretion average time by infectious cats and $1/\theta$ is the parasite's life period on ground.

To consider the transport by water of the parasite, simulating the hypothetical case in which an infected cat places infectious forms on the ground or water and these can be transported through the water, by the rain and/or the rivers, a term that represents this contribution $-\vec{V}.\nabla P$ is added to the equation corresponding to the concentration of parasites P; where $\vec{V}$ denotes water velocity.

According to these considerations, the dynamics is described by the following system of equations, where the subscripts indicate derivatives:

$$\begin{aligned}
S_t(x,y,t) &= \gamma[(S(x,y,t)+I(x,y,t)+R(x,y,t)] - (\lambda P(x,y,t)+\mu)S(x,y,t) \\
I_t(x,y,t) &= -\lambda[P(x,y,t)]S(x,y,t) - (\mu+\eta)I(x,y,t) \\
R_t(x,y,t) &= \eta I(x,y,t) - \mu R(x,y,t) \\
P_t(x,y,t) &= \beta I(x,y,t) - \theta P(x,y,t) + D[P_{xx}(x,y,t) + P_{yy}(x,y,t)] \\
&\quad - \vec{V}.\nabla P(x,y,t)
\end{aligned}$$

Where $N = S + I + R$ is the total population of cats, therefore $N(x,y,t) = exp(\gamma - \mu)t + F(x,y)$ defined in the region $\Omega = (x,y) \in [0,l]x[0,h]$ with $t \in [0,\infty)$.

Boundary conditions are considered (subscripts indicate partial derivatives):

$$\begin{aligned}
P_x(0,y,t) &= kP(x,y,t) \\
P_x(l,y,t) &= P_y(x,0,t) = P_y(x,h,t) = 0
\end{aligned}$$

k a positive constant.

The first condition represents a constant situation of permeability for the local population that crosses the part of the boundary; a proportional passage to the existing population is considered[8]. The derivatives of P with respect to x and to y, equal to zero, represent the parasites no entrance-exit in the respective edges[9]. A homogeneous initial distribution for the populations is considered:

The system models the diffusion of the parasite and the propagation of the infection in the host cats population in a rectangular region of an area $l \times h$; the populations S, I, R and P are functions of x, y and t ; the *T. gondii* difussion is only considered, through the mechanical transportation (insects, birds, rodents and water); the parasite has mobility in the region toward the populations of susceptible cats, that is to say, in regions where there is no cats mobility and *T. gondii* did not exist. $\lambda(P)N/(\mu + \eta)$ is the basic rate of reproduction or epidemic threshold (R_0) and represents the number of secondary infectious produced by the introduction of an infectious cat or an inoculums of parasites in a susceptible host population[7] in a point (x, y); $1/(\mu+\eta)$ is the hope of life of an infectious and $\lambda(P)N$ is the number of new infectious cats. If $R_0 > 1$ then, there is epidemic during a period of time, otherwise the infection is extinguished.

For the construction of the velocity field, the hydrologic map of the department of Quindío-Colombia was taken. Elaborated by the Regional Autonomous Corporation of Quindío, a 12x16 mesh was made and the velocities in each node were calculated. To do this, the measurements in each lower right rectangle of the mesh were taken into account according to the cases further on related. It is deduced from the map that water flows north-south or east-west direction preferentially. For the velocities in some nodes (case v), data from the final report of the project was taken "Modelation of superficial water flowing in the department of Quindío" carried out by the Regional Autonomous Corporation of this department in agreement with the University of Quindío, 2002, where velocities (magnitudes) were measured in some points of main rivers (providing drinkable water to the municipal heads).

- Cases for the angles (measured with the horizontal):
  i)There is no main river in the chart: The average of the angles of affluents in the left vertical exit was taken.
  ii)There is a main river in the chart: The angle of the segment from the entrance to the exit of the river to the rectangle was measured.

- Cases for the slopes:
  iii)There are no contour lines in the rectangle: A 2.5 slope was taken according to the experience had in these cases and to the knowledge of the zones of concern.
  iv)There are contour lines in the rectangle: The slope by the coefficient given by the incline (contour lines difference) and the horizontal distance was calculated. The horizontal distance in the case ii) is the main river length and in the case i) is the average of the affluents lengths inside the chart.

- Cases for the velocities:
  v)There is a main river in the rectangle: The measured velocities were interpolated (CRQ-UQ).
  vi)There is no main river in the rectangle: The velocity magnitude was calculated by the formula:

$$V = \alpha k S^{0.5} \tag{1}$$

Where, V is the velocity, m/s, α is the constant of units conversion, equal to 10 in SI and to 33 in CU (10 in our case), k is an adimensionalized function that depends on the type of land cover and S is the slope, m/m[10].

According to this, the velocities field given in the Figure 1 was obtained (zero was the velocity assigned to the nodes that are not on the Quindío map).

3. Numerical solution

To find the solution of the system, the explicit numerical method of finite differences was used[11]. To obtain simulation, the mathematical package MATLAB was used. A initial population of 2 susceptible cats evenly distributed was placed. The value of the parameter θ can be estimated taking into account that in favorable humidity conditions, temperature, etc., the infecting form can remain viable in the environment (soil) 18 months approximately and thus $\theta = 0.05$. Every female cat has 14 cats in a year, (`www.albaonline.org/este/proyes.htm`), therefore, the birth and mortality rate are estimated to be 0.024 and 0.5 respectively. The coefficient of diffusion is considered to be equal to water diffusion 0,005(cm^2/seg). The minimum quantity of parasites that a susceptible cat should consume to be infectious is 10.000 and to have a sure infection is 1,000.000. The value of

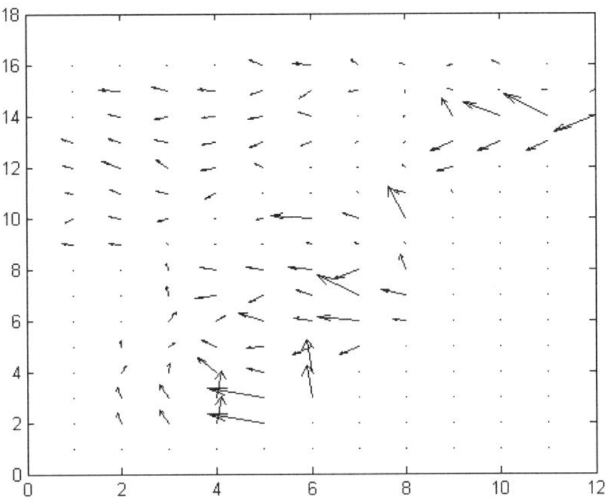

Figure 1. Velocity field.

β, number of oocysts/day excreted by each infectious cat, was calculated taking into account the amount of parasites expelled during this period.

The figures 3 to 5 show the behavior of the populations when an infectious cat is placed in different positions, after some iterations. As it was expected, there are some variations according to the position of the infectious cat, in keeping with the directions of the velocity field.

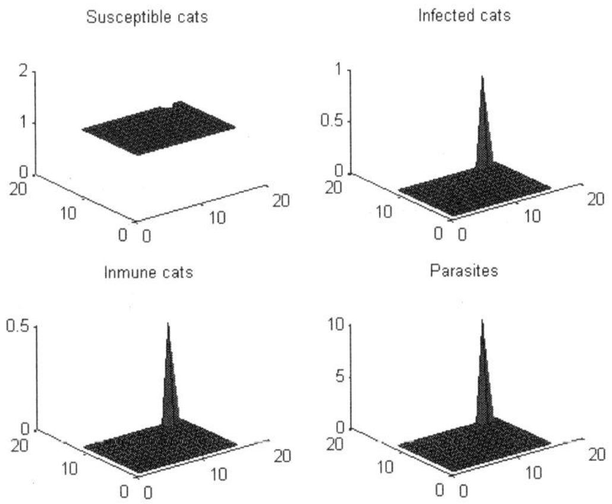

Figure 2. Populations without transportation by water. We ubicated a infected cat at the upper right corner of the map.

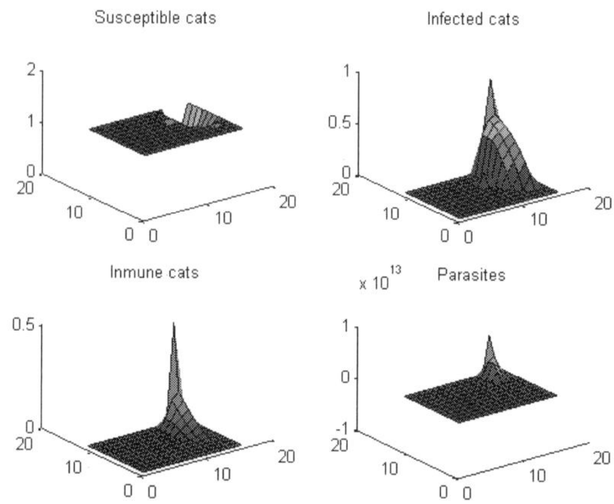

Figure 3. Populations with transportation by water. We ubicated a infected cat at the upper right corner of the map.

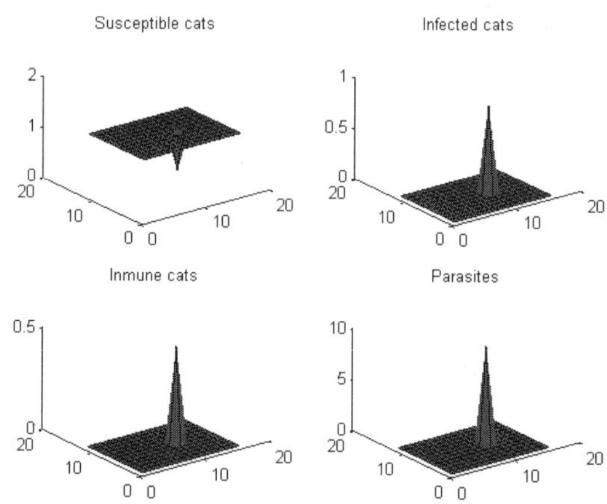

Figure 4. Populations without transportation by water. We ubicated a infected cat at the center of the map.

4. Conclusion

In conclusion, the comparison among the two model simulations, with and without water transportation of the parasite showed that the first one is fundamental in the dissemination of the parasite. That is to say, a hypothetical inoculum definitely contributes to the transmission of the infection

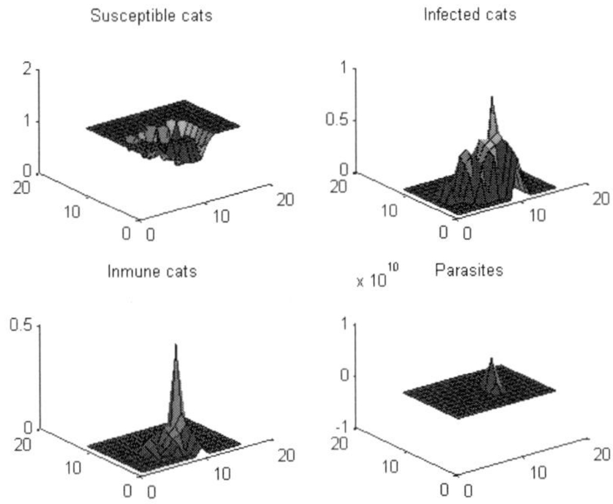

Figure 5. Populations with transportation by water. We ubicated a infected cat at the center of the map.

by this parasite which occurs in some place along the way of the water that finally supplies a population of intermediary or final hosts. As the considered region has an irregular frontier, the method of finite differences used in this work is not the most appropriate one. A proposal for a subsequent work is the use of finite elements to solve the diffusion type model - advection.

Acknowledgments

For their help in carrying out this work, we would like to thank Vicerrectoría of Investigations of Quindío University. In the same way, we thank Doctor Jorge Enrique Gómez Marín from Quindío University for his valuable contribution concerning the epidemiological part, and Doctor João Frederico A. C. Meyer of Campinas University, Brazil, for his help in the area of numerical analysis.

References

1. López CA, Díaz JR, Gómez JE. 2005. Factores de riesgo en mujeres embarazadas, infectadas por *Toxoplasma gondii* en Armenia-Colombia. Revista de Salud Pública, 7(2):180-190.
2. Mandell GL, Bennett JE, Dolin, R. 2000. Principios y Prácticas en Enfermedades Infecciosas. Vol. 2. Editorial Médica Panamericana.
3. Ruiz A, Frenkel JK. 1980. *Toxoplasma gondii* in Costa Rica cats. The American Society of Tropical Medicine and Higiene, 26(6):1150-1160.

4. Isaac JR, Bowie WR, King A, Stewart GI, Ong CS, Fung CP, Shokeir MO, Dubey JP. 1998. Detection of *Toxoplasma gondii* Oocysts in Drinking Water. Applied and Environmental Microbiology, 64(6): 22782280.
5. Eng SB, Werker DH, King AS, Marion SA, Bell A, Issac JR, Stewart GI, Bowie WR. 1999. Computer-Generated Dot Maps as an Epidemiologic Tool: Investigating an Outbreak of Toxoplasmosis. Emerging Infectious Diseases, 5(6):815-819.
6. Trejos AD, Duarte GI. 2005. Un modelo matemático de la propagación de *Toxoplasma gondii* a través de gatos. Actualidades Biológicas, 27(83):143-149.
7. Anderson RM, May RM. 1991. Infectious diseases of humans.dynamics and control. Oxford University Press.
8. Meyer J F, Pregnolatto S A. 2003. Mathematical model and numerical simulation of the population dynamics of Capybaras: an epizootic model with dispersal, migration and periodically varing contagion. Biomatemática. XIII:145-152.
9. Edelstein-Keshet L. 1988. Mathematical Models in Biology. McGraw-Hill, USA
10. McCuen RH, Johnson PA, Ragan, RM. 2002. Highway Hidrology, Hydraulic Design Series Number 2. Second Edition. USA
11. Burden RJ, Faires JD. 1985. Análisis Numérico. Grupo Editorial Iberamericana.

SIS EPIDEMIC MODEL WITH PULSE VACCINATION STRATEGY AT VARIABLE TIMES

F. CÓRDOVA-LEPE*

Instituto de Ciencias Básicas, Universidad Católica del Maule and Depto. de Matemáticas, Universidad Metropolitana de Ciencias de la Educación, Avenida San Miguel 3605, Talca, Chile.
E-mail: `fcordova@ucm.cl`

R. DEL-VALLE

Instituto de Ciencias Básicas, Universidad Católica del Maule, Avenida 3605 San Miguel, Talca, Chile.
E-mail: `rdelvalle@ucn.cl`

G. ROBLEDO

Facultad de Ciencias, Universidad de Chile, Las Palmeras 3425, Ñuñoa, Santiago, Chile.
E-mail: `grobledo@uchile.cl`

An SIS model with pulse vaccination at variable times is presented. The local stability of the disease–free solution is completely studied, whereas some initial results concerning global stability are discussed. By using numerical simulations, the convergence rate toward the disease–free solution is compared with a fixed time vaccination strategy.

*Work partially supported by grant FIBAS–1509 UMCE and GMMRP UCM.

1. Introduction

In this note, we propose a pulse vaccination strategy at variable times for a SIS epidemic model, that is described by the impulsive system of differential equations at variable times:

$$\begin{cases} \left.\begin{aligned} \dot{S}(t) &= m(1-S(t)) - \beta S(t)I(t) + gI(t), \\ \dot{I}(t) &= \beta S(t)I(t) - (g+m)I(t), \\ \dot{V}(t) &= -mV(t), \end{aligned}\right\} \quad \text{if} \quad t \neq \tau_k, \\ \left.\begin{aligned} S(t^+) &= (1-p)S(t), \\ I(t^+) &= I(t), \\ V(t^+) &= V(t) + pS(t), \end{aligned}\right\} \quad \text{if} \quad t = \tau_k, \\ \quad \Delta\tau_k = \tau\big(\beta S(\tau_k)I(\tau_k)\big), \end{cases} \tag{1}$$

where $S(t)$, $I(t)$ and $V(t)$ denote respectively the fraction of susceptible, infected and vaccinated population at time t, and satisfy $S(t)+I(t)+V(t) = 1$ for any $t > 0$.

In addition, there exists an increasing sequence of vaccination times $\{\tau_k\}_{k\geq 0}$, determined by a first vaccination time τ_0 and a decreasing, continuous and differentiable function $\tau\colon [0,1] \mapsto]0,\infty[$,

$$\tau(x) = \frac{\tilde{\tau}}{ax+1}, \quad x \in [0,1], \quad \tilde{\tau} > 0 \quad \text{and} \quad a > 0,$$

which means that at $t = \tau_k$, a fraction $p \in]0,1[$ of the susceptible population is vaccinated.

Our purpose is to generalize the fixed time pulse vaccination strategy developed by Zhou *et.al* [13] by considering time variable impulses. In consequence, $\beta S(\tau_k)I(\tau_k)$, *i.e.*, the incidence at time τ_k will determine the instant of the $(k+1)$–th vaccination. As $\tau(\cdot)$ is decreasing, a bigger incidence implies a nearest date for the next vaccination.

1.1. *The model and its motivations*

The system (1) is a generalization of the classical model:

$$\begin{cases} \dot{S}(t) = m\,(1-S(t)) - \beta S(t)I(t) + gI(t), \\ \dot{I}(t) = \beta S(t)I(t) - (m+g)I(t), \end{cases} \tag{2}$$

that considers vital dynamics, but the death rate $m \in]0,1[$ coinciding with birth rate. Between S and I, the transition rate is βI, with $\beta > 0$. In this case, contagion does not confer immunity and the recovery rate is given by $g \in]0,1[$.

The asymptotic behavior of (2) is determined by the constant

$$R_0 = \frac{\beta}{m+g}, \tag{3}$$

that is called the basic reproductive number in the literature (see *e.g.* Diekmann *et. al* [5] and MacDonald [10]). It can be proved (see *e.g.*, Heathcote [6]) that the disease–free equilibrium ($S = 1$ and $I = 0$) is globally asymptotically stable (GAS) if $R_0 < 1$. On the other hand, if $R_0 > 1$, then the disease–free equilibrium becomes unstable and a GAS endemic equilibrium ($S = 1/R_0$ and $I = 1 - 1/R_0$) appears.

The global asymptotic stability of the endemic equilibrium has stimulated a considerable amount of theoretical and practical research. The main goal is to avoid the stability of the endemic equilibrium of (2): a first way is to consider β and g as control variables by reducing β and/or increasing g such that the disease-free equilibrium becomes a GAS solution.

A second way is to introduce the vaccination on the system: a constant fraction $p \in]0,1[$ of the susceptible population is vaccinated per unit time, the immune (vaccinated) population at time t will be denoted by $V(t)$. Hence, the system (2) with vaccination becomes:

$$\begin{cases} \dot{S}(t) = m\big(1 - S(t)\big) - \beta S(t)I(t) + gI(t) - pS(t), \\ \dot{I}(t) = \beta S(t)I(t) - (m+g)I(t), \\ \dot{V}(t) = pS(t) - mV(t). \end{cases} \tag{4}$$

As above, the behavior of (4) is determined by the reproductive number

$$R_1 = R_0 \frac{m}{m+p} < R_0. \tag{5}$$

Indeed, the disease-free equilibrium of the system (4) is given by $(R_1/R_0, 0, 1 - R_1/R_0)$, which is GAS if $R_1 < 1$. On the other hand, if $R_1 > 1$ then this equilibrium becomes unstable and a GAS endemic equilibrium $(1/R_0, 1 - 1/R_1, 1/R_1 - 1/R_0)$ appears.

The continuity of the vaccination process in (4) is dropped by Zhou *et. al* [13], where a time scheduled vaccination (at each time $\tau_k = k \geq 0$) is introduced. Hence, the model (4) becomes the impulsive system:

$$\begin{cases} \left.\begin{aligned} \dot{S}(t) &= m(1 - S(t)) - \beta S(t)I(t) + gI(t), \\ \dot{I}(t) &= \beta S(t)I(t) - (g+m)I(t), \\ \dot{V}(t) &= -mV(t), \end{aligned}\right\} \quad \text{if} \quad t \neq k, \\ \left.\begin{aligned} S(t^+) &= (1-p)S(t), \\ I(t^+) &= I(t), \\ V(t^+) &= V(t) + pS(t), \end{aligned}\right\} \quad \text{if} \quad t = k. \end{cases} \tag{6}$$

In this impulsive framework (we refer the reader to the classical works of Bainov *et. al* [1,2] and Samoilenko *et. al* [11]) the disease–free equilibrium takes the form of a periodic function $\big(\bar{S}(\cdot), \bar{I}(\cdot), \bar{V}(\cdot)\big)$ defined as follows:

$$\bar{S}(t) = 1 - \bar{V}(t), \quad \bar{I}(t) = 0, \quad \bar{V}(t) = \frac{pe^{-m(t-k)}}{1-(1-p)e^{-m}}, t \in]k, k+1], k \geq 0.$$

The asymptotic behavior of (6) is determined by the reproductive number:

$$R_2 = R_0 \left[\frac{1-e^{-m}}{1-(1-p)e^{-m}} \right]. \tag{7}$$

It was proved by Zhou *et. al* [13] that the periodic solution $(\bar{S}, \bar{I}, \bar{V})$ of (6) is globally attractive if $R_2 < 1$. Note that $R_1 < R_2 < R_0$.

Notice that if in (6) the instants of pulse vaccination $\{k\}_{k\geq 0}$, are replaced by a sequence $\{\tau_k\}_{k\geq 0}$, we can add an evolution law relating $\Delta\tau_k$ with the state variable. So (6) is a particular case of (1) with $\tau(x) = 1$, for any $x \in [0, 1]$. Can we improve the control? That is the question conducing our work.

1.2. *Novelty of this work and outline*

The dynamics of (1) is governed by an impulsive system of new type, that was presented by Córdova–Lepe [3], where an introduction to existence and representation of solutions theory was proposed. To the best of our knowledge, this work shows the first epidemic model incorporating pulse vaccination with variable time. Notice that other related impulsive models with fixed time approach are *e.g.* Shulgin *et. al* [12], Hin *et. al* [7], Li *et. al* [8,9] and references therein. It is interesting for us to determine if a time variable approach leads a faster convergence to the periodic disease-free solution.

The paper is organized as follows. In Section 2, firstly, we study the invariant subsystem with disease–free solutions and the existence of a GAS periodic solution is proved. Second, a necessary and sufficient condition ensuring its local stability is obtained by introducing a new reproductive number. In Section 3, a result about global stability is presented. Finally, several numerical simulations are shown in Section 4.

2. Pulse vaccination strategy at variable times

2.1. *Initial conditions with non infected population*

First, we will study the disease–free solution $\big(\tilde{I}(t), \tilde{S}(t), \tilde{V}(t)\big)$ of (1) with initial condition $(\tilde{S}_0, 0, \tilde{V}_0)$ at τ_0 satisfying $\tilde{S}_0 + \tilde{V}_0 = 1$. Moreover, for any

$k \geq 0$ we have $I'(t) = 0$ if $t \neq \tau_k$, and $I(\tau_k^+) = I(\tau_k)$. In consequence, we can deduce that $\tilde{I}(t) = 0$ for any $t \geq \tau_0$ and $\tau_k = k\tilde{\tau}$, $k \geq 0$. So, the disease–free solution satisfies the system:

$$\left\{\begin{array}{ll} \left.\begin{array}{ll} \dot{S}(t) & = m(1 - S(t)), \\ \dot{V}(t) & = -mV(t), \end{array}\right\} & \text{if } t \neq k\tilde{\tau}, \\ \left.\begin{array}{l} S(t^+) = (1-p)S(t), \\ V(t^+) = V(t) + pS(t), \end{array}\right\} & \text{if } t = k\tilde{\tau}. \end{array}\right. \tag{8}$$

The asymptotic behavior is described by the following result:

Theorem 2.1. *The system* (8) *has a periodic solution* $\big(\bar{S}(\cdot), \bar{V}(\cdot)\big)$ *defined as:*

$$\bar{S}(t) = 1 - \bar{V}(t), \quad \bar{V}(t) = \widehat{V}e^{m[(k+1)\tilde{\tau}-t]}, \quad t \in]k\tilde{\tau}, (k+1)\tilde{\tau}], \quad k \geq 0, \tag{9}$$

where $\widehat{V} = pe^{-m\tilde{\tau}}/[1-(1-p)e^{-m\tilde{\tau}}]$, *which is GAS.*

Proof. As (8) is non–coupled in $]k\tilde{\tau}, (k+1)\tilde{\tau}]$, we can verify that

$$\begin{cases} \tilde{S}(t)e^{m(t-k\tilde{\tau})} - \tilde{S}(k\tilde{\tau}^+) = e^{m(t-k\tilde{\tau})} - 1, \\ \tilde{V}(t)e^{m(t-k\tilde{\tau})} - \tilde{V}(k\tilde{\tau}^+) = 0. \end{cases}$$

The discrete part of (8) implies

$$\begin{cases} \tilde{S}(t) = 1 - [1-(1-p)\tilde{S}(k\tilde{\tau})]e^{-m(t-k\tilde{\tau})}, \\ \tilde{V}(t) = [p + (1-p)\tilde{V}(k\tilde{\tau})]e^{-m(t-k\tilde{\tau})}. \end{cases}$$

Let us denote $\tilde{S}(j\tilde{\tau})$ and $\tilde{V}(j\tilde{\tau})$ respectively by $\tilde{S}_j$ and $\tilde{V}_j$ ($j = 0, 1, \ldots.$). Hence, we obtain the discrete one–dimensional map:

$$\begin{cases} \tilde{S}_{k+1} = 1 - [1-(1-p)\tilde{S}_k]e^{-m\tilde{\tau}} = f\big(\tilde{S}_k\big), \\ \tilde{V}_{k+1} = [p + (1-p)\tilde{V}_k]e^{-m\tilde{\tau}} = g\big(\tilde{V}_k\big). \end{cases} \tag{10}$$

It is straightforward to verify that (10) has a unique positive equilibrium $\widehat{E} = (\widehat{S}, \widehat{V})$, where $\widehat{S} = 1 - \widehat{V}$, which defines a unique periodic solution of (8), of period $\tilde{\tau}$, given in (9). In addition, by using the inequalities

$$0 < f'(\widehat{S}) = g'(\widehat{V}) = (1-p)e^{-m\tilde{\tau}} < 1,$$

we can conclude that $\widehat{E}$ is a GAS equilibrium of (10). Then $\big(\bar{S}(\cdot), \bar{V}(\cdot)\big)$ is a GAS periodic solution of (8) and the theorem follows. □

2.2. *Initial conditions with infected population*

Let us consider a solution of (1) with an initial condition $I(\tau_0) > 0$. Let (S_k, I_k, V_k) be the solution at $t = \tau_k$, with $k \geq 0$. By (1), it follows that

$$\left(S(\tau_k^+), I(\tau_k^+), V(\tau_k^+)\right) = \left((1-p)S_k, I_k, 1 - I_k - (1-p)S_k\right).$$

Notice that if $t \in (\tau_k, \tau_{k+1}]$, then

$$I'(t) = \alpha(t)I(t) - \beta I^2(t), \quad \text{with} \quad \alpha(t) = (\beta - g - m) - \beta V(t),$$

and it follows that

$$I(t) = \frac{I_k \pi(t)}{1 + \beta I_k \displaystyle\int_{\tau_k}^{t} \pi(s)\, ds} \quad \text{with} \quad \pi(t) = \exp\Big(\int_{\tau_k}^{t} \alpha(s)\, ds\Big). \tag{11}$$

Let $\lambda = \beta - g - m$ and notice that

$$\pi(t; \tau_k, I_k, V_k) = \exp\left(\int_{\tau_k}^{t} \alpha(s)\, ds\right)$$

$$= \exp\left[\lambda(t - \tau_k) - \frac{\beta}{m} V_k^+ (1 - e^{-m(t-\tau_k)})\right],$$

where $V_k^+ = (1-p)V_k + p(1 - I_k)$. Moreover,

$$\int_{\tau_k}^{t} \pi(s)\, ds = \int_0^{t-\tau_k} \exp\left[\lambda u - \frac{\beta}{m} V_k^+ (1 - e^{-mu})\right] du.$$

Finally, we obtain the discrete map

$$\begin{cases} I_{k+1} = F(I_k, V_k) = \dfrac{I_k \exp\left[\lambda\tau(I_k, V_k) + \dfrac{\beta}{m} V_k^+ \left(e^{-m\tau(I_k,V_k)} - 1\right)\right]}{1 + \beta I_k \displaystyle\int_0^{\tau(I_k,V_k)} \exp\left[\lambda u + \dfrac{\beta}{m} V_k^+ \left(e^{-mu} - 1\right)\right] ds}, \\ \\ V_{k+1} = G(I_k, V_k) = V_k^+ e^{-m\tau(I_k,V_k)}, \end{cases} \tag{12}$$

with $\tau(I, V) = \tau(\beta I(1 - I - V))$.

Theorem 2.2. *The equilibrium* $(0, \hat{V})$ *of map* (12) *is locally stable if:*

$$0 < R_3 = R_0 \left[1 - \frac{p}{\tilde{\tau} m} \frac{1 - e^{-m\tilde{\tau}}}{1 - (1-p)e^{-m\tilde{\tau}}}\right] < 1. \tag{13}$$

Proof. Notice that $R_3 > 0$ is a consequence of the inequality $p(e^x - 1)/[(e^x-1)+p] < x$ for any $x > 0$. Now, we compute the Jacobian matrix associated to (12). As $\frac{\partial F}{\partial V}(0,\hat{V}) = 0$, we have to prove that eigenvalues $\frac{\partial F}{\partial I}(0,\hat{V})$ and $\frac{\partial G}{\partial V}(0,\hat{V})$ are inside the unit disk.

Notice that

$$\frac{\partial G}{\partial V}(I_k,V_k) = (1-p)e^{-m\tau(I_k,V_k)} - [(1-p)V_k + p(1-I_k)]e^{-m\tau(I_k,V_k)} m\frac{\partial \tau}{\partial V}(I_k,V_k).$$

Evaluating at $(0,\widehat{V})$, we have

$$\frac{\partial G}{\partial V}(0,\hat{V}) = (1-p)e^{-m\tilde{\tau}} - \hat{V}m\frac{\partial \tau}{\partial V}(0,\hat{V}) = (1-p)e^{-m\tilde{\tau}} < 1.$$

On the other hand, as $F(0,V) = 0$, it follows by (11) that

$$\frac{\partial F}{\partial I}(0,\hat{V}) = \lim_{I\to 0^+} \frac{\pi(\tau_k + \tau(\hat{I},\hat{V};\tau_k,I,\hat{V}))}{1+\beta I\int_{\tau_k}^{\tau_k+\tau(I,\hat{V}}\pi(s;\tau_k,I,\hat{V})\,ds} = \pi(\tau_k+\tilde{\tau};\tau_k,0,\hat{V})$$

and

$$\pi(0,\hat{V}) = \exp\left([\beta - m - g]\tilde{\tau} - \frac{\beta}{m}\frac{p(1-e^{-m\tilde{\tau}})}{1-(1-p)e^{-m\tilde{\tau}}}\right).$$

Notice that $|\frac{\partial F}{\partial I}(0,\hat{V})| < 1$ if and only if

$$(\beta - m - g)\tilde{\tau} < \frac{\beta}{m}(1-\hat{V})p,$$

which is equivalent to (13) and the local stability follows. □

3. Global Stability (preliminary results)

Lemma 3.1. *Assume that $R_0 > 1$ and that $(S(t), I(t), V(t))$, $t \geq 0$, is a solution of (1) such that $S(0) \in [0, 1/R_0[$. If $\tilde{\tau}$ satisfies*

$$R_4 = R_0\frac{m}{p}\tilde{\tau}e^{(\beta-m)\tilde{\tau}} < 1, \tag{14}$$

then $S(t) < 1/R_0$, for any $t \geq 0$.

Proof. From now on, we assume $\tau_0 = 0$. Let τ_k be an impulse time of a solution $(S(t), I(t), V(t))$ satisfying $S(0) \in [0, 1/R_0[$. Suppose that $S(\tau_k) \in [0, 1/R_0[$ and that $t \in]\tau_k, \tau_{k+1}]$. Since

$$S'(t) = m(1 - S(t) - I(t)) + \beta\left[\frac{1}{R_0} - S(t)\right] I(t),$$

and $S(\tau_k^+) = (1-p)S(\tau_k)$, by continuity of $S(\cdot)$ on $]\tau_k, \tau_{k+1}]$, there exists $\delta > 0$ such that $S(t) \in [0, 1/R_0[$ and

$$0 < S'(t) \leq mV(t) + \beta\left[\frac{1}{R_0} - S(t)\right], \tag{15}$$

for $t \in]\tau_k, \tau_k + \delta] \subset]\tau_k, \tau_{k+1}]$.

Considering that $V(t) = V(\tau_k^+)e^{-m(t-\tau_k)}$, $t \in]\tau_k, \tau_{k+1}]$. By integrating (15), we have

$$S(t) \leq S(\tau_k^+)e^{-\beta(t-\tau_k)} + \frac{mV(\tau_k^+)}{\beta - m}\left[e^{-m(t-\tau_k)} - e^{-\beta(t-\tau_k)}\right]$$
$$+\frac{1}{R_0}\left[1 - e^{-\beta(t-\tau_k)}\right],$$

for $t \in [\tau_k, \tau_k + \delta]$.

Since $S(\cdot)$ is an increasing function in $]\tau_k, t^*]$, $t^* = \tau_k + \delta$, its values in this interval are bounded by

$$\frac{1}{R_0}\left[1 - pe^{-\beta\delta}\right] + \frac{m}{\beta - m}V(\tau_k^+)\left[e^{-m\delta} - e^{-\beta\delta}\right]. \tag{16}$$

By using $e^{-m\delta} - e^{-\beta\delta} \leq (\beta - m)\delta\, e^{-m\delta}$, we obtain an upper bound of (16):

$$\frac{1}{R_0} - e^{-m\delta}\left[\frac{p}{R_0}e^{-(\beta-m)\delta} - m\delta\right]. \tag{17}$$

By using (14) combined with the fact that $\delta < \tilde{\tau}$, it follows that (16) is upperly bounded by $1/R_0$. Hence, for any $t \in]\tau_k, t^*[$ it follows that $S(t)$ is upperly bounded for a constant S^* inferior to $1/R_0$.

Hence, by continuation of the argument from time t^*, where $S(t^*) < 1/R_0$, we conclude that $S(t) < 1/R_0$ for any $t \in]\tau_k, \tau_{k+1}]$, because there is not accumulation to the right before time τ_{k+1}. If we suppose $\delta > 0$ a maximal value satifying $S(t) < 1/R_0$, $t \in]\tau_k, \tau_k + \delta]$, $\tau_k + \delta < \tau_{k+1}$, then $S(\tau_k + \delta) < S^* < 1/R_0$ and we can prolong the solution under $1/R_0$, which contradicts the definition of δ. □

Lemma 3.2. *If*

$$R_5 = R_0\frac{1 - e^{-m\tilde{\tau}}}{1 - (1-p)e^{-m\tilde{\tau}}},$$

then the periodic solution $(\bar{S}(\cdot), 0, \bar{V}(\cdot))$ *satisfies the condition* $\bar{S}(t) < 1/R_0$, *for any* $t \geq 0$.

Proof. The maximum value of the periodic solution is $\hat{S}$. The condition $\hat{S} < 1/R_0$ is equivalent to $R_5 < 1$. □

Theorem 3.1. *Suppose* $\max\{R_4, R_5\} < 1$. *If* $(S(\cdot), I(\cdot), V(\cdot))$ *is a solution such that* $S(0) < 1/R_0$, *then* $I(t) \to 0$ *as* $t \to \infty$.

Proof. Let us consider an initial condition of (1) with $S(0) < 1/R_0$. It is straightforward to verify that $I(t) > 0$ for any finite time $t > \tau_0$. By (1) and lemmas above, for a solution $(S(\cdot), I(\cdot), V(\cdot))$, with impulses at $\{\tau_k\}_{k\geq 0}$, we can deduce that

$$\frac{I'(t)}{I(t)} = -\beta\left[\frac{1}{R_0} - S(t)\right] < 0, \tag{18}$$

for any $t \neq \tau_k$, $k \geq 0$. Since I is continuous, strictly decreasing and a nonnegative function, it is deduced that $I(t) \to 0$ if $t \to \infty$. □

4. Numerical Simulations

We will consider three groups of initial conditions:

Initial Condition	Case I	Case II	Case III
$S(0)$	0.08	0.33	0.82
$I(0)$	0.90	0.33	0.13
$V(0)$	0.02	0.34	0.05

Also, we will consider the following non-dimensional parameters:

Parameter	m	β	g	p	$\tilde{\tau}$
Value	1/70	9/70	1/10	5/100	1

Hence, we compute the following reproductive numbers

$$R_3 = 0.24, \quad R_4 = 0.36, \quad R_5 = 0.25, \quad \text{and} \quad R_0 = 1.125.$$

which implies a GAS endemic equilibrium in abscence of vaccination.

The first simulation (carried by modifying an algorithm developed in Ref. [4]) is shown in the Figure 1 and presents the solutions of the model (6), which deals with a fixed vaccination time. We can observe that the solutions are asymptotically convergent toward a periodic disease–free solution.

The second Figure, shows the solutions of the system (1), which deals with variable vaccination times (In this case, $a = 15$). As before, the convergence toward the periodic disease–free solution can be observed.

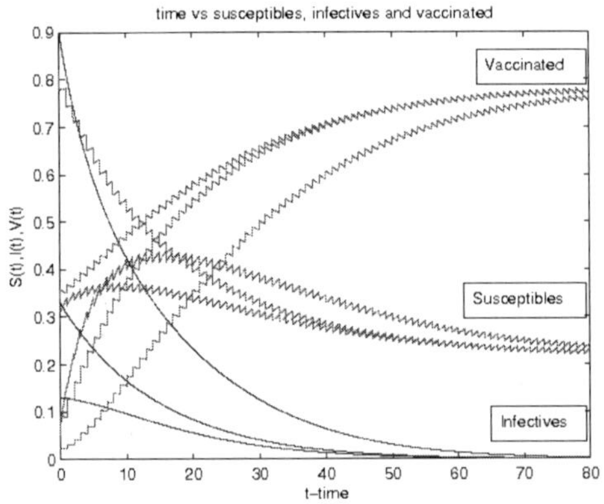

Figure 1. Strategy at fixed times by using (6).

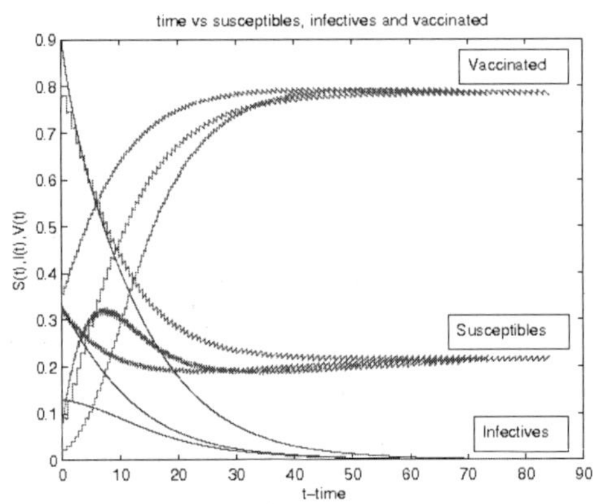

Figure 2. Strategy at variable times by using (1)

To illustrate the effectiveness of a variable time strategy, the previous solutions are presented together in the Figures 3 and 4. It can be observed that a variable vaccination time strategy has a faster convergence toward the periodic disease–free solution. These simulations provide some justification for the use of a time variable vaccination strategy in preference to a fixed time approach in some cases.

We point out that, the simulations suggest us that for a wide set of initial conditions, the condition $R_3 < 1$ for local stability, it is also true for the global stability of the periodic solution $(\bar{S}, \bar{I}, \bar{V})$. Nevertheless, this remains an open question that will require a more accurate analysis.

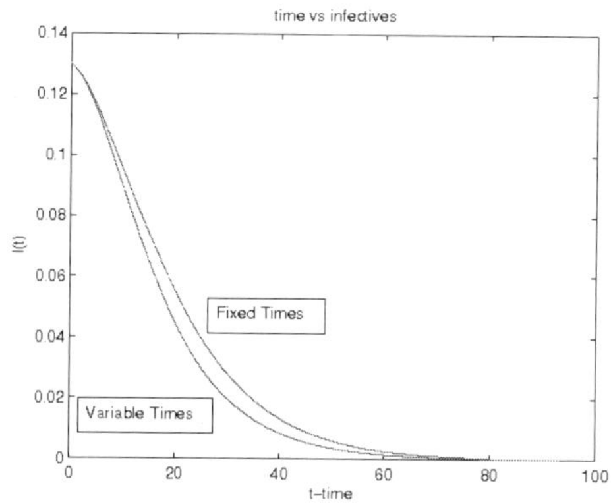

Figure 3. Comparison between both strategies for the infective group. Case II.

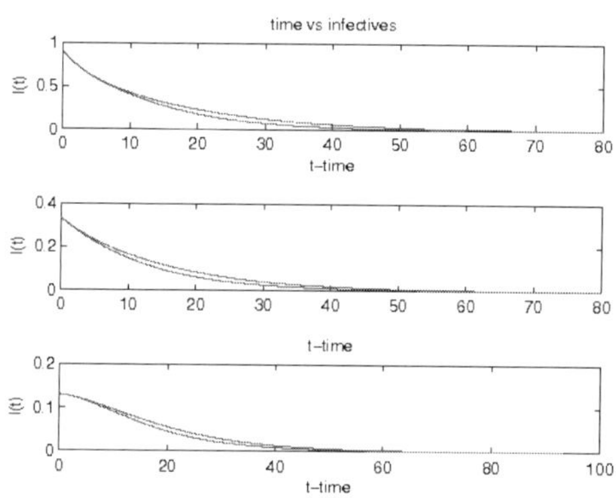

Figure 4. Comparison between both strategies Cases I, II and III.

References

1. D.D. Bainov and P.S. Simeonov, *Systems with impulse effect.* Ellis Horwood Series in Mathematics and its Applications. Ellis Horwood Limited. (1989).
2. D.D. Bainov and P.S. Simeonov, *Impulsive differential equations: periodic solutions and applications.* Logman Scientific and Technical. (1993).
3. F. Córdova-Lepe, *Advances in a Theory of Impulsive Differential Equations at Impulse Dependent Times, with Applications to Bio-economics*, Biomat 2006, Int. Symp. on Math. and Comp. Bio., R. Mondaini & R. Dilao Eds. World Scientific (2006), 343-358.
4. R. Del–Valle. *Algoritmo de discretizaciones para un problema de control optimal* Master Thesis, Universidad Católica del Norte, 2004.
5. O. Diekmann and J.A.P. Heesterbeek, *Mathematical Epidemiology of Infectious Diseases: Model Building Analysis and Interpretation*, Wiley, New York,

2000.
6. H.W. Heathcote, in *Applied Mathematical Ecology*, Edited by S.A. Levin, T.G. Hallam and L.J. Gross, pp. 119–144, Springer–Verlag, Berlin, 1989.
7. Z. Hin and M. Haque, *The SIS Epidemic Model with Impulse Effects*, Proceeding Eight ACIS International Conference on Software Engineering, Artificial Intelligence, Networking and Parallel/Distributed Computing, 505–507.
8. Y. Li, C. Ma & J. Cui, *The effect of constant and mixed impulsive vaccination on SIS epidemic models incorporating media coverage*, Rocky Mountain J.Math. (2008) 38, 1437–1455.
9. Y. Li & J. Cui, *The effect of constant impulse and pulse vaccination on SIS epidemic models incorporating media coverage*, Commun. Nonlinear Sci.Numer.Simulat. (2009) 14, 2353–2365.
10. G. Macdonald, *The Epidemiology and Control of Malaria*, Oxford University Press, 1957.
11. A.M. Samoilenko and N.A. Perestyuk, *Impulsive differential equations.* World Scientific Series on Nonlinear Science, Series A, vol. 14. (1995).
12. B. Shulgin, L. Stone & Z. Agur, *Pulse Vaccination Strategy in the SIR Epidemic Model*, Bulletin of Mathematical Biology (1998) 60, 1123-1148.
13. Y. Zhou & H. Liu, *Stability of Periodic Solutions for an SIS Model with Pulse Vaccination*, Mathematical and Computer Modelling (2003) 38, 299-308.

INDEX